Seismic Activity in Western Europe

with Particular Consideration to the Liège Earthquake of November 8, 1983

NATO ASI Series

Advanced Science Institutes Series

A series presenting the results of activities sponsored by the NATO Science Committee, which aims at the dissemination of advanced scientific and technological knowledge, with a view to strengthening links between scientific communities.

The series is published by an international board of publishers in conjunction with the NATO Scientific Affairs Division

A	Life Sciences	Plenum Publishing Corporation
B	Physics	London and New York
C	Mathematical and Physical Sciences	D. Reidel Publishing Company Dordrecht, Boston and Lancaster
D	Behavioural and Social Sciences	Martinus Nijhoff Publishers
E	Engineering and Materials Sciences	The Hague, Boston and Lancaster
F	Computer and Systems Sciences	Springer-Verlag
G	Ecological Sciences	Berlin, Heidelberg, New York and Tokyo

Series C: Mathematical and Physical Sciences Vol. 144

Seismic Activity in Western Europe

with Particular Consideration to
the Liège Earthquake of November 8, 1983

edited by

Paul Melchior

Royal Observatory of Belgium
Brussels, Belgium

D. Reidel Publishing Company

Dordrecht / Boston / Lancaster

Published in cooperation with NATO Scientific Affairs Division

Proceedings of the NATO Advanced Research Workshop on
Seismic Activity in Western Europe
Brussels, Belgium
20-22 March 1984

Library of Congress Cataloging in Publication Data

NATO Advanced Research Workshop on Seismic Activity in Western Europe (1984 : Brussels,
 Belgium)
 Seismic activity in Western Europe.

 (NATO ASI series. Series C, Mathematical and physical sciences; vol. 144)
 "Proceedings of the NATO Advanced Research Workshop on Seismic Activity in
Western Europe, Brussels, Belgium, March 20–22, 1984''—T.p. verso.
 "Published in cooperation with NATO Scientific Affairs Division."
 Bibliography: p.
 Includes index.
  1. Seismology—Europe—Congresses. 2. Liège (Belgium)—Earthquake, 1983—
Congresses. I. Melchior, Paul J. II. North Atlantic Treaty Organization. Scientific
Affairs Division. III. Title. IV. Series: NATO ASI series. Series C, Mathematical and
physical sciences; no. 144
QE536.2.E85N38   1984   551.2'2'094   84-18240
ISBN-13: 978-94-010-8829-9   e-ISBN-13: 978-94-009-5273-7
DOI: 10.1007/978-94-009-5273-7

Published by D. Reidel Publishing Company
P.O. Box 17, 3300 AA Dordrecht, Holland

Sold and distributed in the U.S.A. and Canada
by Kluwer Academic Publishers,
190 Old Derby Street, Hingham, MA 02043, U.S.A.

In all other countries, sold and distributed
by Kluwer Academic Publishers Group,
P.O. Box 322, 3300 AH Dordrecht, Holland

D. Reidel Publishing Company is a member of the Kluwer Academic Publishers Group

TABLE OF CONTENTS

Recommendation adopted by the participants
to the NATO Workshop 243/84

SEISMIC ACTIVITY
IN WESTERN EUROPE

with particular Consideration to the Liège Earthquake
of November 8, 1983

This Workshop, meeting in Brussels, 20-22 March, 1984,

recognising the difficulty of establishing precise patterns of
seismicity and formulating effective earthquake counter-measures
in North-West Europe,

taking note of the disastrous effects that followed the occurrence
of a moderate-sized earthquake at Liège on 8 November, 1983,

appreciating the valuable results that have already been achieved
with existing facilities and

conscious of the present vulnerability of this region of dense
population and high industrial development to further earthquake
damage

recommends that the following steps be taken to establish the
natural seismic hazard in Belgium and neighbouring countries :

1. The delineation of active seismic zones should be investigated
   by studies of lithospheric structure and presently-active geological processes, if possible in the framework of the International Lithosphere Programme.

2. Permanent national seismograph networks should be upgraded and
   maintained at an adequate level, incorporating modern techniques
   of digital recording and data transmission.

3. There should be continuing support for studies of historical
   seismicity and re-evaluation of early instrumental data.

4. Studies of seismicity should be complemented by investigations
   of other related geophysical parameters, including monitoring
   of vertical and horizontal earth deformation (e.g. satellite
   geodesy, gravity, earth tides, levelling) geomagnetism and
   radon emanation.

5. There should be adequate provision for portable recording networks of seismographs to monitor regions of special interest
   and take part in field studies following strong earthquakes.

further recommends that the following additional steps be taken
to establish the characteristics of strong ground motion and the
measures necessary to reduce its effects :

1. Adequate permanent networks of strong-motion seismographs
   should be established in each country to record detailed characteristics of ground motion in large earthquakes.

2. There should also be adequate reserves of portable strong-
   motion instruments to be deployed following major earthquakes
   and to be used for other special studies.

3. Studies should be made of near-surface geological and topographic conditions that are likely to affect strong ground motion.

4. Field studies should be made of all damaging earthquakes to
   establish the vulnerability of various types of building and
   construction practice to earthquake damage.

5. The characteristics of strong ground motion, as determined
   above shall be applied to formulate adequate anti-seismic design codes and standards of construction.

makes the following general recommendations :

1. It is important that all possible steps be taken to improve the
   technical communication between seismologists and engineers,
   and in particular seismologists should endeavour to provide
   engineers with the measurements of ground parameters that they
   need.

2. International cooperation is essential for efficient regional
   programmes in seismology and earthquake engineering, and should
   be fostered by :

(a) Free exchange of all data, including if possible the direct transmission of computer-readable information.

(b) The co-ordinated planning of national networks. In parti- cular, the region around Liège, being close to several countries, is especially suitable for detailed international cooperation in the carying out of seismological and related studies.

(c) The support of international seismological agencies which collect and collate seismological readings from European countries, and compile overall data bases for the region.

and finally thanks most sincerely the NATO Science Committee for its support, and the Local Organizing Committee for the provision of excellent facilities and generous hospitality.

PROBLEMES ACTUELS DE LA SEISMOLOGIE EN BELGIQUE

Paul Melchior

Observatoire Royal de Belgique

ABSTRACT

The moderate sized earthquake at Liège, on 8 november 1983, has caused disastrous effects, demonstrating the vulnerability of this area of dense population and high industrial development.

The present belgian seismographic network is not appropriate and must be upgraded and equiped with modern techniques of data transmission. Proposals are introduced to achieve such a goal.

Four precursor earthquakes of magnitude around 2.2 to 2.4 have been detected in october 1983 which shows the importance of a continuous and real time watch of all kinds of precursors.

Il m'est agréable de saluer ici la présence de M. le Chef de Cabinet adjoint de M. le Ministre de l'Education Nationale et de M. le Directeur Général de l'Enseignement universitaire et de la Recherche Scientifique.

Leur présence témoigne du souci qu'ont nos autorités de tutelle de trouver rapidement une solution aux difficultés qui entravent le bon fonctionnement de la Séismologie en Belgique.

Je tiens à saluer aussi nos nombreux collègues d'Allemagne, de France, d'Italie, du Luxembourg, des Pays-Bas, du Royaume Uni et de Suisse et tout particulièrement le Dr. Han Shou Liu du NASA Goddard Space Flight Center aux U.S.A. qui nous assurent, pendant ces trois jours, l'apport inestimable de leur compétence scienti-

1

P. Melchior (ed.), Seismic Activity in Western Europe, 1–17.
© *1985 by D. Reidel Publishing Company.*

fique et de leur expérience professionnelle dans l'analyse et la
compréhension des phénomènes séismiques.

Cet "Advanced Research Workshop" n'aurait pas pu avoir lieu sans
le support généreux de la Division des Affaires Scientifiques de
l'OTAN. Je salue ici son représentant le Dr. Luis da Cunha et lui
exprime toute notre gratitude. Le Ministère de l'Education Nationale
de Belgique nous apporte également son soutien et j'en remercie
vivement Monsieur le Chef de Cabinet.

Quand en décembre dernier nous avons envisagé d'organiser ce
"Workshop" nous n'avons pas pensé attirer, dans un délai aussi
bref, une audience aussi nombreuse et aussi concernée. Lorsque le
nombre de participants nous est apparu dépasser nos prévisions,
nous avons été heureux de pouvoir bénéficier de l'hospitalité de
cette salle des séances de l'Institution soeur qu'est pour nous
l'Institut Royal Météorologique de Belgique. J'en remercie cordia-
lement Monsieur Quinet, chef de Département et ses collaborateurs.

La nuit du 8 novembre 1983 a été une dure épreuve pour nos compa-
triotes de Liège et notre première pensée, en ouvrant ce Workshop,
sera certainement pour ceux qui ont souffert et souffrent encore
de la catastrophe.

L'événement séismique est en effet brutal, catastrophique, lorsque
l'épicentre se situe au coeur d'une grande ville. Il nous faut,
désormais, nous rendre à l'évidence : la Belgique est un pays à
séismicité modérée mais nullement négligeable. Les experts le
savaient mais ne pouvaient, ou peut-être, ne désiraient pas en
apporter de preuves convaincantes tant est délicate cette matière
et aussi parce que l'alarmisme n'est jamais opportun.

La mémoire de nos concitoyens est fréquemment et à juste titre
braquée sur le séisme des Flandres, le 11 juin 1938 dont la magnitude 5.9 a été bien supérieure à celle du récent séisme de Liège
(4.9) mais dont le foyer était nettement plus profond (20 Km au
lieu de 4) et l'épicentre en région relativement peu habitée.

Or il nous faut rappeler que, depuis 1900, quelque 70 séismes ont
été ressentis en Belgique par la population et la mémoire ne fera
pas défaut si nous rappelons quelques dates récentes (cf. figure 1)

9 août 1983	Charleroi	magn. 3.3
4 août 1983	Charleroi	3.2
14 et 15 septembre 1982	La Louvière	3.4 (5 répliques)
20 décembre 1970	La Louvière	3.5
3 novembre 1970	Marchienne	3.9
16 janvier 1970	Fontaine l'Evêque	2.8
27 décembre 1968	Coxyde	2.5
août-septembre 1968	Haine St Pierre	4.1 (6 répliques)

28 mars 1967	Carnières	magn. 3.3 (5 répliques)
11 mars 1966	Gosselies	3.2 (4 répliques)
16 janvier 1966	Chapelle-lez-Herlaimont	4.4 (4 répliques)
21 décembre 1965	Ans-Vottem	4.3

Nous ne citons ici que les répliques importantes.

Le réseau actuel de stations séismologiques belges ne peut faire face correctement aux implications de ces manifestations séismiques. Nous verrons ici comment il convient de le développer.

La surveillance de la séismicité du territoire national incombe à l'Observatoire Royal de Belgique, Institution scientifique de l'Etat dépendant des deux Ministères de l'Education Nationale. Au sein de l'Observatoire, la section de Géodynamique est chargée des recherches dans les domaines de la Gravimétrie, de la Séismologie et des Marées Terrestres et, pour ce qui concerne la Séismologie, cette section peut tirer parti des enregistrements acquis en quatre stations : Uccle (dont l'environnement n'est pas favorable), Dourbes (près de Couvin), qui relève de l'Institut Royal Météorologique, Membach (près de la Gileppe) et Walferdange au Grand Duché de Luxembourg (grâce aux Accords Culturels qui nous lient).

Cette distribution est loin d'être satisfaisante. Les instruments acquis à des époques différentes sont disparates. En outre aucune des trois stations extérieures n'est reliée par câble téléphonique à Uccle et les enregistrements en sont acheminés par la poste ou prélevés lorsqu'on envoie une voiture à la station.

Le personnel qualifié n'est plus suffisant en nombre pour assurer le dépouillement des enregistrements des quatre stations si bien que certains d'entre eux sont simplement classés en attendant des jours meilleurs.

Ceci a abondamment été décrit par la presse le 8 novembre dernier et je ne reviendrai pas sur ces points.

Or, à juste titre inquiet de cette situation, j'ai présenté en mars 1983 un plan de restructuration prévoyant la création à l'Observatoire Royal d'un petit Centre de Géophysique Interne destiné à assumer notamment le travail d'analyse séismologique.

Mon inquiétude a été partagée par nos autorités de tutelle et, dès avant le séisme de Liège, les dispositions étaient prises pour assurer le fonctionnement de ce Centre en octroyant des contrats à trois jeunes géophysiciens formés et compétents. Le Centre est entré en activité le 1 janvier 1984. Il a déjà à son actif des travaux et analyses importants dont il sera fait rapport pendant le Workshop.

C'est dans ce contexte que je dois aborder ici la question qui m'a

été sans cesse posée depuis cinq mois : peut-on, pourra-t-on pré-
voir les tremblements de terre ?

La réponse est bien hasardeuse. Il est vrai que des efforts consi-
dérables sont actuellement consentis dans beaucoup de pays du
monde, et en particulier chez nos voisins européens les plus proches.

Il y a vingt ans on aurait sans doute répondu à cette question par
la négative.

Aujourd'hui quelques succès indéniables de prévision - dont le
plus spectaculaire en Chine - (Haicheng, 4 février 1975) ont été
suivis d'échecs hélas aussi spectaculaires - en Chine aussi et
dans la même région (Tangshan, 27 juillet 1976, 650.000 morts).

Comment relever le défi qui nous est ainsi lancé, sinon en vous
montrant maintenant ce qui s'est passé pendant les trois semaines
qui ont précédé le séisme de Liège ...

Car il y a eu des précurseurs indéniables sous la forme de quatre
petites secousses prémonitoires qui ont été parfaitement enregistrées à la station de Membach. Mais si les enregistrements y ont
été poursuivis grâce au dévouement du personnel du service des
barrages à Verviers, ils n'avaient pas pu être dépouillés pour les
raisons que je viens de dire et que l'on ne connait que trop bien
lorsque l'on parle de la grande misère des Institutions Scientifiques de l'Etat. Depuis la création du Centre de Géophysique
interne, ce dépouillement a repris et nous avons aussitôt décou-
vert que quatre séismes "alertes" s'étaient effectivement produits
à Liège : deux le 20 octobre, un le 24 et un le 25 octobre. Leur
magnitude ne dépasse pas 2.4 (figure 2) et ils ne pouvaient être
ressentis par une population qui n'était pas motivée comme elle
l'est aujourd'hui.

Une station séismologique comme celle de Membach qui, au point de
vue de la qualité du site, est la meilleure des stations belges,
capte quelque 3000 séismes par an.

Il faut donc un personnel très compétent, plein d'expérience, pour
dépouiller les kilomètres de courbes enregistrées, mesurer les
temps d'arrivée des signaux, analyser le contenu de chaque signal
séismique, discriminer les tremblements de terre lointains, proches, très proches, les explosions nucléaires souterraines, les
tirs de carrières, ... (cf. fig. 6, 7).

On peut alors logiquement se demander ce que nous aurions pu faire
si les enregistrements de Membach avaient été dépouillés au jour
le jour. Objectivement peu de choses car une seule station ne permet pas de localiser exactement l'épicentre. Si nous avions, par

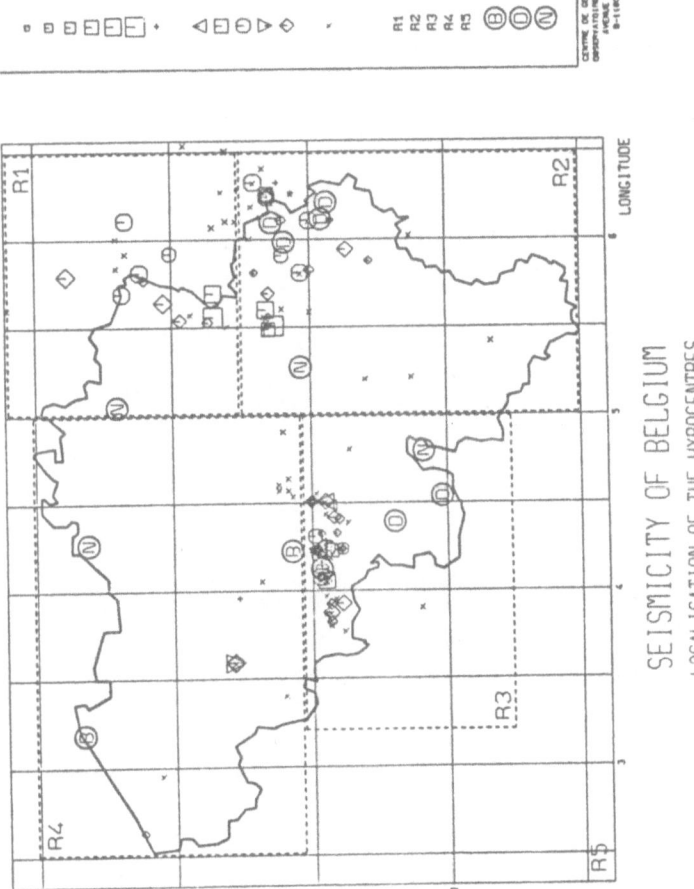

Figure 1.

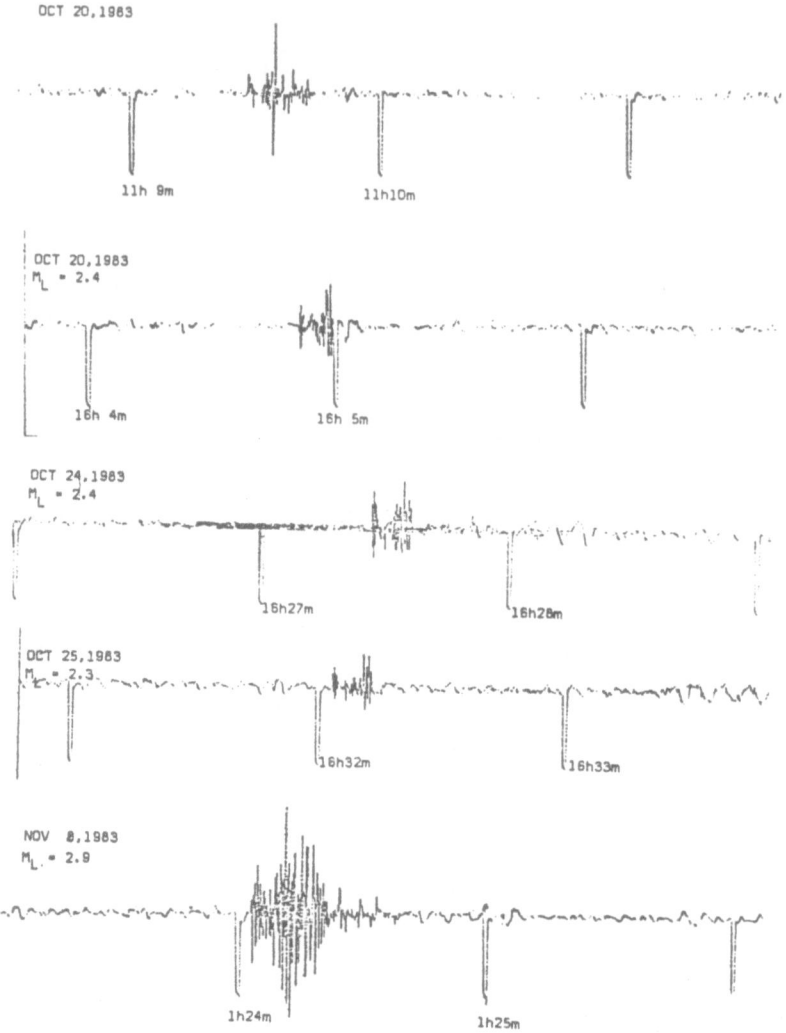

STATION:MEMBACH

SEISMOMETER S-10 (JOHNSON-MATHESON) VERT.

OCT 20,1983

11h 9m   11h10m

OCT 20,1983
M_L = 2.4

16h 4m   16h 5m

OCT 24,1983
M_L = 2.4

16h27m   16h28m

OCT 25,1983
M_L = 2.3

16h32m   16h33m

NOV 8,1983
M_L = 2.9

1h24m   1h25m

Figure 2. Quatre petites secousses prémonitoires les 20, 24 et
25 octobre 1983 comparées à la réplique du 8 novembre.

Réplique du Tremblement de Terre de Liège, 7 mars 1984, 17h 0m 23s5, m = 2.6

FIGURE 3.

contre, disposé d'un petit réseau de stations dans cette même région nous aurions pu situer les quatre épicentres successifs et
compte tenu des résultats et de ce qu'il s'agissait d'une séquence
exceptionnelle d'événements, avertir. Mais avertir de quoi, sans
risquer de provoquer un affolement d'autant plus inconsidéré que
rien ne nous aurait garanti qu'il ne s'agissait pas d'une fausse
alerte ? La séismologie ne pouvant prédire actuellement le lieu
et l'instant précis d'un séisme, il ne peut être question d'arrêter
ou de perturber les activités économiques et humaines pendant des
jours ou des semaines. Et d'ailleurs qui avertir ? Nous l'ignorons.

Nous tirerons la leçon de cette expérience.

Quoiqu'il en soit, dans un domaine aussi complexe, seules des
observations diversifiées, des mesures nombreuses et répétées peuvent nous guider vers la construction d'un modèle mathématique sur
lequel nous pourrons opérer des simulations de situations et esti-
mer correctement les risques.

Les observations à faire se rapportent à des phénomènes très différents, où l'on peut déceler des précurseurs d'un séisme à condition de poursuivre ces observations avec tenacité pendant de
très longues années. Les précurseurs que l'on soupçonne sont nombreux et ceci ne facilite pas notre tâche. D'autre part les expé-
riences antérieures montrent qu'ils ne se manifestent pas toujours
clairement, que certains sont parfois totalement absents et qu'aucun des types de précurseurs n'apparait plus décisif qu'un autre.

Il y a donc un problème d'évaluation, d'analyse critique, de la
signification des symptômes relevés.

Brièvement, les mesures qui devraient être faites relèvent des
diverses disciplines de la géophysique. Elles sont :

de caractère géodésique par contrôles *répétés*, à intervalles fixes,
essentiellement nivellement et gravimétrie, mais aussi par trila-
tération à l'aide des excellents systèmes de mesure électronique
de distance dont on dispose aujourd'hui;

de caractère géodynamique par contrôle *continu* des déformations à
l'aide d'extensomètres, clinomètres et gravimètres et par étude
du flux de chaleur interne;

de caractère hydrologique et géochimique par analyse des eaux
souterraines :
(a) niveau des eaux dans les puits : grâce à une coopération avec
le Service Géologique de Belgique, l'Observatoire Royal dispose
depuis un mois d'un puits profond où 3 nappes superposées peuvent
être contrôlées en continu avec un dispositif électronique nouveau
conçu chez nous.

(b) concentration en Radon dont l'importance en tant qu'indicateur a été mise en évidence en Chine dès 1968, en URSS dès 1971 ainsi qu'en Californie, avec des résultats très encourageants. Un programme important est développé présentement à l'Université de Mons et il en sera question au cours de ce Workshop;

de caractère électromagnétique par la surveillance de la résistivité électrique du sol et des composantes du champ magnétique local qui peuvent être altérée à la suite d'une dilatance des terrains engendrant des effets piézoélectriques ou piézomagnétiques ou par une variation du degré de saturation du sol;

et, évidemment, de caractère séismologique : variation du taux d'activité séismique, petits séismes avertisseurs comme ceux relevés - trop tard - à Liège, migration lente et systématique des foyers, variation du rapport de la vitesse des ondes transversales à la vitesse des ondes longitudinales, variation des paramètres de la marée terrestre locale (observée par plusieurs chercheurs au Canada et en Norvège).

Il semble bien que tous ces précurseurs sont liés à un même phénomène inhérent à la mécanique des roches lorsqu'elles tendent vers la rupture et cette constatation doit permettre de construire le modèle mathématique dont nous avons besoin pour progresser.

En faisant une analyse *a posteriori* des précurseurs enregistrés nous pourrons aussi évaluer les chances réelles d'une prédiction.

Un travail essentiel qui ne peut s'effectuer qu'*a posteriori* est l'étude du *"mécanisme au foyer"* c'est-à-dire la description aussi précise que possible de la distribution des compressions et dilatations subies. Cette étude se base sur les enregistrements obtenus dans le plus grand nombre possible d'observatoires séismologiques. Elle permet de comprendre ce qui s'est passé et contribuera de manière fondamentale à construire ce modèle mathématique nécessaire.

Pour l'analyse du séisme de Liège, nous avons reçu les enregistrements de pas moins de 151 observatoires séismologiques européens situés en Bulgarie, Espagne, Finlande, France, Irlande, Norvège, Pays-Bas, Pologne, République Fédérale d'Allemagne, République Démocratique Allemande, Royaume Uni de Grande Bretagne, Suisse et Tchécoslovaquie.

A l'étude du mécanisme au foyer on associera évidemment la détermination de la magnitude dans l'échelle de Richter et corrélativement l'énergie libérée par le séisme.

L'étude a posteriori des séismes encourus constitue donc une information fondamentale pour l'avenir et, s'il faut espérer que

ces séismes soient aussi rares que possible, il n'en reste pas
moins qu'ils constituent la source essentielle de notre information.

C'est la raison pour laquelle il faut absolument les étudier de
la manière la plus approfondie possible. Il nous faut pour cela
de bons instruments et surtout le personnel hautement qualifié en
nombre suffisant. C'est à cette tâche que le Centre de Géophysique
Interne de l'Observatoire Royal se consacrera dans les mois qui
viennent.

Mais l'étude a posteriori peut remonter bien au delà de l'époque
à laquelle ont été installés les premiers séismographes.

Cette étude se base alors sur les textes les plus anciens, documents, lettres, archives et journaux et permet d'enrichir consi-
dérablement la connaissance de la géographie et de l'histoire
séismique c'est-à-dire de la répartition dans l'espace et dans le
temps de l'activité séismique d'une région (cf. figure 1).

Ceci ne peut cependant pas être ramené à une simple tâche de compilation.

Une sévère critique historique des documents utilisés est nécessaire. Elle doit encore être faite en ce qui concerne la Belgique
et, comme il n'est pas probable qu'un historien de formation puisse
être intégré à la Section de Géodynamique de l'Observatoire Royal,
il nous faudra sensibiliser nos Instituts Universitaires d'Histoire
à ce problème et susciter ici une recherche interdisciplinaire.

Une fois le catalogue des séismes historiques établi et contrôlé
sur ces bases, les méthodes statistiques les plus fiables et les
plus significatives pourront lui être appliquées.

Quoiqu'il en soit, donner une alerte préséismique reste actuellement une démarche extrêmement hasardeuse et ne peut être confiée
qu'à un organisme prudent et compétent.

Pour reprendre une remarque faite en 1982, à l'Académie des
Sciences à Paris par le Professeur Jacques Blamont : "L'information
doit provenir d'une source dont le public soit certain qu'elle
n'est pas polluée par des intérêts politiques, économiques, finan-
ciers ou militaires".

L'Observatoire Royal, Institution scientifique de l'Etat répond
certainement à ces critères et le public pourra lui faire confiance.

De fait, nous avons récemment vécu les retombées d'une information
répandue par un irresponsable annonçant, le 12 février dernier, un
tremblement de terre à Liège pour la nuit du week-end suivant :
ceci nous a contraints à diffuser à plusieurs reprises un démenti
formel.

Au début de cette année, nous avons donc présenté à nos autorités de tutelle un plan de redéveloppement du réseau de surveillance séismologique de la Belgique (fig. 4 et 5) qui s'inspire directement de toutes ces constatations :

1) un réseau fondamental de 5 stations Tournai - Gent - Membach - Dourbes et Walferdange au Grand Duché avec transmission des mesures en temps réel au Centre de Géophysique Interne de l'Observatoire Royal où un petit ordinateur assurera la gestion du réseau.

Nous estimons que ce réseau doit être complété en *imposant* l'installation d'une station séismologique simple mais efficace en chaque site d'implantation d'ouvrages d'art à haut ou moyen risque. Ces stations doivent être reliées à l'Observatoire Royal.

C'est ce que le Ministère des Travaux Publics avait parfaitement compris lors des travaux de surhaussement du barrage de la Gileppe et c'est ce qui est actuellement en discussion avec le même Ministère à propos d'autres travaux.

D'autres sites doivent être surveillés de la même manière (cf. fig. 1).

2) un ensemble de six stations mobiles à déployer dans la province de Liège et un ensemble similaire à déployer dans la province de Hainaut afin d'y surveiller l'activité séismique de niveau imperceptible à la population. Il ne faut pas cependant dissimuler qu'une telle surveillance est difficile dans des régions fortement peuplées et où l'activité industrielle est importante (cf. fig. 6,7).

La mobilité de ces stations permettra de les redéployer dans les régions où se produira ultérieurement un séisme, éventualité que nous ne pouvons malheureusement pas écarter dans un avenir plus ou moins proche.

Il nous faut rappeler à cet égard que nous ne disposons actuellement d'aucun équipement de ce genre. Lors du séisme de Liège nous sommes redevables à nos collègues britanniques (Cambridge University et Imperial College of Technology London) et français (Institut de Protection et de Sûreté nucléaire) d'avoir immédiatement déployé leurs propres équipements dans la région liégeoise. Ils nous feront part de leurs constatations au cours de ce Workshop.

3) une station à hautes performances à Membach, site de qualité exceptionnelle. Un équipement de haute sensibilité permettra de résoudre trois problèmes pour l'avenir :
    - détermination des paramètres dynamiques des séismes importants en Belgique (Liège - Hainaut - Massif du Brabant - Luxembourg - Hautes Fagnes)
    - étude de la sismicité dans la région liégeoise avec couverture de la zone des barrages et de la Centrale de Tihange.

- détection de *toutes* les explosions nucléaires souterraines, travail que l'Observatoire Royal effectue déjà à l'intention du Ministère des Relations Extérieures (Comité du Désarmement). On doit en effet noter qu'avec nos équipements actuels les explosions les moins puissantes échappent à nos capacités de détection.

En terminant je dois relever la contribution efficace que nous apportent, dans l'évaluation des effets macroséismiques, la population, les administrations communales, la presse écrite et radio-télévisée - il m'est agréable de rappeler les excellents contacts que nous avons eus avec elles au lendemain du séisme de Liège.

La section de Géodynamique diffuse, après chaque séisme, auprès des administrations communales, un questionnaire circonstancié et détaillé. Les réponses reçues permettent le tracé d'une carte macroséismique qui est ensuite confrontée aux enregistrements instrumentaux.

Nous ne pouvons trop insister sur l'importance de cette démarche et nous tenons à remercier tous ceux qui ont concouru à établir les réponses.

La presse, pour sa part, remplit ici un rôle d'éducation du public qu'on ne peut sousestimer. La radio, de son côté, peut atteindre le plus rapidement les populations touchées, leur diffuser des avertissements et surtout concourir à calmer leurs appréhensions.

L'Observatoire Royal se propose de travailler en étroite coopération avec le groupe de l'Université de Mons dont le programme d'analyse du contenu en Radon doit être associé à un programme de détection de microsecousses.

Nous sommes, par vocation, ouverts à d'autres coopérations avec les Universités belges aussi bien qu'avec nos collègues des pays voisins qui nous ont apporté une aide très appréciée lors du tremblement de terre du 8 novembre.

Sur le plan européen, la Belgique doit affirmer sa solidarité envers les pays généralement plus frappés par les séismes en s'associant au travail des deux centres d'analyse situés en Europe : le Centre Sismologique Euro-Méditerranéen à Strasbourg en France et le Centre Séismologique International à Newsbury, Royaume Uni de Grande-Bretagne.

TABLE 1
Les tremblements de terre récents à Liège

		M_L	DOU	MEM	
DEC 21, 1965	10h 0m	4.3			ressenti
OCT 20, 1983	11h 9m				
	16h 4m	2.3		2.3	
OCT 24, 1983	16h 27m	2.3		2.3	
OCT 25, 1983	16h 32m	2.2		2.2	
NOV 8, 1983	0h 49m	4.9			ressenti
	0h 55m	2.7	2.7		ressenti
	1h 25m	2.8	2.8	2.8	ressenti
	2h 13m	3.5			ressenti
NOV 10, 1983	12h 52m	pas certain			
NOV 16, 1983	13h 28m	pas certain			
NOV 20, 1983	8h 3m	2.2	2.1	2.3	ressenti
DEC 1, 1983	21h 53m	2.7	2.7	2.8	ressenti
DEC 4, 1983	17h 41m	enregistré par CEA (France) à Cointe			
DEC 15, 1983	15h 17m	2.1	2.0	2.3	
JAN 5, 1984	12h 38m	2.6	2.6	2.7	
JAN 15, 1984	13h 22m	2.2	2.2	2.2	ressenti
MAR 7, 1984	16h 0m	2.6	2.6	2.7	ressenti

TABLE 2
Nombre de stations étrangères
ayant fourni les paramètres d'enregistrement du séisme de Liège

BULGARIE	3
ESPAGNE	4
FINLANDE	3
FRANCE	26
IRLANDE	8
NORVEGE	3
PAYS-BAS	4
POLOGNE	2
REPUBLIQUE DEM. ALLEMANDE	1
REPUBLIQUE FED. ALLEMANDE	14
ROYAUME UNI	78
SUISSE	7
TCHECOSLOVAQUIE	2
TOTAL :	151

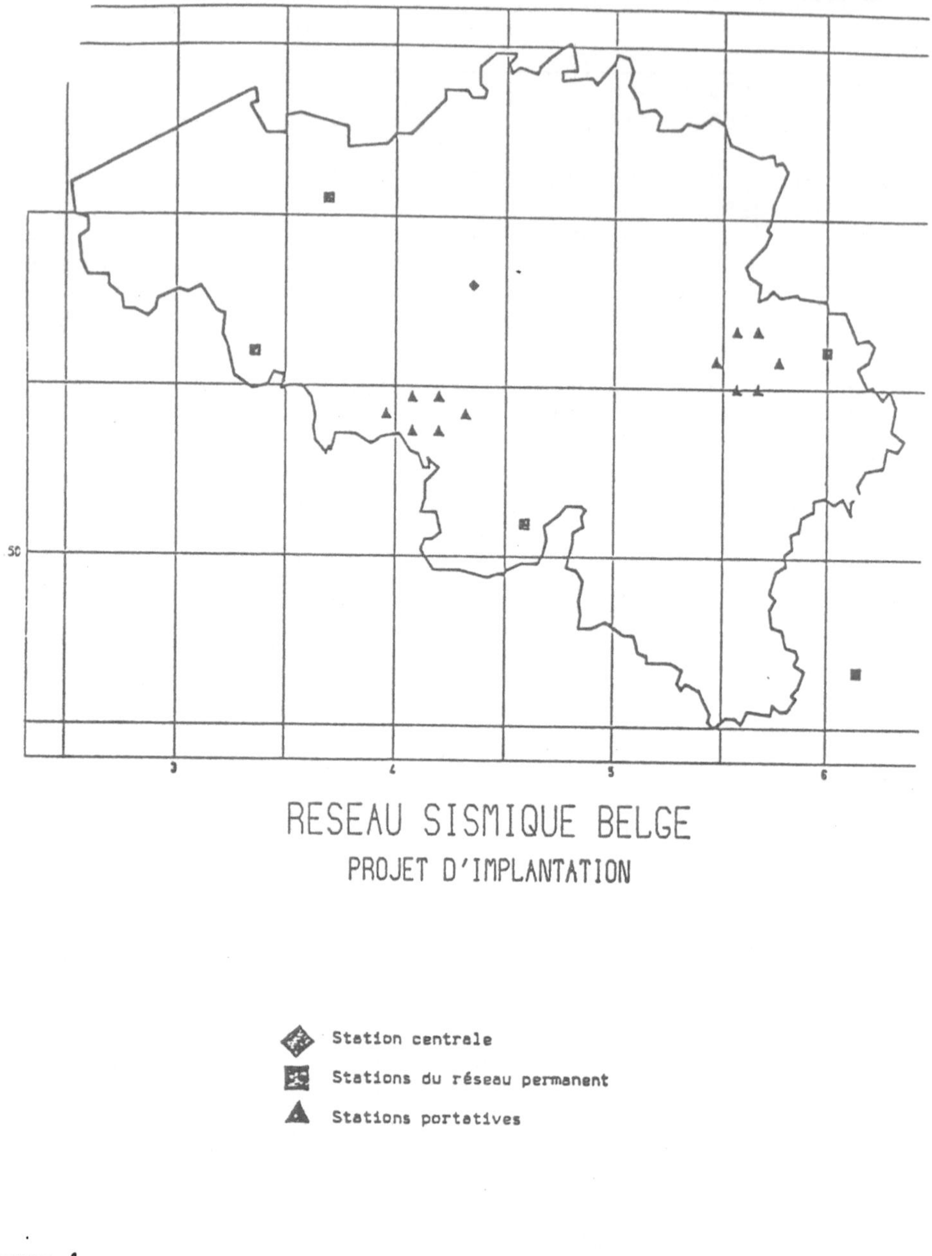

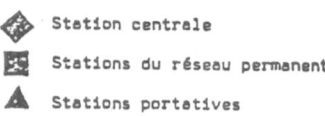

RESEAU SISMIQUE BELGE
PROJET D'IMPLANTATION

◆ Station centrale
▨ Stations du réseau permanent
▲ Stations portatives

Figure 4.

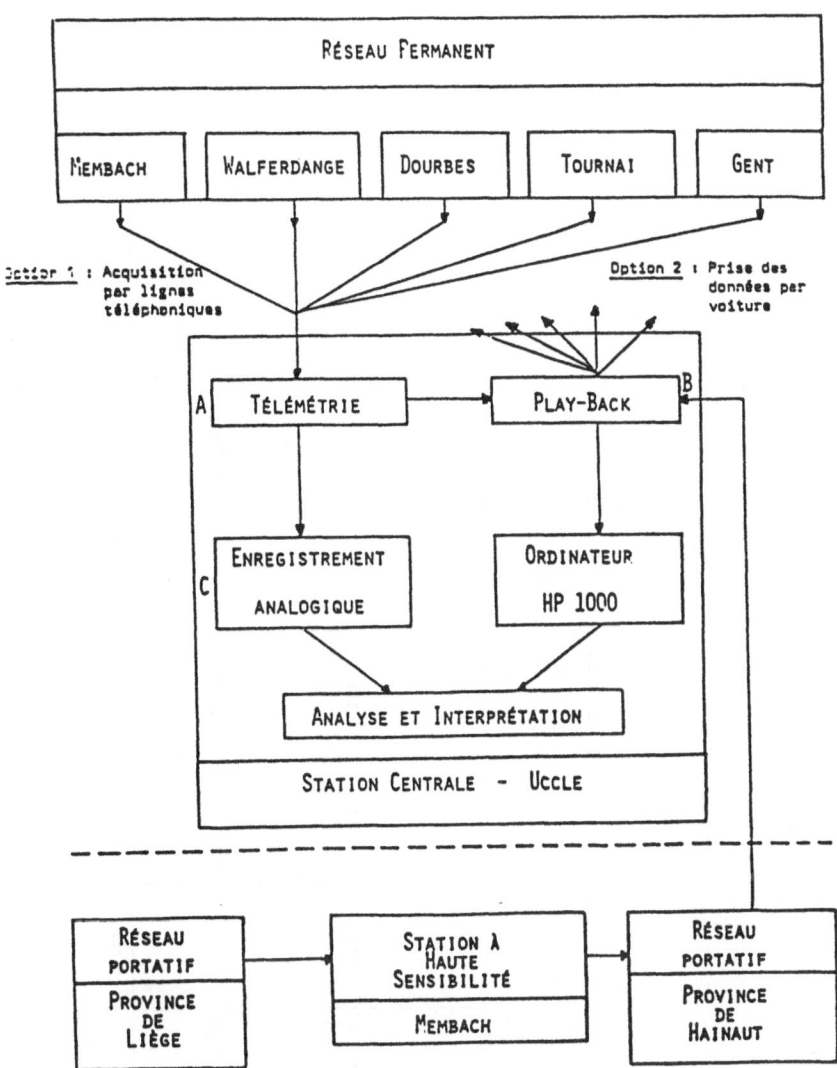

Figure 5.

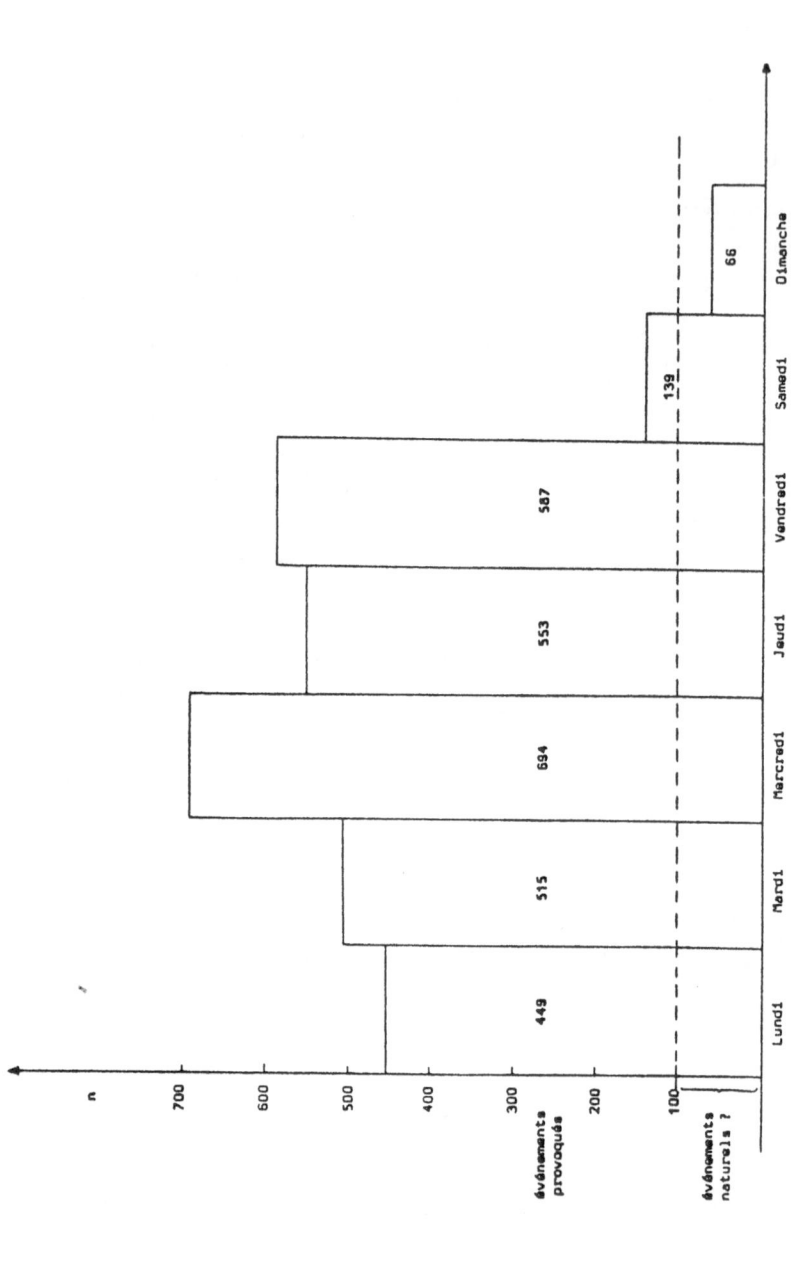

Figure 6. *Station de Membach* : Nombre n d'événements artificiels et de séismes naturels locaux répartis en fonction du jour de la semaine : 3020 événements enregistrés de juin 1977 à janvier 1982. *700 événements (100 par jour) peuvent être considérés comme naturels pendant les 1706 jours concernés soit environ 3 par semaine.*

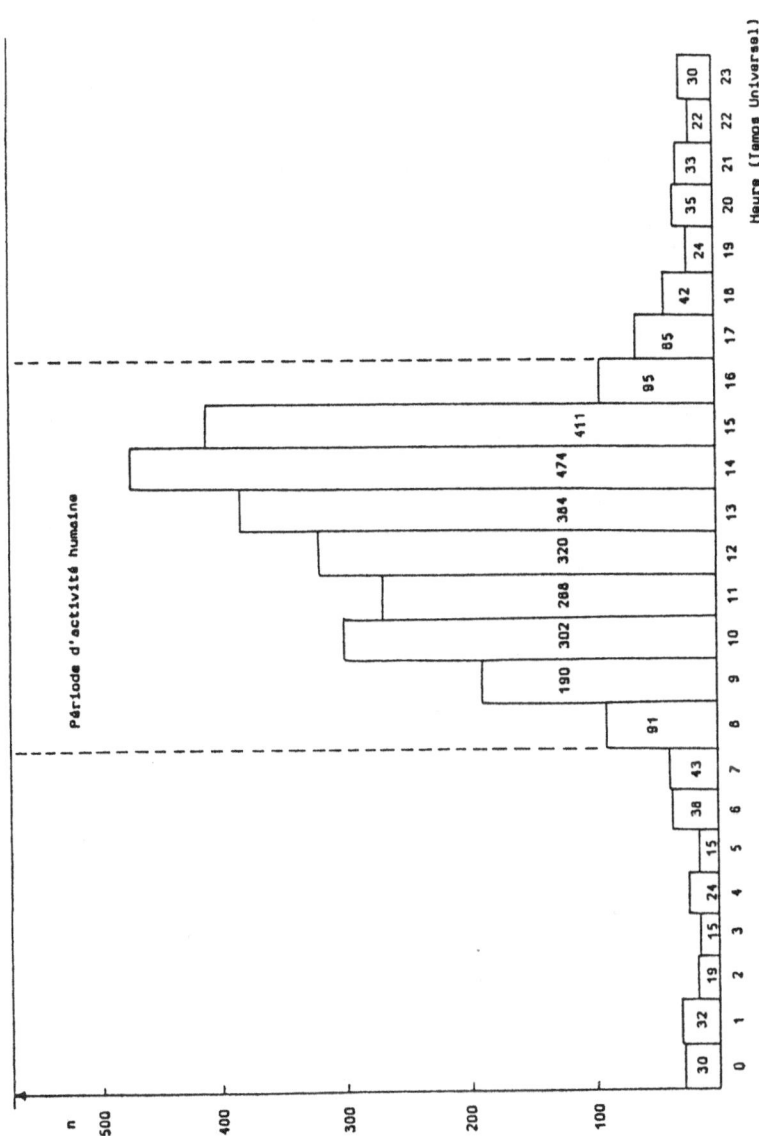

Figure 7. *Station de Membach* : Nombre n d'événements artificiels et de séismes naturels locaux répartis en fonction de l'heure de la journée : 3020 événements enregistrés de juin 1977 à janvier 1982. *700 événements (30 par heure) peuvent être considérés comme naturels pendant les 1706 jours concernés soit environ 3 par semaine.*

SATELLITE-DETERMINED STRESSES IN THE CRUST OF EUROPE

Han-Shou Liu

Geodynamics Branch, Laboratory for Earth Sciences
NASA/Goddard Space Flight Center
Greenbelt, MD 20771, U.S.A.

ABSTRACT

General tectonic frame and seismotectonic dislocations indicate
the kinematics and neotectonic movements in Western Europe. In order to understand the dynamics of the generating forces, a crustal
stress field computed from satellite-derived gravity data is presented. The stresses in the crust of Europe are obtained in terms
of a series involving spherical harmonics of the geopotential. The
computed stress directions are in good agreement with fault-plane
solutions from earthquakes and in situ stress measurements. With
the articulation of the dynamical principles that govern crustal
deformations of a much smaller scale, this study provides a geodynamical basis for intraplate deformations and seismotectonic block
movements in Central Europe.

INTRODUCTION

Sixty-one years ago, Lord Kelvin developed the theory of
spherical shells and obtained general solutions for the equilibrium
and stresses in an elastic shell (1). Based on this theory,
Bijlaard (2) and Liu (3) discussed the stresses in the Earth's
crust due to shift of the axis of rotation and Love (4) obtained
special solutions for tidal deformation on the surface of the
Earth due to the Moon's gravity (Figure 1). It is highly desirable
to extend this theory for investigating the deformation of the
crustal shell of the Earth.

The fundamental problem of the deformation of the crust involves stress conditions on the base of the crust caused by mantle

P. Melchior (ed.), Seismic Activity in Western Europe, 19–36.
© *1985 by D. Reidel Publishing Company.*

convection. Jeffreys (5), Runcorn (6) and Hide and Horai (7) have made harmonic analyses of the stresses in the interior of the Earth. Recently, Liu (8 - 18) has demonstrated that a certain band of the high degree harmonics of the geopotential is related to the subcrustal stresses under Africa, Asia, Australia and North and South America. Therefore it is possible to calculate the stresses in the crust by applying the theory of shells with subcrustal stresses as boundary conditions. The purpose of this paper is to present the calculated stresses in the crust of Europe as inferred from satellite gravity harmonics. The resultant stress patterns will be compared with the observed seismicity and stress data in Central Europe.

PRINCIPAL STRESSES AND MAXIMUM SHEAR STRESS

Stresses acting on a shell element are shown in Figure 2. The boundary conditions on the base of the crust are (see figure 1) :

$$F_E(\zeta,\lambda) = \sum_{n=13}^{\infty} \sum_{m=0}^{m=n} \frac{Mg}{4\pi a^2} \left(\frac{a_e}{a}\right)^{n+1} \cdot \frac{2n+1}{n+1} \cdot \frac{1}{\sin\zeta} \cdot \frac{\partial}{\partial\lambda} \phi_n^m (\zeta,\lambda) \quad (1)$$

$$F_N(\zeta,\lambda) = \sum_{n=13}^{\infty} \sum_{m=0}^{m=n} \frac{Mg}{4\pi a^2} \left(\frac{a_e}{a}\right)^{n+1} \cdot \frac{2n+1}{n+1} \cdot \frac{\partial}{\partial\zeta} \phi_n^m (\zeta,\lambda) \quad (2)$$

In equations (1) and (2) M is the mass of the Earth, g the acceleration due to gravity, a_e the radius of the Earth, a the radius of the outer spherical surface of the flowing region, λ the longitude, ζ the co-latitude, and the surface spherical harmonic

$$\phi_n^m (\zeta,\lambda) = P_n^m (\cos \zeta) \left[\overline{C}_{n,m} \cos(m\lambda) + \overline{S}_{n,m}(m\lambda)\sin(m\lambda) \right] \quad (3)$$

where $\overline{C}_{n,m}$ and $\overline{S}_{n,m}$ are the harmonic coefficients of the geopotential which can be determined by satellite measurements and P_n^m (Cos ζ) is the associated Legendre polynomial of degree n, order m and argument Cos ζ.

The crust may be considered a homogeneous elastic spherical shell. Solutions for the equilibrium of and stresses in the spherical shell of the crust subject to purely convection-generated traction on the boundary can be obtained in terms of series involving spherical harmonics. Kelvin and Love (4) have obtained general and special types of solutions leading to results with regard to the equilibrium of a spherical shell, and forming an introduction to the application of the theory of elasticity to geophysics. We now proceed to apply Kelvin's general solutions and Love's

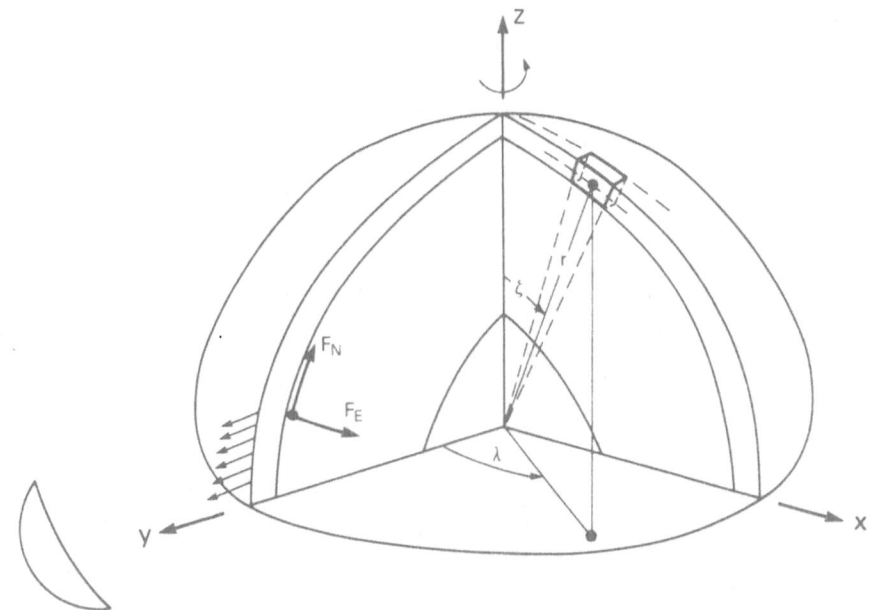

Figure 1. Coordinate system and boundary conditions of a spherical shell.

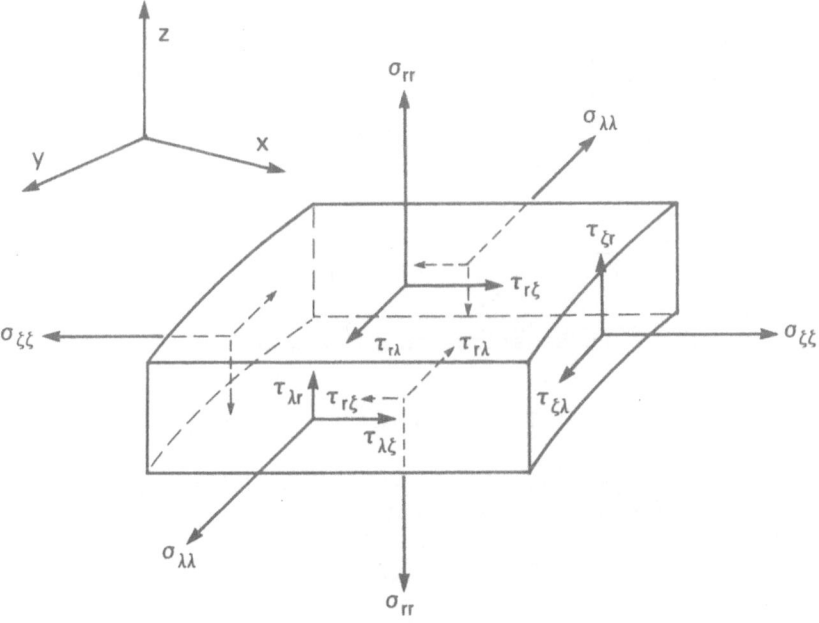

Figure 2. Stresses acting on a shell element.

special solutions to the problem of the spherical shell and to
calculate the orientations of the principal stresses and maximum
shear stress in the crust under Central Europe.

The equations of equilibrium for a homogeneous elastic medium
subject to forces applied on the boundary are

$$(\gamma + \mu) \, (\partial\Delta/\partial x_i) + \mu \nabla^2 u_i = 0 \tag{4}$$

or

$$(\gamma + \mu) \left(\frac{\partial}{\partial x}, \frac{\partial}{\partial y}, \frac{\partial}{\partial z} \right) \Delta + \mu\nabla^2 \, (u, \, v, \, w) = 0 \tag{5}$$

where Δ is the divergence of the displacement $(u, \, v, \, w)$, and γ and
μ are Lamé's constants. Equation (5) can be solved by putting

$$u_i = (\partial\phi/\partial x_i) + v_i$$
$$\Delta' = \Sigma(\partial v_i/\partial x_i) \tag{6}$$

with

$$(\gamma+2\mu) \, \nabla^2 \phi = 0 \tag{7}$$

and

$$(\gamma+\mu) \, (\partial\Delta'/\partial x_i) + \mu \, \nabla^2 v_i = 0 \tag{8}$$

Appropriate solutions of equation (8) can be built up from
the types (4)

$$(u, \, v, \, w) = r^2 \left(\frac{\partial}{\partial x}, \frac{\partial}{\partial y}, \frac{\partial}{\partial z} \right) \omega_n + a_n \, (x, \, y, \, z) \, \omega_n \tag{9}$$

$$(u, \, v, \, w) = \left(\frac{\partial}{\partial x}, \frac{\partial}{\partial y}, \frac{\partial}{\partial z} \right) \phi_n \tag{10}$$

$$(u, \, v, \, w) = \left(y \frac{\partial}{\partial z} - z \frac{\partial}{\partial y}, \, z \frac{\partial}{\partial x} - x \frac{\partial}{\partial z}, \, x \frac{\partial}{\partial y} - y \frac{\partial}{\partial x} \right) \psi_n \tag{11}$$

where ω_n, ϕ_n and ψ_n are spherical solid harmonics of degree n, and

$$a_n = - \, 2 \, [n\gamma + (3n + 1)\mu \,] / [\, (n + 3) \, \gamma + (n + 5) \, \mu] \tag{12}$$

The component tractions X_r, Y_r and Z_r across any spherical
surface $r = $ constant may be expressed as

$$rX_r/\mu = (\gamma/\mu) \, x\Delta + (\partial\xi/\partial x) + r(\partial u/\partial r) - u \tag{13}$$

$$rY_r/\mu = (\gamma/\mu) \, y\Delta + (\partial\xi/\partial y) + r(\partial v/\partial r) - v \tag{14}$$

$$rZ_r/\mu = (\gamma/\mu) \, z\Delta + (\partial\xi/\partial z) + r(\partial w/\partial r) - w \tag{15}$$

where $\xi = xu + yv + zw$. The boundary conditions are

$$
\begin{aligned}
X_{r=a} &= \sigma_{rr} = 0 & X_{r=a_e} &= \sigma_{rr} = 0 \\
Y_{r=a} &= F_N & Y_{r=a_e} &= 0 \\
Z_{r=a} &= F_E & Z_{r=a_e} &= 0
\end{aligned}
\tag{16}
$$

Under the assumption $\gamma=\mu$, which is nearly true for the material of the crust, the six stress elements in the elastic shell of the lithosphere in the polar coordinate system are determined by :

$$
\sigma_{rr} = \sum_{n=13}^{\infty} \sum_{m=0}^{m=n} \frac{\mu}{r} (K_1 + 2K_2) \phi_n^m (\zeta,\lambda)
\tag{17}
$$

$$
\sigma_{\zeta\zeta} = \sum_{n=13}^{\infty} \sum_{m=0}^{m=n} \frac{\mu}{r} \left[(K_1 + 2K_3) \phi_n^m (\zeta,\lambda) + 2K_4 \frac{\partial^2}{\partial\zeta^2} \phi_n^m (\zeta,\lambda) \right]
\tag{18}
$$

$$
\sigma_{\lambda\lambda} = \sum_{n=13}^{\infty} \sum_{m=0}^{m=n} \frac{\mu}{r} \left[(K_1 + 2K_3)\phi_m^m (\zeta,\lambda) + 2K_4 \cot \zeta \frac{\partial}{\partial\zeta} \phi_n^m(\zeta,\lambda) \right.
$$

$$
\left. + 2K_4 \frac{1}{\sin^2\zeta} \cdot \frac{\partial^2}{\partial\lambda^2} \phi_n^m (\zeta,\lambda) \right]
\tag{19}
$$

$$
\tau_{r\zeta} = \sum_{n=13}^{\infty} \sum_{m=0}^{m=n} \frac{\mu}{r} (K_3 + K_4 - K_5) \frac{\partial}{\partial\zeta} \phi_n^m(\zeta,\lambda)
\tag{20}
$$

$$
\tau_{\zeta\lambda} = \sum_{n=13}^{\infty} \sum_{m=0}^{m=n} \frac{\mu}{r \sin \zeta} K_5 \left[\cot \zeta \frac{\partial}{\partial\lambda} \phi_n^m(\zeta,\lambda) + 2\frac{\partial^2}{\partial\lambda^2} \phi_n^m(\zeta,\lambda) \right]
\tag{21}
$$

$$
\tau_{\lambda r} = \sum_{n=13}^{\infty} \sum_{m=0}^{m=n} \frac{\mu}{r \sin \zeta} (K_3 + K_4 - K_5) \frac{\partial}{\partial\lambda} \phi_n^m (\zeta,\lambda)
\tag{22}
$$

in which K_1, K_2, K_3, K_4 and K_5 are constants to be determined by boundary conditions (18). Subject to the boundary conditions in equation (1) and (2), the normal and shear stresses at any point in the elastic shell of the lithosphere are given by equations

(17) to (22). The input gravity data in the derivation of the
stress elements are made up of 840,000 optical, electronic and
laser observations on 27 satellites obtained by Lerch, et al.(19).
By applying the theory of elasticity, the principal stresses and
the maximum shear stress in the crust were determined. Determination of the preferred orientations of these stresses in a test
area covered by $15° \times 20°$ was carried out by a statistical procedure using 300 grid points. For $a_e = 6371$ km, $a = 6341$ km and
$\mu = 6 \times 10^{11}$ dyn cm^{-2}, the calculated directions of the principal
compressional stress σ_1, the principal tensional stress σ_2, and
the maximum shear stress τ_{max} in Central Europe are shown in
Figure 3. Figure 3 shows that the crust of Central Europe is under
NW-SE compression and NE-SW extension. The maximum shear τ_{max}
inclines to the E-W and N-S directions at about 2 degrees.

Calculation of the stresses in the spherical shell of the crust
is a typical mechanics problem. The formulations derived here are
important because they contribute directly to a mathematical understanding of the geophysical problem. In particular, since the
calculation contains virtually no assumptions other than classical
theories of shells plus the satellite gravity data, the stress
solutions must be considered fundamental.

3. PRESENT-DAY STRESS FIELDS

The present-day stress fields in Central Europe can be obtained
from fault-plane solutions of earthquakes, orientation of geological joints, and in situ stress measurements. Fault-plane solutions
of earthquakes in the Rhine Graben system and its surrounding
regions have been obtained by Ahorner (20). Preferred joint in
Europe has been discussed by Scheidegger (21). Several groups of
investigators (22-26) have performed in situ stress measurements
in Europe. The results of these studies are displayed in Figure 4.
Figure 4 indicates that the crust of Central Europe is under NW-
SE compression. The nearly perfect agreement in the directions of
P-axes from 30 fault-plane solutions and σ-axes from 31 in situ
stress measurements (Figure 5) in Central Europe makes this region
a key area for testing the reliability of the satellite-determined
stresses in the crust of Europe. The present study has determined
the orientations of principal stresses in the crust of Central
Europe. This satellite-determined stress field removes some critical objections in (27) from theoretical considerations against
the applicability of the relation between fault-plane solutions
and the directions of principal stresses if earthquakes occur on
pre-existing faults in a heterogeneous (as opposed to homogeneous)
medium. While there is a nearly perfect agreement between the directions of the maximum horizontal compressive stress (28) and the
principal compressional stress (Figure 3), a detailed verification
of seismic events and stress measurements has yet to be achieved.

Another challenge in both theory and observation is to establish

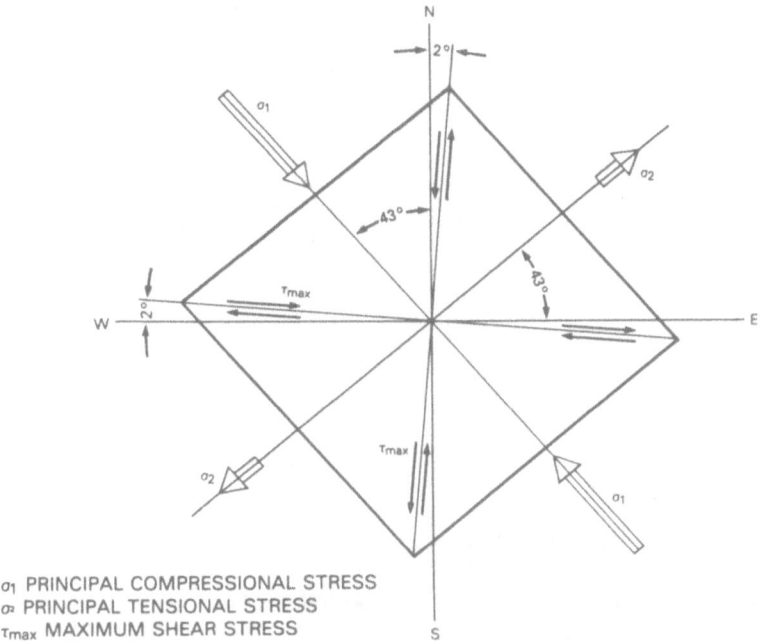

σ₁ PRINCIPAL COMPRESSIONAL STRESS
σ₂ PRINCIPAL TENSIONAL STRESS
τmax MAXIMUM SHEAR STRESS

Figure 3. Orientations of the principal stresses in the crust of Central Europe as inferred from satellite gravity data.

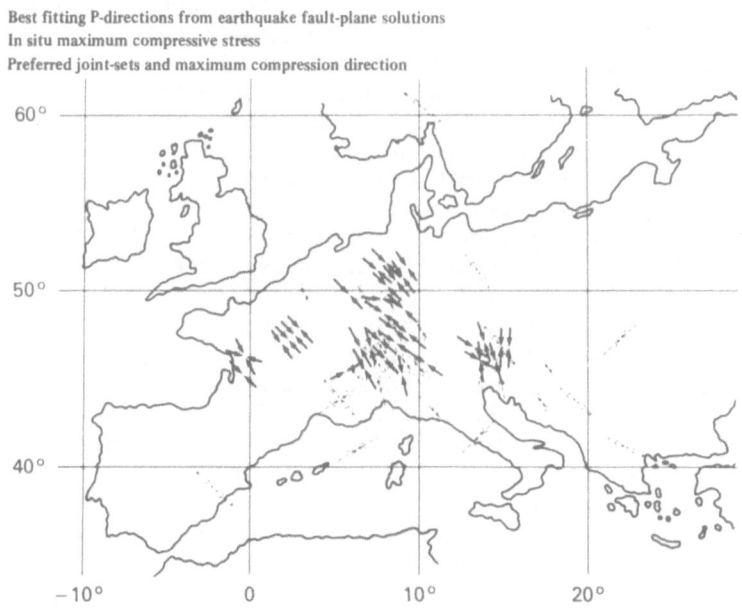

Best fitting P-directions from earthquake fault-plane solutions
In situ maximum compressive stress
Preferred joint-sets and maximum compression direction

Figure 4. Present-day stresses in Europe deduced from seismic and geological phenomena.

fault-plane solutions   in situ stress measurements

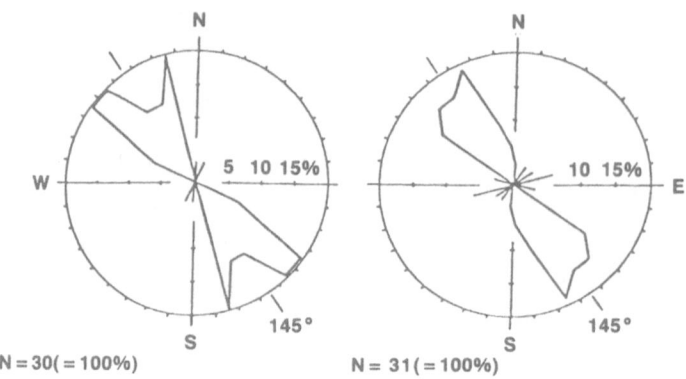

in a definite way the possible connection between the present-day
stress field deduced from seismic activity and the sub-crustal
stresses as inferred from satellite gravity data. The neo-tectonic
movements in Central Europe may be related to the present-day
intraplate stresses (20). However, the causes for the present-day
stresses in this region are as yet unknown. Interpretations of the
origin of the present-day stress fields in the eastern part of the
Eurasian plate have been given previously by Liu (15, 17). Liu (17)
has shown that subcrustal stress pattern has revealed an anomalous
lens of upwelling mantle material under the Baikal rift region
which causes earthquakes and fault motions. The fault-plane solutions and observed seismotectonic block movements along the Baikal-
Stanovoy seismic belt in Siberia are in good accord with the subcrustal stresses inferred from satellite gravity data. Furthermore,
from the satellite-determined subcrustal stress patterns, Liu (10)
was able to develop a seismogenic model for the disastrous 1976
Tangshan earthquake near Peking in China. The similarity of the
intraplate seismicity between the Brussels-Liège region of Belgium
and the Peking-Tangshan region of China suggests that crustal deformations in both regions may be caused by the same geodynamical
processes of the convection-generated subcrustal stress system.
    Stress equations (1) and (2) are derived from a laminar viscous
mantle flow model with a Newtonian viscosity developed by Runcorn
(6). Liu (8-18) has shown that these equations are approximately
valid for computation of subcrustal stress patterns.
    The successively higher degree harmonics which have been placed
in the model of the Earth's gravity field have been one of the most
impressive series of measurements in the space program. The harmonic coefficients $C_{n,m}$ and $S_{n,m}$ have been determined by satellite
and gravity measurements of the geopotential (19). The low-degree

harmonics for n $\leqslant$ 12 may reflect a large-scale mantle flow system
(6). The high degree harmonics for n $\geqslant$ 13 may result from a short
wave-length convection system (8 - 18). The present-day stresses
in Central Europe are probably related to the small-scale subcrustal stress field which corresponds to the high-degree harmonics.
By introducing the high-degree harmonic coefficients of the geopotential (19) into the stress equations (1) and (2), data can be
processed by a computer to produce the results of the magnitudes
and directions of the resultant stress. The magnitudes and directions of the resultant stresses under the crust associated with
mantle convection currents can be expressed by:

$$\sigma(\zeta,\lambda) = \left[F_E^2 (\zeta,\lambda) + F_N^2 (\zeta,\lambda) \right]^{1/2} \tag{23}$$

and

$$\theta = \tan^{-1} \left[F_N(\zeta,\lambda)/F_E (\zeta,\lambda) \right] \tag{24}$$

The numerical methods of stress calculations have been developed by Liu (8-18). For a more detailed computation of the magnitudes and directions of stresses exerted by the convection currents
under the crust of Europe, the following recursive Legendre functions were applied:

$$P_n^{m+2} (\cos \zeta) = -2(m+1) \cot \zeta P_n^{m+1} (\cos \zeta) - (n-m)(n+m+1)P_n^m(\cos \zeta) \tag{25}$$

$$P_{n+2}^m (\cos \zeta) = \frac{2n+3}{n-M+2} \cos \zeta P_{n+1}^m (\cos \zeta) - \frac{n+m+1}{n-m+2} P_n^m (\cos \zeta) \tag{26}$$

$$P_n^m(u) = (-1)^m (1-u^2)^{m/2} \frac{d^m}{du^m} \left[P_n^{(u)} \right] \tag{27}$$

$$\frac{d}{d\zeta} \left[P_n^{m+2} (\cos \zeta) \right] = -2(m+1) \left\{ \cot \zeta \frac{d}{d\zeta} \left[P_n^{m+1} (\cos \zeta) \right] \right.$$

$$-cs^2 \zeta P_n^{m+1} (\cos \zeta) \right\} - (n-m)(n+m+1) \frac{d}{d\zeta} \left[P_n^m (\cos \zeta) \right] \tag{28}$$

$$\frac{d}{d\zeta} \left[P_{n+2}^m (\cos \zeta) \right] = \frac{2n+3}{n-m+2} \left\{ \cos \zeta \frac{d}{d\zeta} \left[P_{n+1}^m (\cos \zeta) \right] \right.$$

$$- \sin \zeta P_n^m (\cos \zeta) \right\} - \frac{n+m+1}{n-m+2} \frac{d}{d\zeta} \left[P_n^m(\cos \zeta) \right] \tag{29}$$

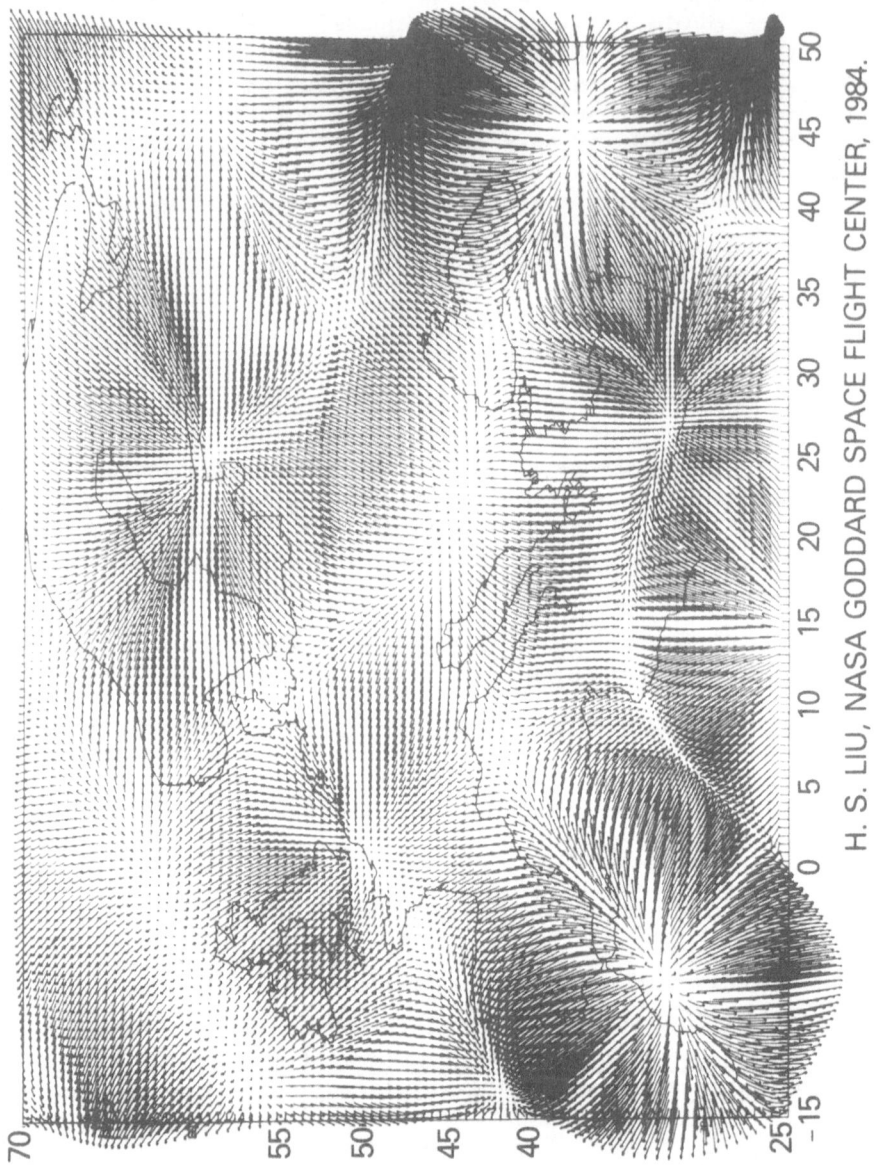

H. S. LIU, NASA GODDARD SPACE FLIGHT CENTER, 1984.

Figure 6. Satellite-determined subcrustal stresses under Europe.

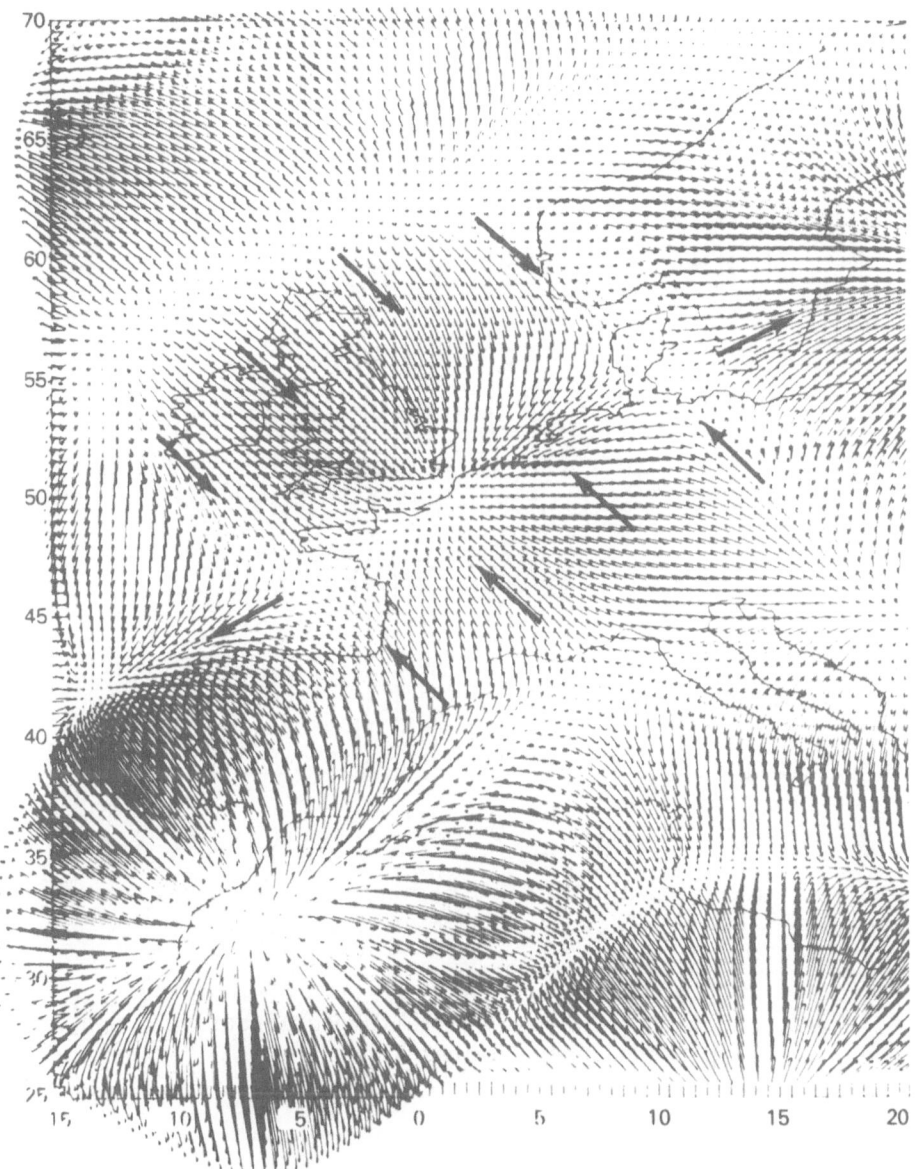

Figure 7. Subcrustal stress pattern under Western Europe as inferred from satellite gravity data.

MATERIAL TESTS: CRYSTAL DEFORMATION

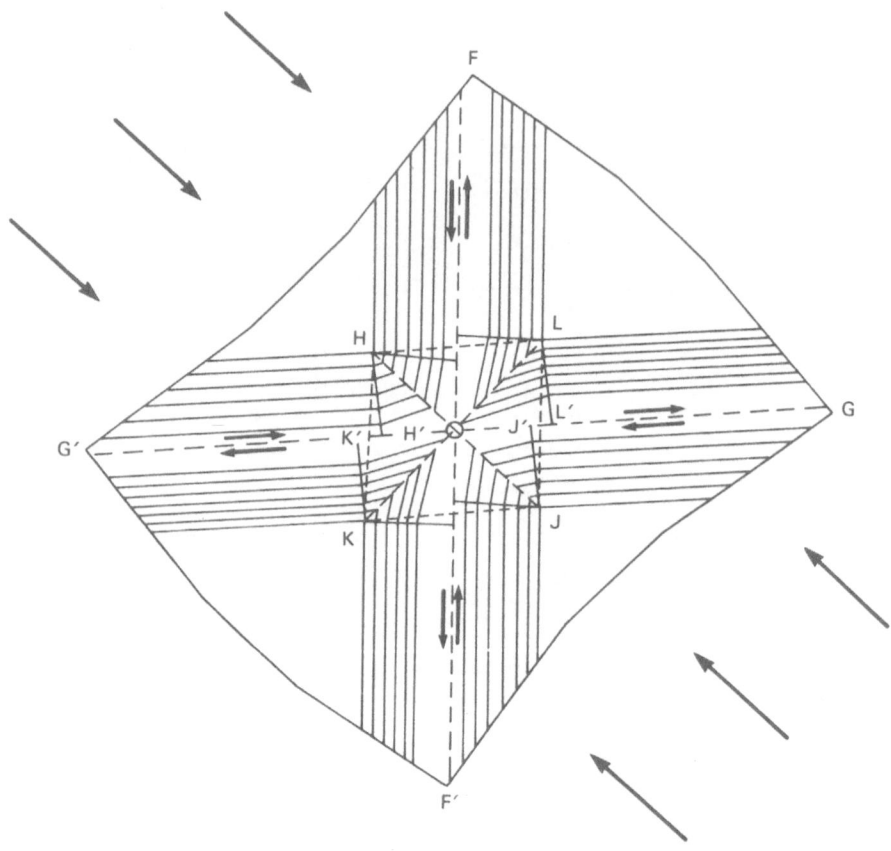

Figure 8. Accommodation of two simultaneously active conjugated
          slip bands under compression.

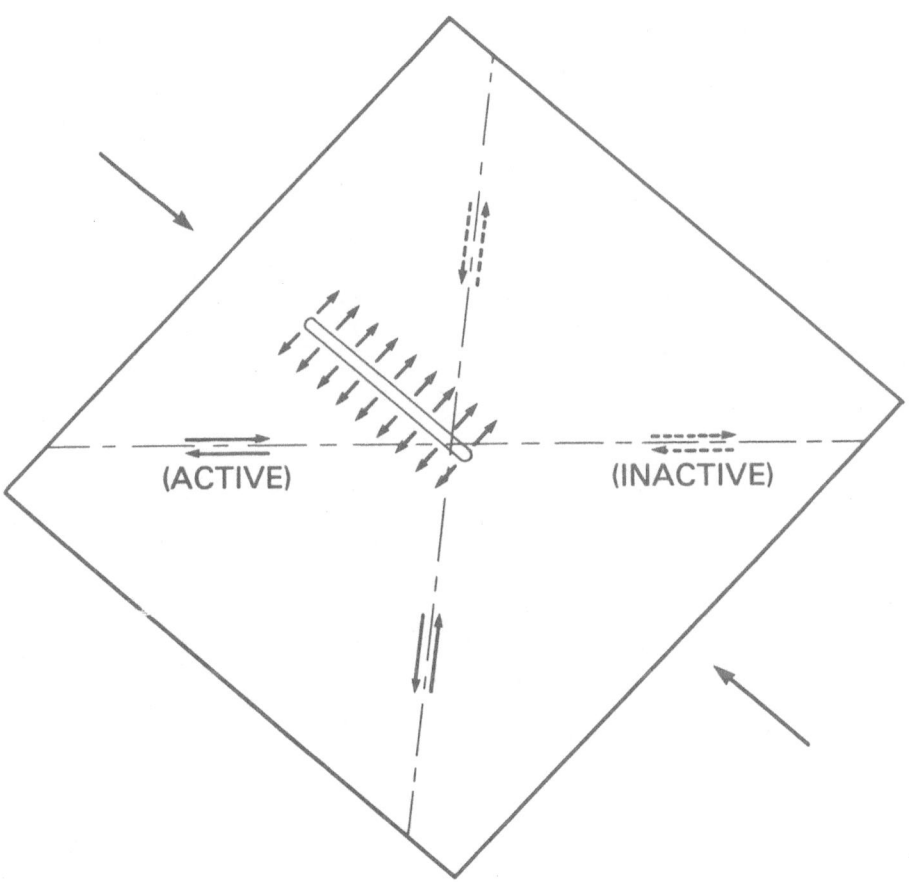

Figure 9. Shear and tensional stress in a fractured crystal under compression.

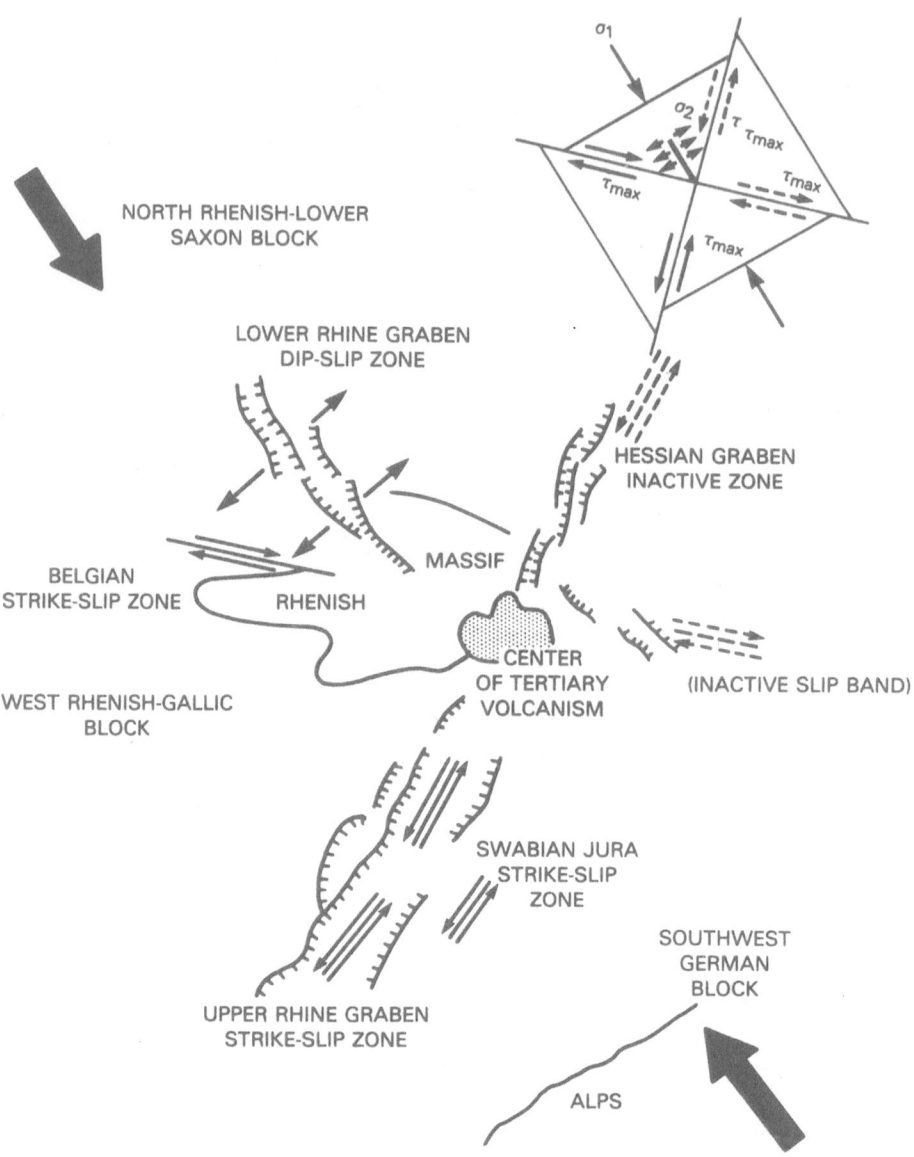

Figure 10. Seismotectonic scheme of the Rhine Graben system with the calculated crustal stress components and possible horizontal crustal block movements.

The coefficients of $C_{n,m}$ and $S_{n,m}$ (19) were introduced into
equations (1) and (2) and processed by a computer. Results from
equations (1), (2), (23) and (24) were obtained at points having
coordinates $25° \leqslant \zeta \leqslant 70°$ and $-15° \leqslant \lambda \leqslant 50°$. A map of the subcrustal stresses exerted by mantle convection under Europe taken
over half-degree grid and synthesized for $13 \leqslant n \leqslant 25$ is shown in
Figure 6. The subcrustal pattern under Western Europe is enlarged
in Figure 7. The arrows in Figures 6 and 7 indicate the stress
vectors, and the magnitudes of these vectors are of the order of
10^8 dyn cm^{-2}, which is the same order of magnitude as for stresses
associated with the fracture of continents (8). It can be seen
that the stress pattern in Figure 7 is characterized by a major
bilaterally opposed compressive stress pattern in the NW-SE direction and a minor bilaterally opposed tensional stress pattern
in the NE-SW direction, as developed by Liu (12). An important
question to be addressed is that of the seismic consequence of
this subcrustal stress pattern, as we shall discuss presently.

4. MODEL OF INTRAPLATE DEFORMATIONS AND MOVEMENTS

Argon and Orowan (29) have shown that a small crystal plate
under compression will be deformed into two active conjugate
slip bands (Figure 8). According to the principles of experimental mechanics (30) the conjugate slip bands may become inactive.
Consequently, a plate with a fracture under NW-SE compression can
be modeled by the diagram as shown in Figure 9 (18). This diagram
shows that two branches of the conjugate slip bands are active
while two other branches become inactive. Tensional stresses will
be developed perpendicular to the fracture line. Geodetic as well
as seismotectonic studies (20) have shown that the crustal deformation and seismotectonic block movements in Central Europe can be
represented as summarized in Figure 10. The two active strike-
slip zones, Belgian zone, and upper Rhine Graben-Swabian Jura
zone, correspond to the computed maximum shear stress planes while
the Hessian Graben zone is an inactive slip zone which corresponds
to the inactive conjugate slip band. The Lower Rhine Graben dip-
slip zone is under tension which agrees with the tensional feature
in our model. Thus, based on our model, the differing seismotectonic regimes of the Upper and the Lower Rhine Grabens, as well
as the Belgian and Swabian Jura zones, are easy to understand.
Tensional crustal deformations occur under the influence of the
subcrustal stress field in a distinct manner only in the lower
Rhine Graben, whereas in the other seismic zones shear movements
are typical. Seismic activity in Central Europe may be attributed
to the secondary adjustive movements of small crustal units within
the Eurasian plate. As indicated in Figure 10, the North Rhenish-
Lower Saxon and S-W German Seismotectonic Block movement is consistent with the computed subcrustal stress fields. Since crustal
stress fields computed on the basis of satellite gravity data

resemble closely the experimental pattern of single-crystal deformation, much of the intraplate deformation in Central Europe seems
to be explainable by the dynamical processes that govern single
crystals which are a billion times smaller in size than seismotectonic blocks.

5. CONCLUSION

   Despite efforts to investigate seismic activity in Western
Europe, the origin of the generating forces or stresses which cause
seismotectonic dislocations and neo-tectonic block movements in
Central Europe remains a mystery. By modeling the stress pattern
derived from satellite gravity data, the result of this study
seems to provide an avenue for interpretation of the present-day
crustal stresses in Central Europe that are amenable to experimental inquiry. As the general stress field in Western Europe
becomes better understood from the analysis of satellite gravity
observations, the seismological features of Central Europe find a
coherent, unified explanation.

REFERENCES

1. Thompson, W. and Tait, P.G. 1923,
   "Treatise on Natural Philosophy",Cambridge University Press,
   Cambridge.

2. Bijlaard, P.P. 1936,
   Union of Geod. and Geophys. Symp., Edinbourg.

3. Liu, H.S. 1974,
   J. Geophys. Res. 19, pp. 2568-2572.

4. Love, A.E.H. 1944,
   "A Treatise on the Mathematical Theory of Elasticity",Dover,
   New York.

5. Jeffreys, H. 1962,
   "The Earth", 4th ed.,p.328, Cambridge University Press,London.

6. Runcorn, S.K. 1967,
   Geophys. J.R. Astron. Soc. 14, pp. 375-384.

7. Hide, R. and Horai, K. 1968,
   Phys. Earth Planet. Inter. 1, pp 305-308.

8. Liu, H.S. 1977,
   Phys. Earth Planet. Inter. 15, pp. 60-68.

9. Liu, H.S. 1978,
    Phys. Earth Planet. Inter. 16, pp. 247-256.

10. Liu, H.S. 1979a,
    Phys. Earth Planet. Inter. 19, pp. 307-318.

11. Liu, H.S. 1979b,
    Modern Geology 7, pp. 29-36.

12. Liu, H.S. 1980a,
    Tectonophysics 65, pp. 225-244.

13. Liu, H.S. 1980b,
    Modern Geology 7, pp. 81-93.

14. Liu, H.S. 1980c,
    Modern Geology 7, pp. 161-169.

15. Liu, H.S. 1981a,
    Phys. Earth Planet. Inter. 25, pp. 64-70.

16. Liu, H.S. 1981b,
    Modern Geology 8, pp. 23-26.

17. Liu, H.S. 1983a,
    Phys. Earth Planet. Inter. 31, pp. 77-82.

18. Liu, H.S. 1983b,
    Phys. Earth Planet. Inter. 32, pp. 146-159.

19. Lerch, F.J., Klosko, S.M., Labscher, R.E. and Wagner, C.A.
    1979, J. Geophys. Res. 84, pp. 3897-3916.

20. Ahorner, L. 1975,
    Tectonophysics 29, pp. 233-249.

21. Scheidegger, A.E. 1980,
    Rock Mechanics, Suppl.9, pp. 109-124.

22. Bruckl, E., Roch, K.H. and Scheidegger, A.E. 1975,
    Tectonophysics 29, pp. 315-322.

23. Greiner, G. and Illies, J.H. 1977,
    Pageophysics 115, pp. 11-26.

24. Greiner, G. and Lohr, J. 1980,
    Rock Mechanics, Suppl. 9, pp. 5-15.

25. Gysel, M. 1975,
    Tectonophysics 29, pp. 301-314.

26. Hast, N. 1973,
    Philos. trans. R. Soc. (London), Ser.A 274, pp. 409-419.

27. McKenzie, D.P. 1969,
    Bull. Seismol. Soc. Am. 59, pp. 591-601.

28. Ahorner, L., Naier, B., and Bonjer, K.P. 1983,
    in Fuchs, K. et al. (ed.), "Plateau Uplift", pp. 187-197.

29. Argon, A.S. and Orowan, E. 1964,
    Philos. Mag., 9: pp. 1003-1021.

30. Briggs, A., Clarke, F.J.P. and Tattersall, H.G. 1964,
    Philos. Mag., 9: pp. 1041-1055.

GENERAL TECTONIC FRAME OF CENTRAL EUROPE WITH EMPHASIS GIVEN TO NEOTECTONIC MOVEMENTS

N. Pavoni

Eidgenössische Technische Hochschule Zürich
Institut für Geophysik
ETH-Hönggerberg - CH-8093 Zürich

The geological evolution of Western and Central Europe is characterized by a sequence of tectonic cycles. Orogenic events associated with suturing of continents and continental fragments were followed by periods of predominant crustal extension and rifting. Since the Silurian three major cycles are generally recognized : The Caledonian, the Hercynian and the Alpine cycle.

In early Jurassic, about 180 million years ago, sea floor spreading started in the North Atlantic region. The sequence of magnetic anomalies allows a rather detailed reconstruction of the tectonic evolution of the Atlantic and its different segments. A comparison shows that the Mesozoic/Cenozoic tectonic evolution of Central Europe is closely related with the spreading history of the Atlantic and the drift of the African plate relative to the Eurasian plate.

Vertical crustal movements are documented by accumulation of thick series of sediments in distinct sedimentary basins. The geological history of Central Europe makes clear that sedimentary basins are formed by two fundamentally different processes, either by crustal stretching or by crustal loading in the foreland of an advancing thrust belt.

In a phase of crustal extension sedimentary basins are formed by rifting and graben formation. The crust is divided in a mosaic of blocks with different rates and amounts of vertical movements. The crust below the basin is thinned by stretching and probably also by "subcrustal erosion". Examples of major sedimentary basins formed by crustal extension are the Polish Trough, the North Danish Basin, the Horn Graben, the Glückstadt Trough,

37

P. Melchior (ed.), Seismic Activity in Western Europe, 37–39.

the Lower Saxony Basin and the West Netherland Basin, all of them
active during the Mesozoic, as well as the North Sea Basin, the
Ruhr Graben and the Rhine Graben, which were active mainly during
the Cenozoic. Volcanism may or may not be associated with the rifting process.

It is of interest to note that certain crustal blocks, the
so called highs or positive areas, kept a relative elevated position during long periods of time. Their sedimentary cover - if
ever - is incomplete, fragmentary and of reduced thickness. Denu-
dation and erosion were particularly active in these areas. Major
positive areas in Central Europe are the Rhenish Massif, the
London-Brabant High, the Armorican Massif and the Bohemian Massif.

An actively prograding thrust belt produces a tectonic loading
of the crust of the foreland and a migration of the axis of the
foredeep toward the foreland. This type of basin formation is well
illustrated in the evolution of the Molasse Basin in the Alpine
foreland and the basin formation in the Carpathian foreland. With
the termination of thrusting activity and subduction the crust
reacts isostatically, the fold belt and the foredeep basins are
subjected to uplift and erosion.

Strike-slip faulting, giving way to lateral horizontal movements, is of upmost importance in the structural evolution of
Europe. Extensive wrench-faulting occured during continental suturing. The resulting fault systems transsect the fold belts as
well as the adjacent foreland. Old fault zones are often reacti-
vated in later phases. In cases where the fold belt shows an ar-
cuate structure, e.g. the arc of the Western Alps, the arrangement
of strike-slip faults in the outer parts and in the adjacent foreland of the belt displays a fan-like pattern. Strike-slip faulting
also plays an important role in the formation of local fold structures.

In Central Europe strike-slip deformation and compression due
to plate collision is observed even at considerable distances
(1000 km) from the collision front. Such remote action of collision may cause up-thrusting of major basement blocks (e.g. Rhenish
Massif) and partial inversion of rifts (e.g. Rhine Graben).

The wide distribution of faults and fault systems of similar
trend and sense of displacement in Central Europe proves the existence of large-scale regular tectonic stress fields in the upper
crust for long periods of time. An analysis of the fault systems
of the Variscan fold belt in Europe e.g. indicates that the maximum horizontal compression in this fold belt was consistently
oriented N-S from France to Poland. The late-Cenozoic and neotectonic deformation of Central Europe caused by the latest phase of

Alpine collision is characterized by NNW-SSE oriented maximum
horizontal compression and ENE-WSW extension.

In certain areas, as e.g. in Switzerland, detailed seismotectonic analyses show that the tectonic stress field which causes
the present seismic activity is very similar in its orientation
to the stress field which - during the last 5 - 10 million years-
produced the neotectonic deformation. It is a valuable working
hypothesis to assume that the present tectonic movements, of which
the earthquakes are a clear manifestation, are in direct connection and continuation of late-Cenozoic movements.

REFERENCES

Fuchs, K., von Gehlen, K., Mälzer, H., Murawski, H. and Semmel, A.,
    editors (1983) : Plateau uplift, The Rhenish Shield - A
    case history. 411 p., Springer, Berlin.

Rutten, M.G. (1969) : The geology of Western Europe. XVIII +
    520 p., Elsevier, Amsterdam.

Van Loon, A.J., editor (1978) : Key-notes of the MEGS-II
    (Amsterdam, 1978). Geologie en Mijnbouw, Special Issue,
    57/4, 481-654.

Ziegler, P.A. (1982) : Geological atlas of Western and Central
    Europe. 130 p. and 40 enclosures. Shell I.P.M., The Hague.

THE GENERAL PATTERN OF SEISMOTECTONIC DISLOCATIONS IN CENTRAL EUROPE AS THE BACKGROUND FOR THE LIÈGE EARTHQUAKE ON NOVEMBER 8, 1983

L. Ahorner

Abteilung für Erdbebengeologie, Geologisches Institut,
Universität zu Köln, Federal Republik of Germany

Abstract

The seismotectonic activity of the Central European area between the Alps and the North Sea is controlled by two essential factors. The first is a wide-spread regional stress field within the Earth's crust characterized by a rather uniform trend of the maximum compressive stress in NW-SE direction (145° with respect to North). The second controlling factor is the post-Hercynian block mosaic and the existence of deep reaching major fault structures like the Rhine Graben System which form distinct zones of crustal weakness. Under the influence of the large-scale stress field characteristic seismotectonic block movements are going on along pre-existing zones of crustal weakness which can be described by a rather simple deformation model.

The focal mechanism of the Liege earthquake on November 8, 1983, fits well to the large-scale deformation model. The fault plane solution elaborated on the basis of P wave first motions of 65 seismic stations in Central Europe reveals a prevailing strike-slip dislocation combined with a small thrust component. The fault movement is either right-lateral along a WSW-ENE trending plane, or left-lateral along a NNW-SSE trending plane. From geological evidence it is assumed that the WSW-ENE plane is the true rupture plane. The pressure axis is orien=tated NW-SE and nearly horizontal.

The unusual high stress drop of the Liège mainshock of approximately 156 bar as estimated from S wave spectra points to comparatively high crustal stresses at shallow depth in the northwestern border re=gion of the Rhenish Massif. This is in agreement with the results of a hydraulic fracturing experiment in the research borehole Konzen 1 (near Monschau, 50 km SE of Liège), where a maximum shear stress of 38 bar has been found at a depth of 400 m with increasing tenden=cy to greater depth.

41

P. Melchior (ed.), Seismic Activity in Western Europe, 41–56.
© *1985 by D. Reidel Publishing Company.*

1. Large-Scale Seismotectonic Deformation Model

Seismological research work in the last two decades which has been carried out by many investigators (e.g. Schneider 1968; Ahorner 1970, 1975, 1983; Bonjer 1979) has shown that the seismotectonic activity of the Central European area between the Alps and the North Sea is controlled by two essential factors.

The first controlling factor is a wide-spread regional stress field within the Earth's crust characterized by a rather uniform trend of the maximum horizontal compressive stress in NW-SE direction. This tecto= nic stress field can be inferred from earthquake fault plane solutions (Ahorner, Murawski & Schneider 1972; Pavoni & Mayer-Rosa 1978; Ahorner, Baier & Bonjer 1983), from sigma 1 axes of in-situ stress measurements (Illies & Greiner 1979, Baumann & Illies 1983), and even from satellite gravity data (Han-Shou Liu 1983). Results from different methods agree in indicating a mean direction of the maximum hori= zontal compressive stress of about 145° with respect to North in the total area between the Alps and the North Sea (Figure 1). We can explain the wide-spread tectonic stress field as the result of large-scale plate tectonics. The NW-SE compression within the western part of the Eurasian plate is obvio usly caused by the interaction between diverging convective mantle streams moving from the Mid-Oceanic rift zone in the North-Atlantic towards Southeast, and the collision between the Eurasian plate and the African plate in the Mediterranean region (Ritsema 1968; Ahorner 1975).

The second controlling factor for the seismotectonic activity is the post-Hercynian block mosaic of Central Europe and the existence of deep reaching major fault structures which form distinct zones of crustal weakness. The most important of these fracture zones are the Upper Rhine Graben between Basel and Frankfurt, and its tributary graben systems to the North, which follow the Middle Rhine valley from Mainz to Bonn, and the Lower Rhine Embayment (Lower Rhine Graben) from Bonn to the Netherlands.

Under the influence of the large-scale tectonic stress field cha= racteristic seismotectonic block movements are going on along the pre-existing zones of crustal weakness which can be described by a rather simple deformation model (Figure 2). Fault zones which are arranged parallel to the direction of the maximum shear stress (SSW-NNE and WNW-ESE) of the present-day stress field are affected mainly by hori= zontal strike-slip movements. This is true, for instance, for the Swa= bian Jura earthquake zone south of Stuttgart, and for the main parts of the Upper Rhine Graben (south of Heidelberg). Both structural zones are characterized nearly exclusively by seismotectonic strike-slip dislo= cations with a left-lateral moving sense, as predicted by the large-scale deformation model for NNE trending zones. At the other hand, the WNW trending Belgian earthquake zone between Aachen and Ostend, which follows the conjugate shear stress direction, reveals a clear pre= dominance of right-lateral strike-slip mechanisms. A well documented example for this type of dislocation is given by the Brussels earthquake on June 11, 1938 (M_L=5.9) which shows a distinct elongation of the macroseismic isoseismals along the WNW–ESE trending fault plane (Ahorner, Murawski & Schneider 1972; van Gils & Zaczek 1978).

In contradiction to the above mentioned NNE or WNW trending fault zones stand those structural features which are arranged parallel to the direction of the maximum compressive stress and perpendicular to the direction of the least compressive stress or tensional stress, re= spectively. This is true mainly for the NW–SE trending zone of the Lower Rhine Embayment (or Lower Rhine Graben), the Middle Rhine zone, and the northernmost part of the Upper Rhine Graben (north of Heidelberg), where extensional crustal deformation is going on. A broad zone with still active rifting and prevailing tensional dip-slip focal me= chanisms can be followed over a total length of 350 km from Nijmegen in the North to Heidelberg in the South.

The simple deformation model of Figure 2 which describes the general pattern of seismotectonic dislocations in Central Europe has been established already some 15 years ago (Ahorner 1970), but it is valid in its main features up to now. Improved fault plane solutions of recent earthquakes and numerous in-situ stress measurements which have been carried out in the last years have confirmed the model. This is demonstrated for instance by the new seismotectonic map for the Central European area published recently by Ahorner, Baier & Bonjer (1983).

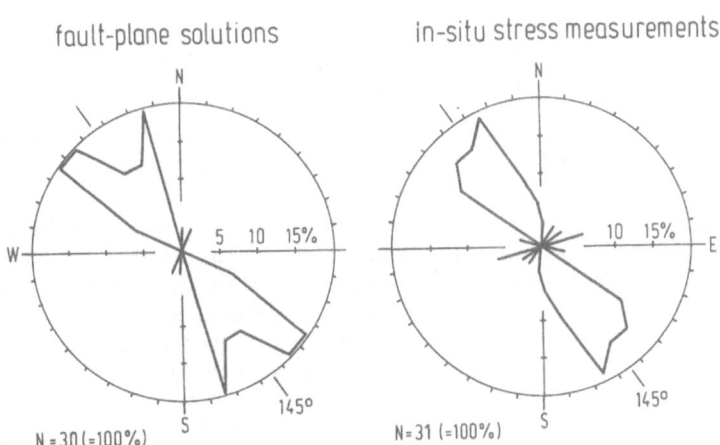

Figure 1. Direction of the maximum horizontal compressive stress in Central Europe (from Ahorner, Baier & Bonjer 1983). The left diagram gives the direction of pressure axes of earth= quake fault plane solutions, the right diagram the direction of sigma 1 axes of in-situ stress measurements (stress data from Baumann & Illies 1983)

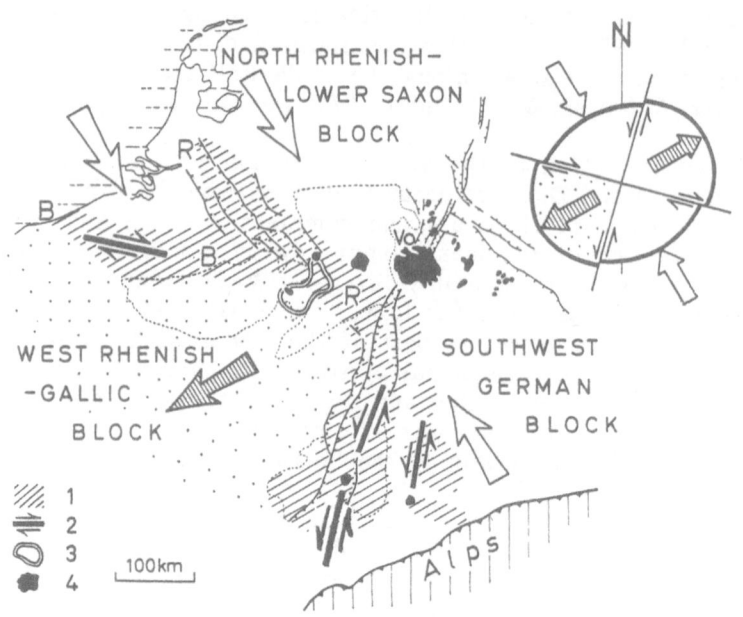

Figure 2. Large-scale seismotectonic deformation model for the Central
European area (from Ahorner 1970, with slight modifications).
1 = earthquake zones (R = Rhenish zone, B = Belgian or Bra=
bant zone), 2 = seismotectonic strike-slip dislocations, 3 =
region with Quaternary volcanism, 4 = main centers of Ter=
tiary volcanism. Contours of the Rhenish Massif are pointed.

2.   Seismotectonics of the Rhenish Massif and its Vicinity

For understanding the seismotectonic processes causing the Liège
earthquake on November 8, 1983, it is useful to look at the general
pattern of seismotectonic dislocations within the Rhenish Massif and
its vicinity, because the Liège focal region is situated at the north=
western border of this massif. The large outcropping basement complex
of the Rhenish Massif (Rheinisches Schiefergebirge, Ardennes) consists
mainly of Paleozoic rocks which have been strongly folded during the
Hercynian orogeny. In the younger geological history (Upper Tertiary
and Quaternary) the Rhenish Massif has undergone some 300 m of pla=
teau uplift, accompanied with neotectonic fault movements and young
volcanism (Fuchs et al. 1983).

The seismicity and the earthquake focal mechanisms of the Rhenish
Massif have been investigated recently by Ahorner (1975, 1983) and
Ahorner, Baier & Bonjer (1983). Figure 3 summarizes the main results.
The maximum compressive stress is orientated within the Rhenish Massif

in the same NW-SE direction as in the whole Central European area. This is evident from the prevailing orientation of the pressure axes (P axes) of more than twenty earthquake fault plane solutions and of the sigma 1 axes of 18 in-situ stress measurements. The most important seismotectonic feature is the still active rift zone, already mentioned, which can be followed in NW-SE direction approximately along the course of the river Rhine from the northern end of the Lower Rhine Graben (near Nijmegen) over the Middle Rhine valley to the northern= most part of the Upper Rhine Graben (near Heidelberg). The Rhenish Massif is bisected by this rift zone into two half parts. Earthquake focal mechanisms within the rift zone are predominantly of the tensional dip-slip type with fault planes striking NW-SE. The rather uniform SW-NE orientation of the tension axes (T axes) of those dip-slip mechanisms suggests that the two crustal blocks bordering the rift zone are moving away from each other in SW-NE direction. This is also the direction of the least compressive stress or tensional stress, respectively, of the present-day stress field.

The Lower Rhine Graben is the most prominent segment of the neotectonic rift zone. Quaternary deposits are dislocated here along NW-SE trending normal faults (e.g. Erft Fault, Peel Boundary Fault) up to 175 m, especially in the western part of the graben, where the seismic activity is concentrated (Figure 4). Strong earthquakes with local magnitudes up to $M_L=6$ have been released in historical time at the western border faults of the Lower Rhine Graben against the Eifel Mountains and the Hohes Venn. The largest shock with intensity VIII of the MSK scale occurred 1756 near the town of Düren. A more recent damaging earthquake with intensity VII-VIII and local magnitude $M_L=5.7$ happened 1951 near the town of Euskirchen (event no.5 on Fi= gure 3). The epicenter of the Euskirchen earthquake is situated within a NW-SE trending zone of strong differential crustal movements which forms the border line between the subsiding Lower Rhine Graben and the rising Eifel Mountains (Ahorner 1983). The present-day vertical strain rate of this zone is in the order of 1 to 2 mm per year, as de= duced from geodetical measurements (Mälzer, Hein & Zippelt 1983). The fault plane solution of the Euskirchen earthquake reveals a tensio= nal dip-slip dislocation along a fault plane striking NW-SE (Figure 5). The tension axis (P axis) is arranged nearly horizontal in SW-NE direc= tion. This focal mechanism is in full agreement with the geological and geodetical situation in the epicentral region.

Focal mechanisms similar to that of the Euskirchen earthquake have been found for many earthquakes occurring within the active rift zone which extends from the Lower Rhine Graben along the Middle Rhine zone to the northernmost part of the Upper Rhine Graben.

Outside of the active rift zone the tensional dip-slip component decreases and the strike-slip component becomes more important in the earthquake focal mechanisms of the Rhenish Massif. In some cases even thrust fault mechanisms have been observed. A typical example for a strike-slip dislocation in the eastern half part of the Rhenish Massif is

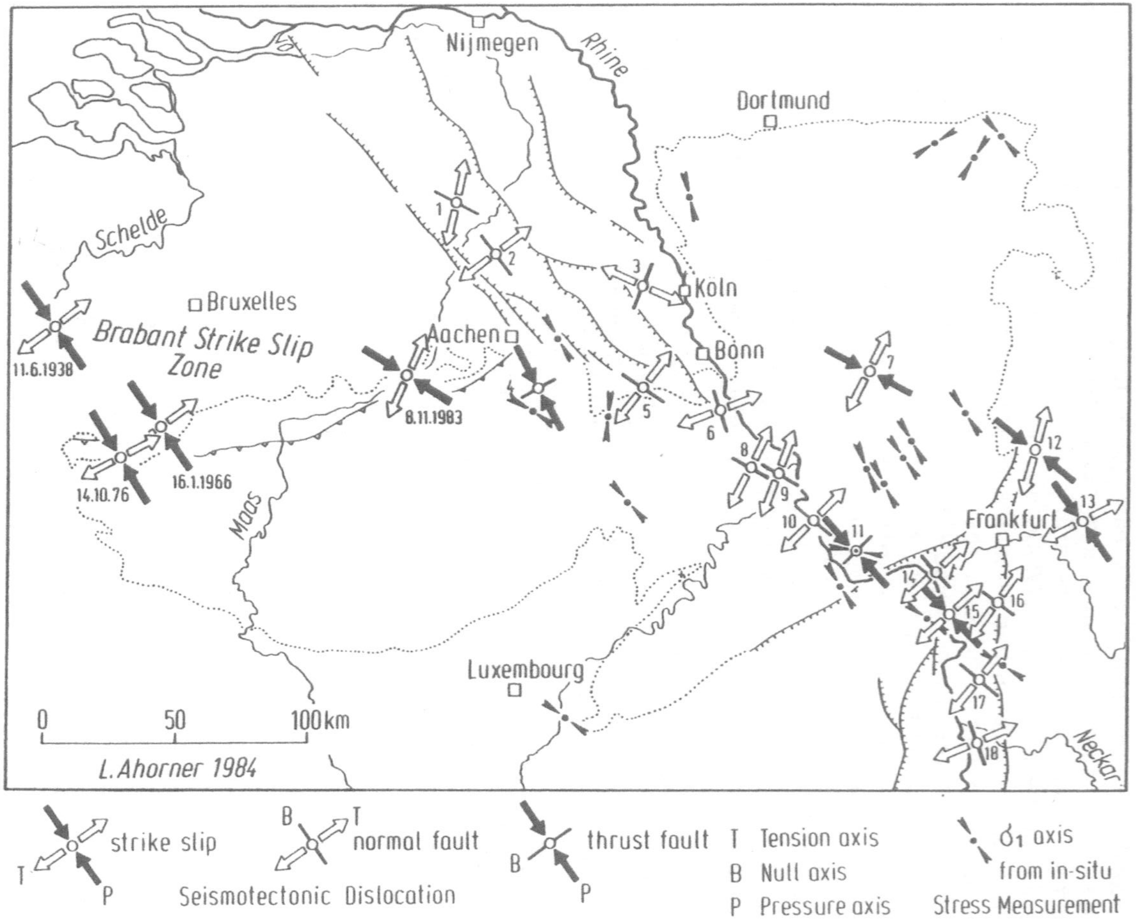

Figure 3.

Seismotectonic map for the Rhenish Massif and its vicinity. Fault plane solutions of earthquakes show strike-slip, normal fault, and thrust fault dislocations. The general trend of the compressive crustal stress is indicated by the sigma 1 axes of in-situ stress measurements. Fault plane solutions are from Ahorner, Baier & Bonjer (1983), stress data from Baumann & Illies (1983). Contours of the Rhenish Massif are dotted. Major Quaternary faults are drawn in the Lower Rhine Graben and the Upper Rhine Graben.

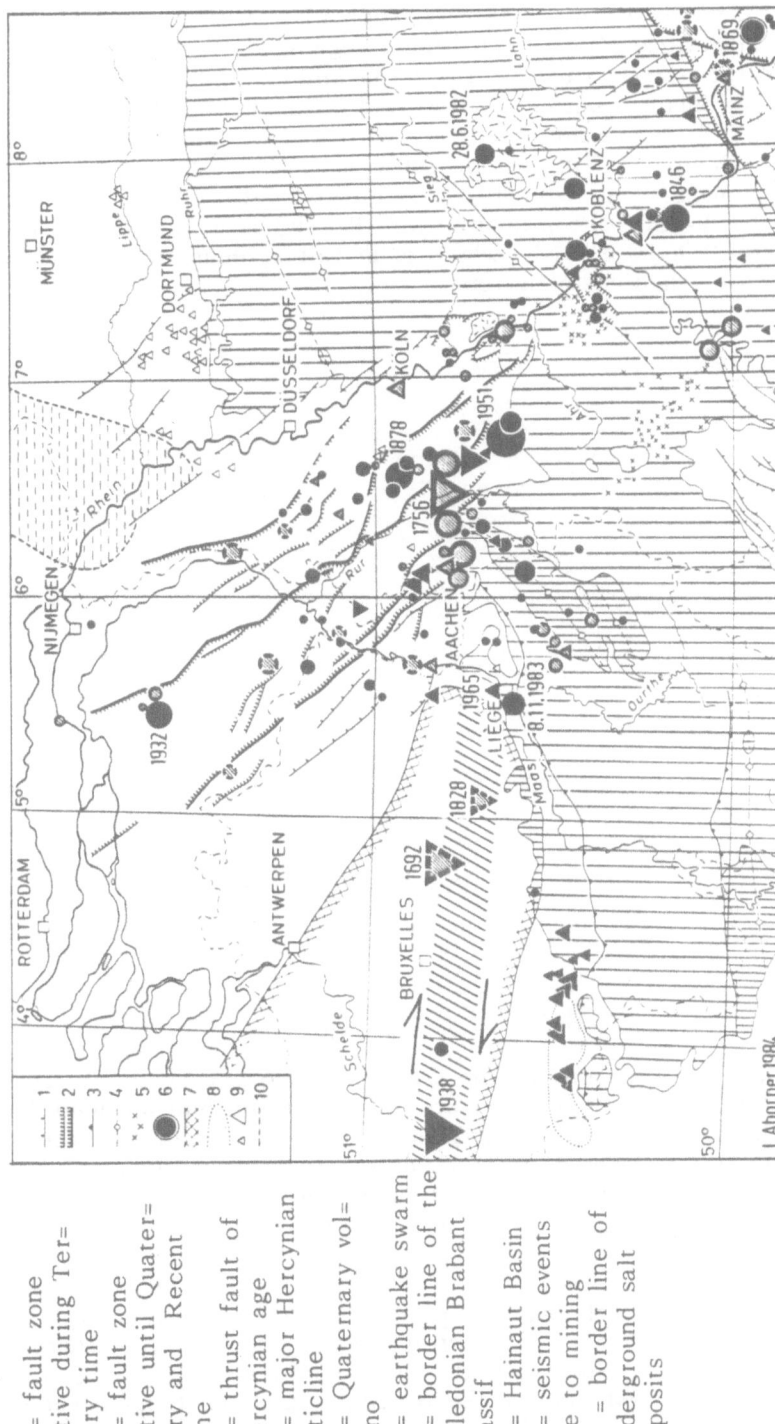

1 = fault zone
active during Ter=
tiary time
2 = fault zone
active until Quater=
nary and Recent
time
3 = thrust fault of
Hercynian age
4 = major Hercynian
anticline
5 = Quaternary vol=
cano
6 = earthquake swarm
7 = border line of the
Caledonian Brabant
Massif
8 = Hainaut Basin
9 = seismic events
due to mining
10 = border line of
underground salt
deposits

Figure 4. Earthquake epicenters 1500-1982 (black dots and triangles) and geological features of the Rhenish Massif and its northern foreland (from Ahorner 1983 with slight modifications). The Rhenish Massif is vertically hatched. The Brabant Strike Slip Zone is diagonally hatched. Triangles represent events with a focal depth slightly deeper than normal (h=15-25 km), if the peak points downward.

the Bad Marienberg, Westerwald, earthquake on June 28, 1982 (event
no.7 on Figure 3). The intensity of this shock was V–VI MSK scale
and the local magnitude M_L=4.7. The focal region is situated about
40 km east of the active rift zone along the Middle Rhine valley. The
well defined fault plane solution for the Bad Marienberg earthquake
reveals a strike-slip dislocation combined with a minor tensional dip-
slip movement (Figure 5). The moving sense is either right-lateral
along a fault plane striking WSW-ENE, or left-lateral along a fault
plane striking NNW-SSE. The pressure axis (P axis) is arranged NW-SE,
the tension axis (T axis) SW-NE, which is in full agreement with the
regional stress field in Central Europe.

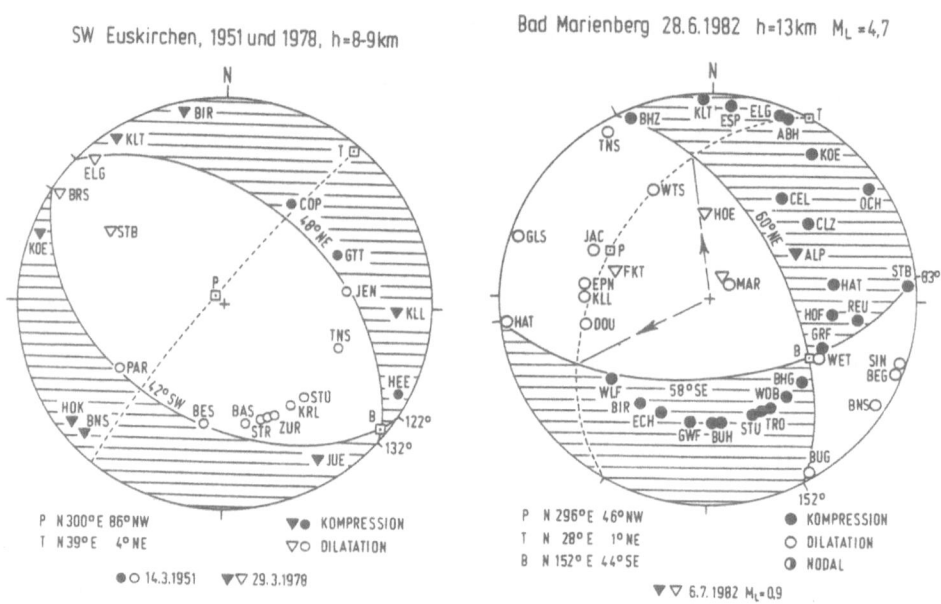

Figure 5. Fault plane solutions of the Euskirchen earthquake on March
        14, 1951 (event no.5 on Figure 3) and of the Bad Marienberg
        earthquake on June 28, 1982 (event no.7 on Figure 3). Both
        solutions are from Ahorner (1983). Equal-area projection of
        the lower focal hemisphere. Azimut and plunge of the pressure
        axes P, tension axis T, and null axis B are given.

Strike-slip mechanisms of the same type as for the Bad Marienberg
earthquake have been found in the eastern border region of the Rhenish
Massif (northeast of Frankfurt) and also in the northwestern border re=
gion of the Rhenish Massif against the Brabant Massif on Belgian terri=
tory. The WNW-ESE trending Belgian or Brabant earthquake zone is

characterized nearly exclusively by strike-slip focal mechanisms, with a few exceptions situated mainly in the Hainaut area, where non-tecto= nic earthquakes due to the outwashing of underground evaporites of Pa= leozoic age possibly occur with extreme shallow focal depths (Ahorner 1983).

Figure 4 shows with hatched signature (diagonally hatched) the po= sition of the supposed Brabant Strike Slip Zone, a broad zone of crustal deformation with a right-lateral moving sense. Large earthquakes with local magnitudes between 5.5 and 6 have been generated by this zone in the past. The focal depths of most shocks are slightly deeper than normal (h=15-25 km). Examples are the Tirlemont earthquake of 1692 (intensity VII-VIII MSK scale), the Tirlemont earthquake of 1828 (inten= sity VII-VIII), and the Brussels earthquake of 1938 (intensity VII, magni= tude 5.9). More examples are described in the publication of Van Gils & Zaczek (1978) which gives an excellent overview of the historical seismi= city on the Belgian territory.

It is evident from Figure 4 that the Liège earthquake on Novem= ber 8, 1983, occurred in the eastern structural prolongation of the Bra= bant Strike Slip Zone. Therefore the focal mechanism of this shock should be in some kind influenced by the right-lateral shear process connected with the Brabant earthquake zone. The following chapter will show if this general idea deduced from the regional pattern of seismotectonic dislocations is confirmed by the actual fault plane so= lution of the Liège earthquake.

3.   Focal Mechanism of the Liège Earthquake

Figure 6 shows a fault plane solution for the Liège mainshock on November 8, 1983, 0:49 GMT, which has been elaborated at the Depart= ment of Earthquake Geology of the Geological Institute at the Univer= sity of Cologne in cooperation with Dr.Pelzing from the Geological Survey at Krefeld. The solution is based on the first motion of short period P waves recorded at 65 seismic stations in Central Europe. First motion readings from stations with a distance less than 200 km from the epicenter were weighted higher in the routine of fixing the spatial position of the two fault planes than the readings from more distant stations which often show a rather bad signal/noise ratio and do not allow in any case to read reliable first motions. The data from the near stations in Belgium, the Netherlands and in West Germany support the finally accepted solution without any inconsistency. In case of the distant stations discrepancies arise only with two stations in Northern France (FLN, LDF) and with several stations in the United Kingdom and in Ireland (DYA, DCO). But in general, if one neglects the few exceptions, the final solution is well defined by the great majority of P wave first motion data delivered by the seismic station network in Central Europe.

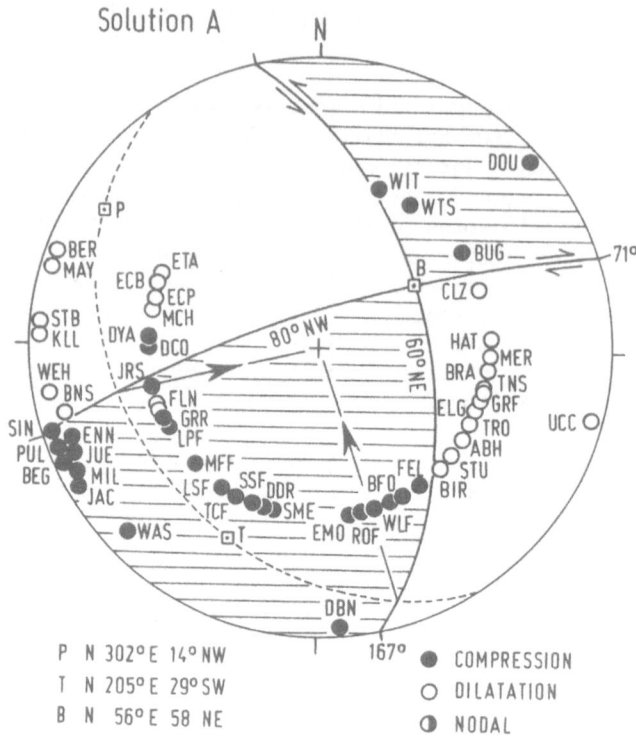

Figure 6. Fault plane solution for the Liège earthquake on November 8, 1983, at 0:49 GMT (mainshock). Equal-area projection of the lower focal hemisphere. Azimut and plunge of the pres= sure axis P, tension axis T, and the null axis B are given. The possible slip vectors on both fault planes are indicated by black arrows.

    The focal mechanism found is a clear strike-slip dislocation com= bined with a minor thrust component. The movement is either right- lateral along a plane striking WSW-ENE and dipping with 80° to the NW, or left-lateral along a plane striking NNW-SSE and dipping with 60° to the NE. The pressure axis (P axis) is arranged nearly horizon= tal in NW-SE direction. The tension axis (T axis) is orientated SW-NE and dips with about 29° to the SW. The P wave first motion data do not allow to decide which of the two planes is the true rupture plane. Judging from the geological situation and the known fault pattern in the epicentral region, however, it is most likely, that the WSW-ENE trending plane might be the true rupture plane. Many fault lines known from geological field work and underground mining are trending in the

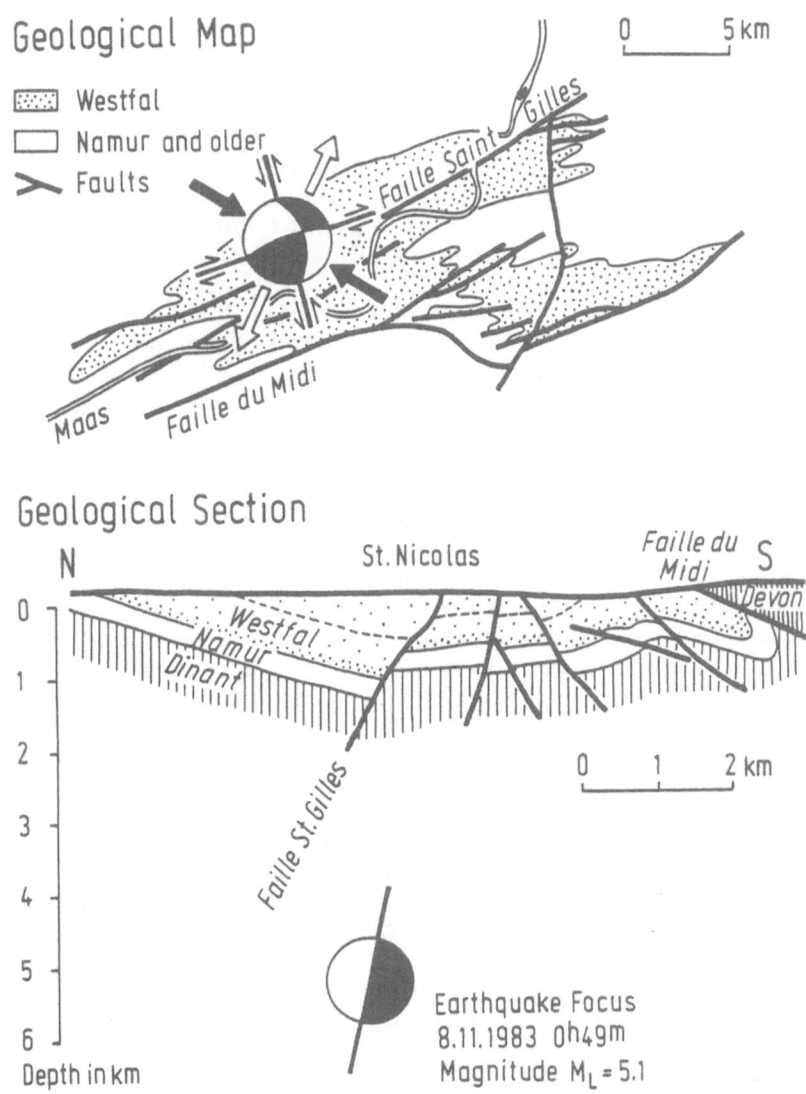

Figure 7. Focal mechanism of the Liège earthquake on November 8, 1983, and the near-surface geological structure. The pressure axis (black arrows) and the tension axis (white arrows) of the strike-slip focal mechanism are plotted. The tectonic fault pattern in the epicentral region speaks for a rupture plane in the earthquake focus trending WSW-ENE. Geology after Fourmarier (1954)

same direction (Figure 7). If we consider the shallow focal depth of
the Liège mainshock, of only 4 km from macroseismic data (De Becker
& Camelbeeck 1985) or 6 km from instrumental data (Ahorner & Pelzing
1985), a close relationship between the seismotectonic dislocation at
depth and the near-surface geology can be expected.

The choice of the WSW-ENE trending plane as true rupture plane
implies for the focal process a right-lateral strike-slip movement, com=
bined with a minor compressional dip-slip (thrust) component. This
means that the northern block was shifted during the earthquake along
the assumed rupture plane to the East and slightly upward with respect
to the southern block. An average amount of focal (block) displacement
in the order of 0.27 m has been calculated by Ahorner & Pelzing (1985)
using S wave displacement spectra and the Brune source model. The
estimation of the source length (length of active rupture plane) with
the same method yields a length of about 1.8 km.

The small thrust component involved in the prevailing strike-slip
mechanism points to compressional forces acting in the direction per=
pendicular to the rupture plane. Those forces are to explain, if we re=
member, that the maximum compressive stress of the regional tectonic
stress field is orientated NW-SE. Numerous near-surface in-situ stress
measurements within the Rhenish Massif have proven a mean direction
of the maximum horizontal compressive stress of about 145° with respect
to North (Baumann & Illies 1983). In addition, a hydraulic fracturing
experiment carried out recently by Rummel & Baumgärtner (1984) in
the 400 m deep research borehole Konzen 1 (situated in the Hohes
Venn near Monschau, about 50 km southeast of Liège; see Figure 8) ,
has proven a local direction of the maximum compressive stress of
about 120° with respect to North. An important result of this experi=
ment was, that the magnitude of the tectonic stress increases with
depth. At the deepest point of the borehole the maximum horizontal
stress exceeds the vertical stress by about 80 bars. The resulting shear
stress is 40 bars. If we extrapolate the observed trend to greater depths,
which are comparable with focal depths of earthquakes, stress values
of some hundred bars are achieved.

This means, that strong compressional forces acting in NW-SE
direction may affect the crustal rocks beneath the Hohes Venn and
the northwestern border region of the Rhenish Massif. In the light
of these findings, the thrust component of the Liège focal mechanism
is well understandable. The same ist true for the unusual high stress
drop of about 156 bars which has been determined by Ahorner & Pel=
zing (1985) for the Liège mainshock.

The causes of the stress concentration in the Hohes Venn and in
the northwestern border region of the Rhenish Massif are not clear at
this time. One thinkable explanation would be that the local stress
concentration is due to the increased uplift movement of the Hohes
Venn and the Northern Eifel Mountains with respect to the northern

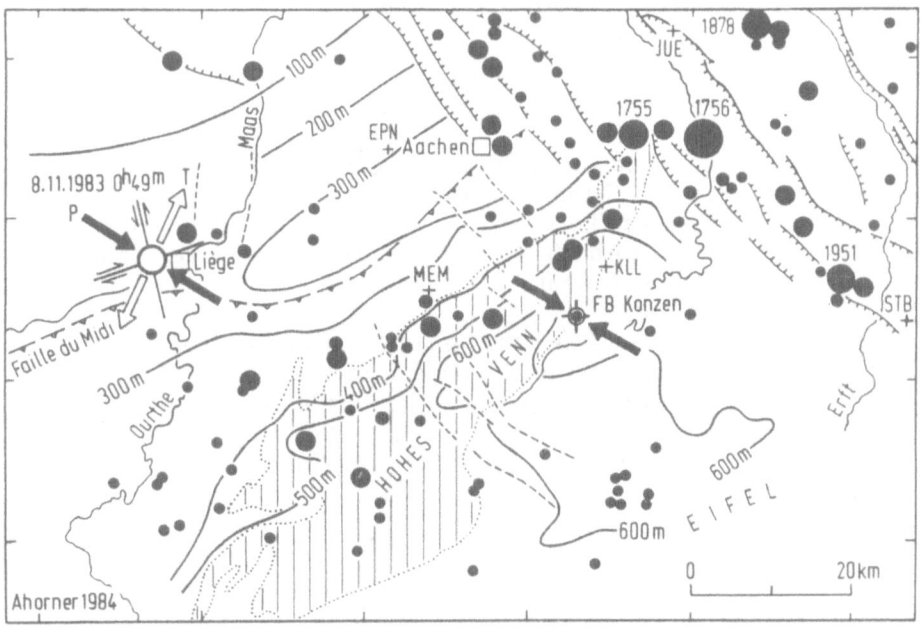

Figure 8. Earthquake epicenters 1500-1982 (black dots) and the tectonic
stress field in the northwestern border region of the Rhenish
Massif. The direction of the maximum compressive stress is
is indicated by black arrows which denote the pressure axis
of the focal mechanism of the Liège earthquake 1983 and the
sigma 1 axis of the hydrofracturing experiment in the research
borehole Konzen 1 (FB Konzen). Isolines give the height of the
base of Tertiary (Oligocene) deposits in meter above sea level.
The post-Oligocene uplift of the Hohes Venn amounts more
than 600 m with respect to the northern foreland. A SW-NE
striking hinge-zone of increased uplift coincides with the epi=
center alignment at the northern flank of the Stavelot-Venn
anticline. From Ahorner (1983, with modifications).

foreland. The relative uplift of this area amounts more than 600 m
since the Oligocene, and is still alive in Recent time. A present-day
uplift rate of 1 to 2 mm per year has been proven for the Northern
Eifel and the Hohes Venn by geodetical measurements (Mälzer, Hein
& Zippelt 1983; Ahorner 1983). It seems possible that uplift-related
crustal stresses are superimposed in the rising areas and especially
along their border regions to the tectonic stress field which is due
to the large-scale plate tectonics as mentioned before.

Summarizing the results of our analysis we can say that the
focus of the Liège earthquake on November 8, 1983, is situated in a
structural zone which is highly capable for generating tectonic earth=
quakes from several reasons.

The most important of these reasons are:

(1) The crustal rocks in the Liège area are strongly fractured by pre-existing faults of Caledonian, Hercynian and post-Hercynian age, which are trending mostly in a W-E or WSW-ENE direction.

(2) The focal region of Liège is situated in the eastern prolongation of the Brabant Strike Slip Zone, which forms a broad band of crustal weakness between the Caledonian Brabant Massif and the Hercynian Rhenish Massif with a still active large-scale right-lateral shear deformation.

(3) The Liège area is situated in a region of high crustal stress concentration which might be due to the Quaternary and Recent uplift of the Hohes Venn and the Northern Eifel. The uplift-related stresses are superimposed to the regional stresses of the Central European area which are caused by large-scale plate tectonics.

Acknowledgements

The author is indebted to colleagues from many seismological observa= tories for supplying him with seismograms and first motion readings of the Liège earthquake.

References

Ahorner, L. (1970): Seismotectonic relations between the graben zones of the Upper and Lower Rhine valley. - In: H.J.Illies & St. Müller (eds.), Graben Problems, p.155-166; Schweizerbart, Stuttgart

Ahorner, L. (1975): Present-day stress field and seismotectonic block movements along major fault zones in Central Europe. - Tec= tonophysics 29: 233-249; Amsterdam

Ahorner, L. (1983): Historical seismicity and present-day microearth= quake activity of the Rhenish Massif, Central Europe. - In: K.Fuchs et al.(eds), Plateau Uplift, p.198-221; Springer, Hei= delberg

Ahorner, L., Baier, B., and Bonjer, K.P. (1983): General pattern of seismotectonic dislocation and the earthquake generating stress field in Central Europe between the Alps and the North Sea. - In: K.Fuchs et al.(eds.), Plateau Uplift, p.187-197; Springer, Heidelberg

Ahorner, L., Murawski, H. and Schneider, G. (1972): Seismotektonische Traverse von der Nordsee bis zum Apennin. - Geol.Rdsch. 61: 915-942; Stuttgart

Ahorner, L. and Pelzing, R. (1985): The source characteristics of the Liège earthquake on November 8, 1983, from digital recor= dings in West Germany. - (this volume)

Baumann, H. and Illies, H.J. (1983): Stress field and strain release in the Rhenish Massif. - In: K.Fuchs et al. (eds.), Plateau Up= lift, p.177-186; Springer, Heidelberg

Bonjer, K.P. (1979): The seismicity of the Upper Rhine Graben. A con= tinental rift system. - In: Petrovski, J. and Allen, C. (eds.), Proc.Int.Res.Conf.Intra-Continental Earthquakes Ohrid/Yugo= slavia 1979, p.107-115

De Becker, M. and Camelbeeck, T. (1985): Macroseismic inquiry of the Liège earthquakes on November 8, 1983. - (this volume)

Fourmarier, P. (1954): Prodrome d'une description géologique de la Belgique. - 826 p.; Liège

Gils van, J.-M. and Zaczek, Y. (1978): La séismicité de la Belgique et son application en génie paraseismique. - Ann.Trav.Publ.Belg. 6: 1-38; Bruxelles

Fuchs, K., von Gehlen, K., Mälzer, H., Murawski, H. and Semmel, A. (1983)(eds.), Plateau Uplift, 411 p.; Springer, Heidelberg

Illies, H.J. and Greiner, G. (1979): Holocene movements and the state of stress in the Rhine graben rift system. - Tectonophysics 52: 349-359; Amsterdam

Liu, Han-Shou (1983): A dynamical basis for crustal deformation and seismotectonic block movements in Central Europe. - Phys. Earth Planet.Inter. 32: 146-159; Amsterdam

Mälzer, H., Hein, G. and Zippelt, K. (1983): Height changes in the Rhenish Massif: Determination and analysis. - In: K.Fuchs et al. (eds.), Plateau Uplift, p.164-176; Springer, Heidelberg

Pavoni, N. and Mayer-Rosa, D. (1978): Seismotektonische Karte der Schweiz 1:750 000. - Eclogae geol.Helv. 71/2: 293-295; Basel

Ritsema, A.R. (1968): Seismo-tectonic implications of a review of European erathquake mechanisms. - Geol.Rdsch. 59: 36-56; Stuttgart

Rummel, F. and Baumgärtner, J. (1985): Hydraulic fracturing in-situ Spannungsmessungen in der 400 m tiefen Erkundungsbohrung Nordeifel (Hohes Venn). - im Druck

Schneider, G. (1968): Erdbeben und Tektonik in Südwest-Deutschland. - Tectonophysics 5: 459-511; Amsterdam

SEISMICITY OF WESTERN EUROPE AS DETERMINED FROM ISC DATA FILES

R. D. Adams

International Seismological Centre, Newbury and
Department of Geology, University of Reading, U.K.

ABSTRACT

The extent of damage caused by the moderate-sized earthquake at Liège on 8 November 1983 is a reminder of the vulnerability of populated areas with relatively low seismic activity
to the effects of close shallow earthquakes. An essential
first step in understanding the seismicity of an area is the
compilation of comprehensive catalogues of earthquakes, and
this'paper presents plots of events in northwest Europe taken
from the files of the International Seismological Centre in
various combinations of area, time-period and magnitude. The
diffuse nature of the activity in such areas makes precise
delineation of active areas difficult, and the information in
such catalogues, although a valuable starting point in the
studies of seismicity, must be supplemented by both historical
investigations of past events, and more detailed investigations
of present events using modern techniques of recording and
analysis. A brief field trip to Liège enabled local effects
to be estimated generally at intensity VII, but reaching VIII
in the worst areas.

INTRODUCTION

The occurrence of an earthquake on 8 November 1983, of
magnitude 4.9, centred within the city of Liège in Belgium
reinforces the contention that this part of northern Europe will
be subjected to periodic shaking from earthquakes of similar
magnitude. Given their infrequency, however, and the consequent
lack of effective earthquake precautions in standards of building

P. Melchior (ed.), Seismic Activity in Western Europe, 57–69.
© *1985 by D. Reidel Publishing Company.*

design and construction, such earthquakes can cause serious damage, and the total cost of damage at Liège due to this event has been put at about £30,000,000.

In determining the pattern of expected earthquakes to help in improving earthquake preparedness and countermeasures, the first step is to establish the past pattern of occurrence, through the compilation of earthquake catalogues. This paper looks in detail at the information available for the Liège region, and surrounding parts of northern Europe, that is held at the International Seismological Centre. This information is restricted to events that have been located instrumentally from early in the twentieth century. To determine the full seismic history of a region, however, the short period of instrumental recording is quite inadequate. This is particularly so in an area of low activity, where earthquakes may occur in cycles, with active periods separated by decades or even centuries of quiescence.

It is essential, therefore, that catalogues of earthquakes be extended backwards in time as far as possible. The first step is the revision of early instrumental locations using modern analysis techniques and earth models. Next, historical records of earthquake felt effects have to be collected and analysed. This will involve studying old documents such as newspapers, diaries, church and civil records and their evaluation by competent historians as well as seismologists. Finally, archeological and geological investigations may reveal evidence of earthquake-associated earth movements extending back further than the historical record. It is only by using all these methods that the full seismic history and potential of a region can be evaluated.

Records of instrumentally determined earthquakes still remain the core of earthquake catalogues, and among agencies collecting this information on a global scale the International Seismological Centre carries out the most comprehensive compilation. Its work and relevant records for northern Europe will now be described.

ANALYSIS AND STORAGE OF EARTHQUAKE INFORMATION AT ISC

The International Seismological Centre's prime task is to collect standard earthquake readings from all available recording stations in the world, to re-evaluate the information, and to publish definitive lists of world earthquakes and the information on which the determinations are based. The aim of the ISC's work is to be as complete as possible; the start of analysis is therefore delayed for nearly two years

to ensure as full reporting from outlying stations as can be
achieved. The requirement for rapid analysis is catered for
elsewhere.

In Europe, national agencies have a need to produce rapid
locations for major events that affect their territory, and
for the region as a whole the Centre Séismologique Europío-
Mediterranéen in Strasbourg carries out a rapid service for
gathering of data and epicentre determinations. The region
covered by CSEM extends from the Mid-Atlantic Ridge to the
Urals and includes the countries on the southern and eastern
borders of the Mediterranean. A rapid "Preliminary Determination
of Epicenter" service is provided on a global scale by the
National Earthquake Information Service of the US Geological
Survey at Boulder, Colorado. The PDE service mails cards giving
epicentral estimates with a delay of about six weeks, and later
issues a printed monthly listing. These rapid preliminary
lists contain about 800 events a month. The ISC's services do
not therefore compete with those of PDE, but complement them,
in providing a final revision of global earthquake analysis,
concentrating on completeness rather than speed.

Analysis Procedures

The procedures used for analysis at ISC are described by
Adams, Hughes and McGregor (1), and follow the scheme shown in
Figure 1. The analysis is done in monthly batches, since 1983
on the Centre's own VAX 11/750 computer. After all available
epicentral estimates and read-
ings of station arrivals have
been stored chronologically,
they are subjected to a "group"
routine, which first combines
origin estimates of the same
earthquake submitted by different agencies. Station readings
and comments about felt effects
are then associated with this
event, and its parameters re-
determined using all available
observations and a version of
the least-squares procedure that
uses weighting according to
Jeffreys' Method of Uniform
Reduction (2). This initial
procedure produces about 1400-
1800 events a month. This
number of events corresponds
to an earthquake on the average
about every 30 minutes, and

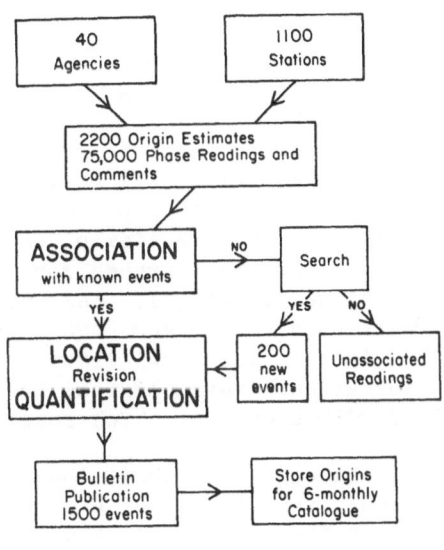

Figure 1. Scheme of
analysis at ISC.

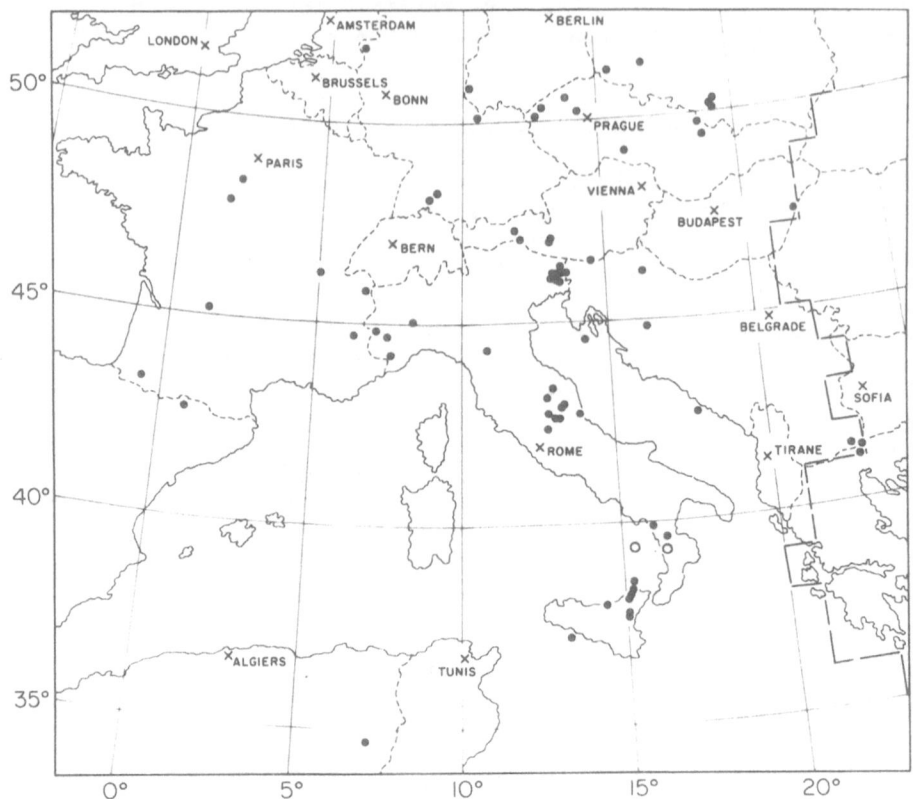

Figure 2. Previously unreported earthquakes found by
ISC in western Europe for 1978.

consequently there will be instances of station readings that
are associated with the wrong origin. After these and other
errors have been rectified, the remaining unassociated phase
observations are searched for new events using a procedure of
Engdahl and Gunst (3). This produces initially about 400 new
events each month, some 200 of which are well enough determined
to retain in the final listings. The monthly <u>Bulletin</u> <u>of</u> <u>the</u>
<u>International Seismological Centre</u> finally contains about
1600-2000 events of which some 10% are previously unreported.

     Although each event found by the computer is checked by a
seismologist, it is not always possible to detect spurious events
in the time available, and some events appear in the catalogue
which on more detailed analysis prove to be due to errors in
phase identification or association. A more detailed study
of previously unreported events in Europe for 1978 by Adams (4)
showed that out of 218 events reported, 12 were false and 10

were substantially mis-located. Figure 2 shows the remaining
well-established new events in western Europe for this year.
Often such events are found close to international boundaries
where no single national agency has enough readings for a
determination, but ISC can successfully combine readings
reported by agencies in neighbouring countries.

ISC's Historical File

    ISC maintains an Historical File of earthquakes dating back
to events of 1904 to which all new events are added. Naturally
the information relating to early years is restricted to large
events only and is poor in quality. There is increasing
improvement in the quality of the determination of parameters,
and the number of events recorded with time. In recent years,
there has been particular effort to ensure that more events have
an estimate of magnitude.

    The main sources of data in the file are:

"Seismicity of the Earth" Gutenberg and Richter	1904-1952
US Coast and Geodetic Survey	1928-1963
International Seismological Summary	1937-1963
Bureau Central International de Séismologie	1950-1963
Sykes (ridge events)	1950-1963
International Seismological Centre	1964-1981
US National Earthquake Information Service	1982-1984

    For the sake of completeness information more recent than
that from ISC analysis is included from the PDE service, about
six months in arrears, and for even more recent major events, very
preliminary information is entered from telexes received from
NEIS within a few hours.

    The increasing completeness of the file is shown by the
total number of earthquake origins to the end of each decade.
To the end of 1963 more than one origin may be present for some
events.

To end of	No of events
1910	123
1920	422
1930	1 354
1940	7 253
1950	16 926
1960	46 174
1970	138 978
1980	288 370
(1983)	317 129

The file may now be searched by any combination of requirements
of time, area, or magnitude, and the output listed or written
on magnetic tape. A simple line-printer plot of the events
found is also available, with a rough outline of coastline.

ISC RECORDS OF EARTHQUAKES IN NORTH-WEST EUROPE

    In north-western Europe most earthquakes are initially
located by a national or regional seismological agency. The
Laboratoire de Détection et de Géophysique (LDG) has reported
events in France down to magnitudes less than three since about
1975; other agencies regularly supplying information in this
area include Gräfenberg Array (GRF), Fürstenfeldbruck (FUR) and
particularly the centre at Strasbourg, known originally as
Bureau Central International de Séismologie (BCIS), and from
1976 as Centre Séismologique Européo-Mediterranéen (CSEM). Many
of the events in the ISC file still retain the origins determined by these agencies, but the larger events for which there
are enough observations have had their locations revised by
ISC using all available data. This revision sometimes results
in a change of position of several tens of kilometres, and
small events cannot be regarded as well-located. Local magnitudes are not determined by ISC, but magnitudes assigned by
local agencies are retained and published. For larger events
determinations of mb and Ms from distant stations may be
available.

    Users of the file should be aware of its deficiences. In
early years only the largest events are detected, and positions
can be unreliable. In cases of doubtful position seismologists
may wish to consult the detailed station readings in the
Bulletin of the International Seismological Centre to assess
the uncertainties for themselves and possibly make some adjustments. As well as omissions the file will contain some spurious
events, (e.g. Tittel and Wendt, 5), a common cause being the mis-
identification of PKP phases from Pacific earthquakes. Also
included in the file will be some artificial explosions such
as quarry blasts, or occasionally explosions used for geophysical
exploration. In later years known explosions are flagged as such
in the file, and are listed separately in the ISC's Regional
Catalogue of Earthquakes. Despite these deficiences the ISC file
remains a full source of information for the study of European
seismicity, provided that it is interpreted with care by
seismologists familiar with the region, and with the location
techniques used.

    Events on the ISC Historical File for western Europe are
shown in Figures 3-6 with a variety of criteria of area, time
and size, to give a background of reported seismicity for earth-

quakes in the Liège region and particularly for that of
8 November 1983. In an attempt to maintain some uniformity
of detection capability in the plots shown, successive plots
are for smaller areas, more recent time intervals, and for
earthquakes with lower magnitudes.

   Figure 3 show all those events on the file for the area
between $45^{\circ}-60^{\circ}$N and 10°W-15°E, with a reported magnitude of
five or more. Some early events of this size may not have been
allocated a magnitude, and may therefore be missed. The first
event on this list is that of 16 November 1911, in southern
Germany, and the next is not until 1927. Concentrations of events
are shown in southern Switzerland (1946 and 1965) and in the
Friuli region at the head of the Adriatic Sea (1976). There

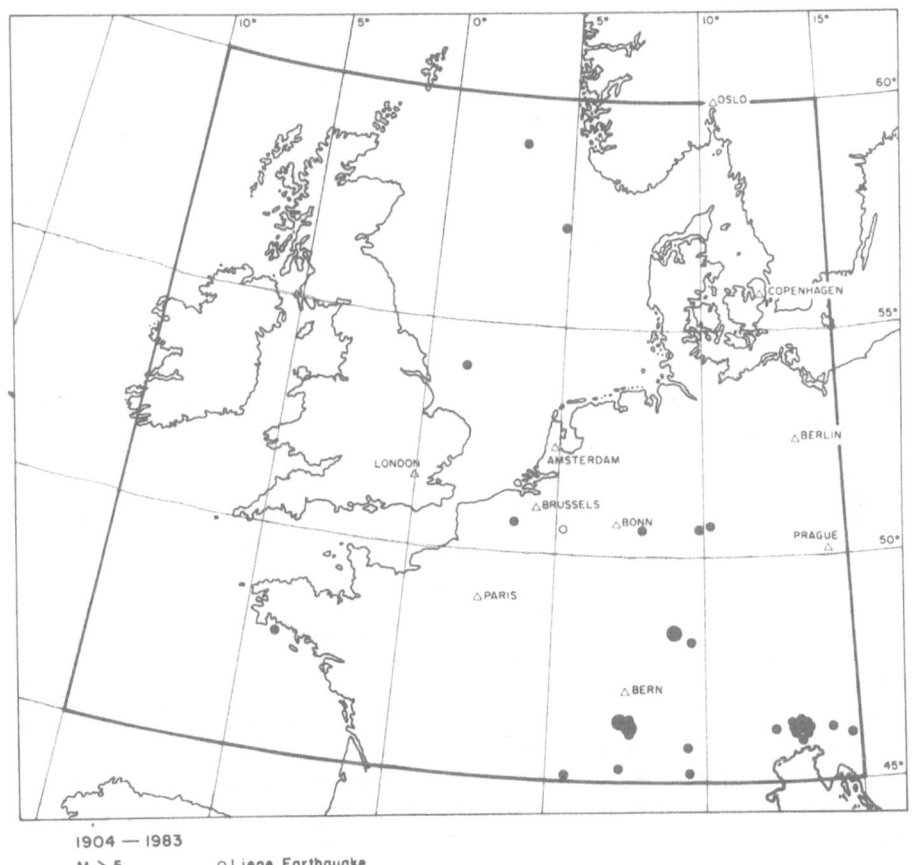

1904 — 1983

M ≥ 5   o Liege Earthquake

Figure 3. Events on ISC Historical File with magnitude
5 or more between $45^{\circ}-60^{\circ}$N and 10°W-15°E.

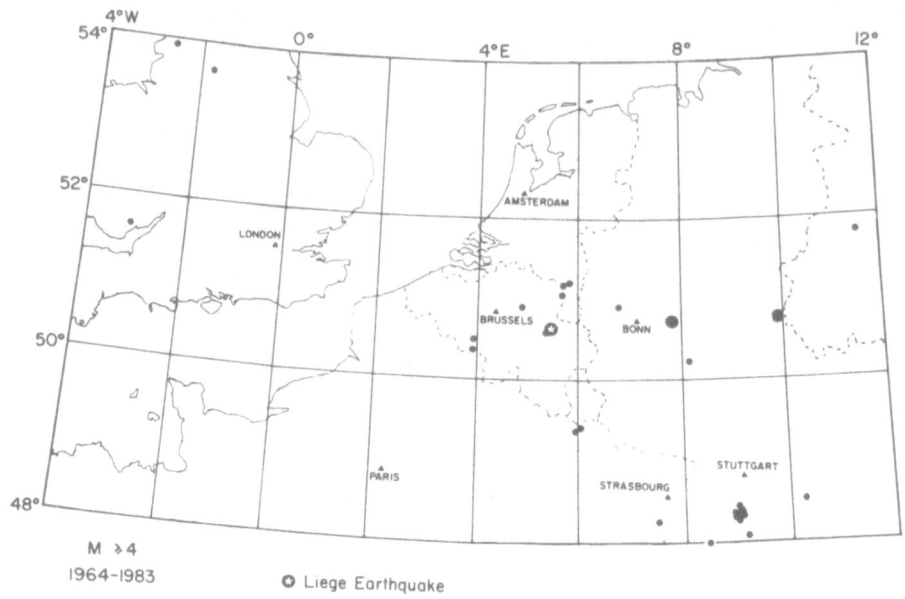

Figure 4. Events on ISC Historical File for region
shown, from 1964 to 1983, with magnitude 4 or more.

is little activity in northern Europe; the only event shown in
the Belgian region is to the south-west of Brussels, and occurred
on 11 June 1938. It is seen that no event in this area between
50° and 60°N has been allocated a magnitude of six or more this
century.

Figure 4 shows earthquakes of magnitude four and greater
for a smaller area from 1964 to the present. A starting date
of 1964 was chosen because of a rise in the number of detected
events from then. Again, some events in this magnitude range may
be missed because they were not allocated magnitudes. The
greatest concentration of activity is seen south of Stuttgart,
and is the Schwabian-Jura sequence of 1978. The two largest
events are those to the east of Bonn, one close to the border
with the German Democratic Republic in June 1975, and the other
in June 1982. This plot includes the event near Liège on
21 December 1965 and also some activity near the Belgian-Netherlands border, and in the region of Mons to the south-west of
Brussels.

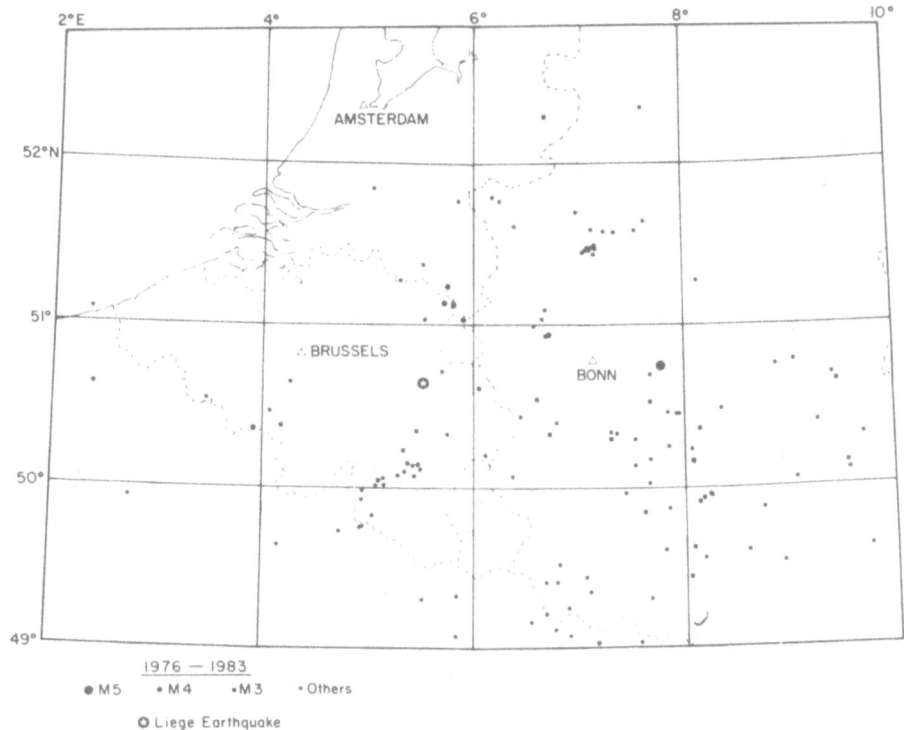

Figure 5. All events on ISC Historical File for region shown, from 1976 to 1983.

All events in the file, irrespective of magnitude, are shown in Figure 5 for the period since 1976, for which many determinations from LDG are available. Again, there is activity in the Limburg region near the Belgian-Netherlands border, and also numerous events reported from the Ardennes to the south of Liège. Many of these are likely to be artificial explosions related to mining and quarrying. This plot shows only little activity to the south-west of Brussels, for the extensive activity near Charleroi in 1948 and 1965 falls outside the chosen time-period. Some of these events are shown, however, in Figure 6, which plots all events since 1960 in a more restricted area. The main feature of this plot is the swarm of activity near Charleroi and Mons which occurred in 1965. The difference shown in the activity of this area between Figures 5 and 6 emphasises the difficulty in describing patterns of activity in areas such as north-western Europe where the level of seismicity is low, and activity is sporadic. Such areas often exhibit bursts of activity in limited areas followed by quiet periods. Assessments of seismicity, therefore can depend strongly on the

period studied. This fact stresses the need to extend records
obtained from instrumental recording as far back as is possible.

These plots show that the most seismically active parts of
Belgium are in the east, along the Rhine Graben and the Ardennes.
Significant earthquakes, particularly in the form of swarms
have occurred in the Hainault region near the south-western
border, but in recent years there appears to have been little
activity in the Flanders region or the Brabant Massif.

THE LIEGE EARTHQUAKE, 8 NOVEMBER 1983

There is a long historical record of earthquakes affecting
the Liège region, and among events reported are those of 1395,
1504, 1640, 1692 and 1755 (van Gils and Zaczek, 6). The most
recent previous earthquake to be centred near Liège was that of
21 December 1965 (ML 4.4) which was felt most strongly in the
north-east part of the city, with an intensity of VII.

The earthquake at Liège on 8 November 1983 was the largest
in Belgium since that south-west of Brussels on 11 June 1938
which had a magnitude of 5.6 (Gutenberg and Richter, 7). The
magnitude of the Liège earthquake is estimated as 4.9 (ML), not
large by world standards, but its position immediately beneath
the city and its shallowness resulted in much more severe
effects than might generally be expected from such an event.
The preliminary estimates of its position are as follows:-

Agency	Origin time h m s	Lat N	Long E	Assumed Depth (km)
Uccle	00 49 34.5	50.64°	5.51°	0
Uccle	00 49 34.2	50.63°	5.38°	4
PDE	00 49 32.0	50.73°	5.35°	10

The position given by Uccle, constrained to a surface focus
is immediately beneath the area of highest damage, and cannot be
greatly in error. Other studies, however, suggest that the focus
is a few kilometres deep. The position given by the PDE service
is about 10 km north-west of Liège, and does not accord with
locally observed effects. If this solution were fixed at the
surface, the epicentre could well move closer to the area of
maximum intensity.

The maximum felt effects were in an area only a few
kilometres across mainly in the suburbs of Glain, St Nicolas
and St Gilles in the hills to the west of the city centre.
This concentration of effects indicates a focus not more than a
few kilometres deep. From a visit about ten days after the

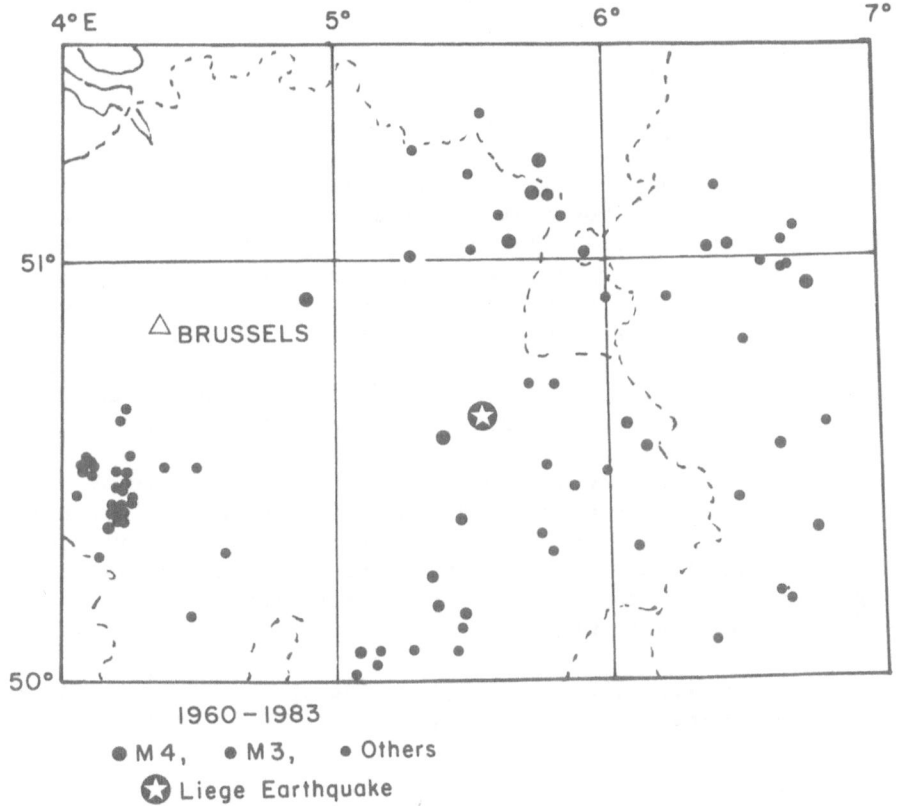

Figure 6. All events in ISC Historical File for region shown, from 1960 to 1983.

earthquake, the intensity in this area was estimated to be generally VII (MM or MSK) but reaching VIII locally, particularly in the cemetery of Montegnée, where damage to monuments showed evidence of strong acceleration, both vertical and horizontal.

This area of strong shaking is several kilometres south-west of that of the 1965 earthquake. For the structures most affected in 1983, this was the first occasion on which they had been shaken with such severity, and the damage was therefore greater than it would have been in a shock of similar size in an area prone to more frequent shaking. A similar situation arose in 1974 in New Zealand, when an earthquake of magnitude 5.0 occurred 10 km from the city of Dunedin, in an area that had not experienced strong shaking since European settlement and the intensity, also rated VII, caused damage far greater than would generally be expected from such an event in a more active part of New Zealand (Adams and Kean, 8).

68   R. D. ADAMS

CONCLUSION

　　The seismic history of northern Europe shows that there is
significant activity, of which the Liège earthquake is the most
recent example. Such events of only moderate magnitude can
cause casualties and damage far beyond what would be expected in
areas of more frequent shaking. The plots of earthquakes presented in this paper show that the activity in northern Europe
is more diffuse and more difficult to localise than in many areas
of higher activity, such as those where inter-plate earthquakes
are clearly defined. The study of seismicity in such areas of
lower activity is therefore a more difficult task, but just
because of its infrequent nature earthquake activity here presents
more of a risk to communities that have short memories, and do
not wish to finance adequate earthquake countermeasures.

　　Despite this difficulty, precise studies of seismicity using
modern techniques of recording and analysis, and backed by full
investigations of past seismicity, including the compilation of
catalogues such as that searched here, can help to specify the
seismic hazard, and to delineate those areas most likely to
experience severe shaking.

　　It is important that the Liège earthquake serve to remind
all those concerned of the possible danger of such earthquakes
in the future.

ACKNOWLEDGEMENTS

　　I thank the staff of the Royal Observatory, Brussels, and
particularly Miss M De Becker and Mr T Camelbeeck for providing
much information about the Liège earthquake, both at the time of
its occurrence and subsequently; Prof N N Ambraseys, Imperial
College, London, for sharing his experience with me during a
visit to Liège on 17 and 18 November; and Messrs A G Cross
and J Watkins of the Department of Geology, University of Reading
for assistance in preparing the figures.

REFERENCES

1. Adams, R.D., Hughes, A.A. and McGregor, D.M. 1982,
       Analysis Procedures at the International Seismological
       Centre. Phys. Earth Planet. Interiors 30. pp 85-93.

2. Jeffreys, H. 1961, The Theory of Probability. Oxford
       University Press, Oxford, 3rd edition.

3. Engdahl, E.R. and Gunst, R.H. 1966, Use of a High Speed
    Computer for the Preliminary Determination of
    Earthquake Hypocentres. Bull. Seismol. Soc. Amer. 56,
    pp 325-336.

4. Adams, R.D. 1981, ISC Determinations of Previously Unreported
    Earthquakes in Europe. Proc. 2nd Internat. Symp. on
    Analysis of Seismicity and on Seismic Hazard. Liblice,
    Czechoslovakia, May 18-23, 1981, pp 132-139.

5. Tittel, B. and Wendt, S. 1982, Vorbehalte gegenueber der
    unkritischen Verwendung wenig gesichter seismologischer
    Daten. Gerlands Beitr. Geophysik 91, pp 271-273.

6. Van Gils, J-M., and Zaczek, Y. 1978. La Séismicité de
    la Belgique et son Application en Génie Parasismique.
    Annales des Travaux Publics de Belgique. No. 6, pp 38.

7. Gutenberg, B. and Richter, C.F. 1954, Seismicity of the
    Earth and Associated Phenomena. Princeton University
    Press, New Jersey.

8. Adams, R.D. and Kean, R.J. 1974, The Dunedin Earthquake,
    9 April 1974. Part 1; Seismological Studies. Bull. N.Z.
    Nat. Soc. Earthquake Engineering 7, pp 115-122.

THE SEISMICITY OF THE RHINEGRABEN RIFT SYSTEM
- SOURCE PARAMETERS, PROPAGATION- AND SITE EFFECTS -

K.-P. Bonjer

Geophysical Institute, University of Karlsruhe
Hertzstrasse 16, D-7500 Karlsruhe 21, FRG

EXTENDED ABSTRACT

In the last two decades seismological investigations of the Rhinegraben Rift System considerably improved the knowledge about the manifold menu presented by the seismotectonic regime acting today. One of the earliest findings was the consistency of the direction of the stress field having a prevailing NW-SE trending strike (Ahorner, 1970). The Lower Rhine Embayment, the Rhenish Massif and the northernmost part of the Upper Rhinegraben - approximately up to the latitude of Heidelberg - still experience active rifting (Ahorner et al., 1983). The middle and southern part of the Upper Rhinegraben dominantly show strike slip dislocations (Schneider et al., 1966; Ahorner, 1970; Ahorner and Schneider, 1974; Bonjer, 1979; Bonjer et al., 1984) as displayed in Fig. 1. However, in the southern Rhinegraben - Black Forest area very recent observations (e.g. in Fig. 2) indicate that normal faulting may play a similar intrinsic role like in the northern Upper Rhinegraben area (Bonjer and Apopei, 1984), as already found and postulated by geological investigations (e.g. Illies, 1981).

As far as the epicenter distribution and the focal depth in the Upper Rhinegraben area are concerned some peculiarities have been found (Bonjer et al., 1984). The comparison of historical and instrumental data (Fig. 3) reveals three facts. Firstly, local source areas exhibit a significant geographical consistency for the past millennium. Secondly, short-time but high sensitivity observations show that most of these focal areas are marked by persistent microseismicity which already reflects the general pattern of the Rhinegraben seismicity.

P. Melchior (ed.), Seismic Activity in Western Europe, 71–83.
© *1985 by D. Reidel Publishing Company.*

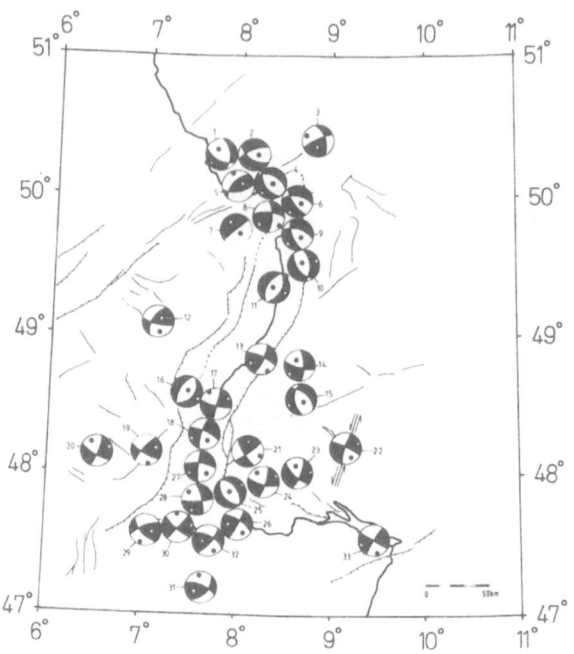

Fig. 1: Fault-plane solutions (Wulff projection) in the Upper
         Rhinegraben area (after Bonjer et al., 1984). Black
         quadrants: compression; white quadrants: dilatation;
         black dots: P-axes; white dots: T-axes.

        Both main border fault-systems create only negligible
seismic activity, if at all. This has already been mentioned by
Hiller et al. (1967). Concerning the instrumental data, this
lack of activity seems to be obvious at the western border
faults from rough inspection of the epicenter map (Fig. 4).
Apparently, such a deficiency of seismicity does not exist at
the eastern border faults, if only the epicenter distribution is
considered. However, it emerges from Fig. 5 that even the
eastern border faults are lacking seismic activity at least down
to a depth of about 10 km. The thickness of the seismogenic part
of the crust varies laterally in the Upper Rhinegraben area. In
the Vosges Mountains and in the graben proper, depths of 13 and
16 km respectively are not exceeded. Maximum depths down to
about 20 km are found in the Black Forest. No earthquake was
reliably detected in the lower crust or in the mantle. The
asymmetry in the maximum focal depth perpendicular to the strike
of the graben has been explained by Bonjer et al. (1984) by

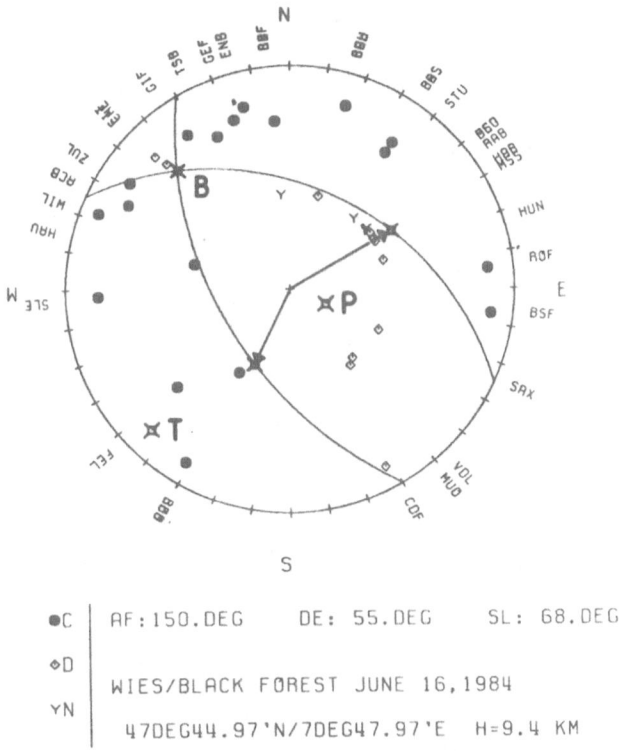

Fig. 2: Fault-plane solution (lower hemisphere Wulff projec-
         tion) of the M_L = 2.9 Wies event (after Bonjer and
         Apopei, 1984). The strike of the B-axes (represent the
         direction of the regional stress field) is 137 degrees.

lateral variations in the temperature-depth distribution using
the model of Brace and Kohlstedt (1980). This result is supported by the findings of Chen and Molnar (1983).

      Due to the increased number of digital recording stations
in the southern Rhinegraben area, studies of spectral source
parameters, propagation- and site effects were initiated (Durst,
1981; Gruber 1983). Following a graphical representation of
spectral source parameters (Hanks and Thatcher, 1972) a
composite Ω_o-f_o-graph of Rhinegraben earthquakes was compiled
(Fig. 6). A cut-off frequency of 19 Hz is clearly visible for
the Sierentz-data (Gruber, 1983) recorded by the station

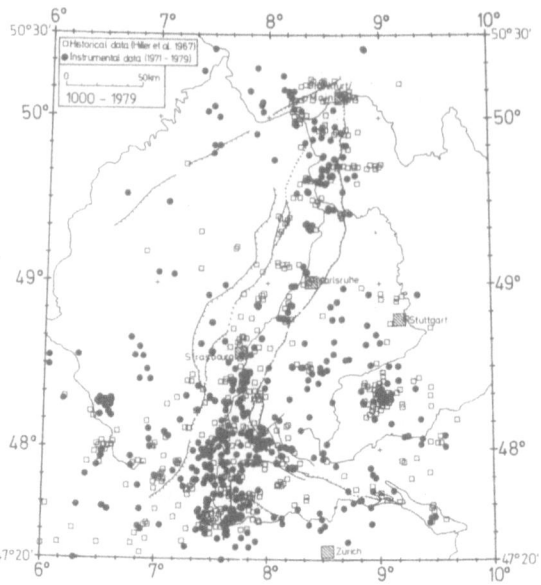

Fig. 3: Comparison between historical and present day seismic
         activity (after Bonjer et al., 1984).
         □ = historical (mainly macroseismic) epicenters
             (1000-1970)
         ● = instrumentally determined epicenters (1971-1980).

Feldberg (FEL) at an epicentral distance of about 46 km. This
cut-off frequency is not a site effect because of the higher
corner frequencies, observed for earthquakes at shorter epicentral distances (Bonjer and Gruber, 1983). To what extent the
low-pass filtering has to be attributed to path or source
effects remains open so far. In Fig. 6 the Mammoth-Lake data of
Archuleta et al. (1982) are also displayed to demonstrate the
existence of the f_{max}-effect (Hanks, 1982), which often is used
as a proof for the violation of the similarity law of Aki
(1967). On the other hand, the Feldberg-data are, when accepting
the constancy of stress drops, at least indicative that in the
moment range $18 < \log M_0$ (dyne cm) < 22 similarity holds (Bonjer
et al., 1982). Another result of the above mentioned studies was
the confirmation of a linear relationship between $\log M_0$ and the
local magnitude M_L (Durst, 1981; Gruber, 1983) - regardless of
the derivation procedure. For the southern graben the following
relationship has been found (Gruber, 1983; Bonjer and Gruber,
1983), when referring to Richter's (1935) California magnitude

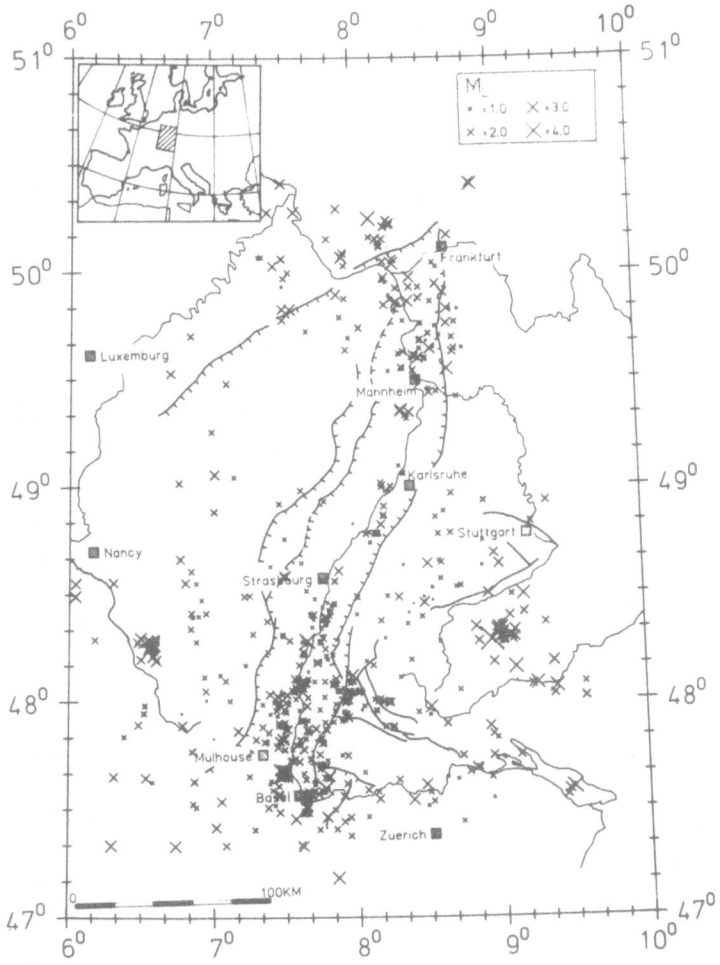

Fig. 4: Epicenters (1971-1980) in the Upper Rhinegraben area
         (after Bonjer et al., 1984).

(i.e. taking the difference in Q_S into account):

$$\log M_0 = (1.02 \pm 0.07) \, M_L + (17.84 \pm 0.13) \qquad (1)$$

The seismotectonic meaning of the intercept term in eq.
(1) - if this exists at all - has not been discussed to our
knowledge up to now, probably due to insufficient data towards
the zero-magnitude. A first attempt has been made by Jung (1983)

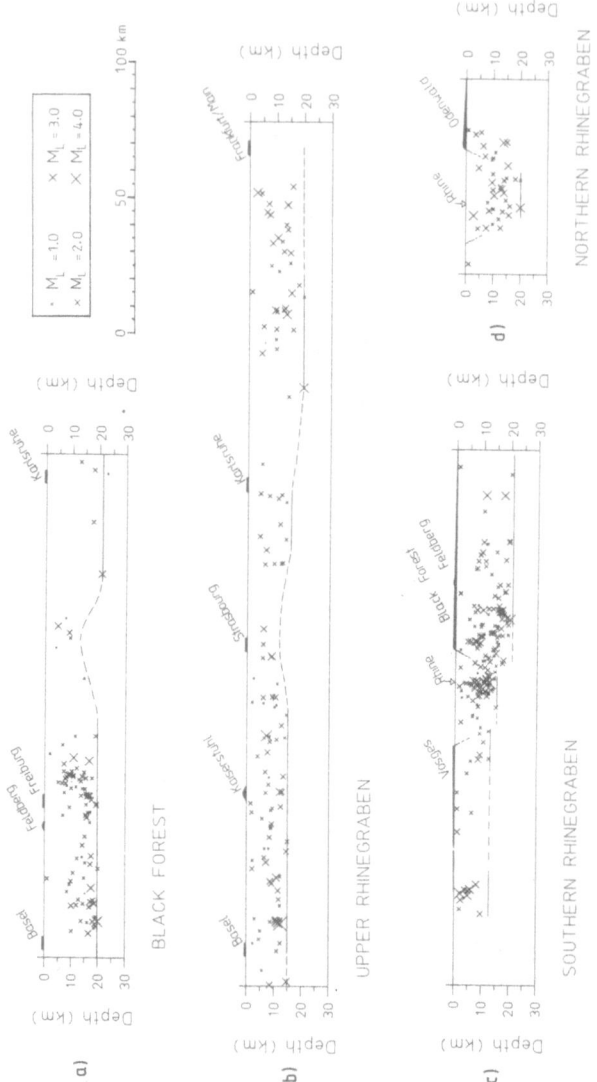

Fig. 5: Depth distribution of foci (1971-1980) after Bonjer et al. (1984). Cross section a: Black Forest, parallel to the strike of the Graben; b: Graben proper, average strike direction N20°E; c: Perpendicular to the strike of the southern Graben. Hypocenters between Strasbourg and Basel were projected; d: Northern Graben, strike direction approximately EW. The boundaries of greatest focal depths are marked by solid lines.

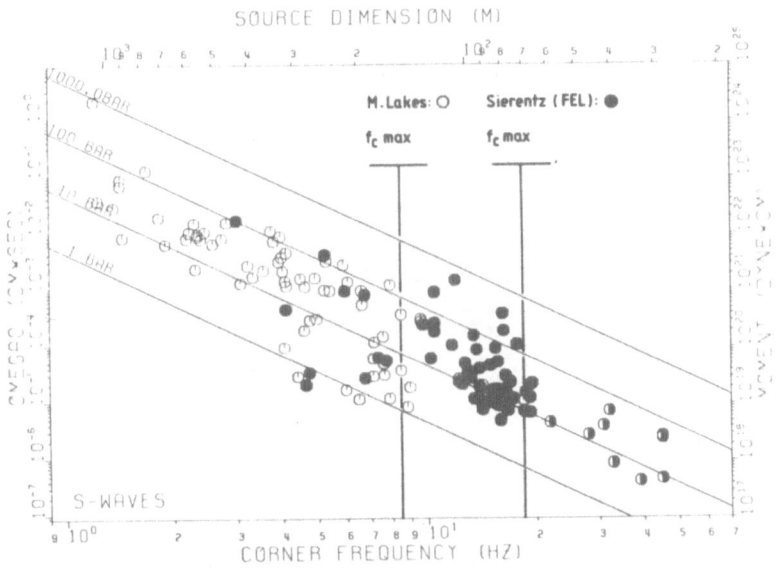

Fig. 6: Graphical representation of the spectral plateau Ω_0 as a function of the corner frequency f_0. Open circle: Mammoth Lakes data; solid circles: Sierentz data recorded at FEL (Black Forest); half filled circles: local events recorded at FEL with epicentral distances less than 30 km.

comparing the crustal and intermediate depth seismicity of the Vrancea focal area in Romania. Although differences in the elastic parameters are not sufficient to account for the offset in crustal and intermediate data (Apopei and Bonjer, 1984) we still expect, that the intercept term may serve as a tool to distinguish between different seismotectonic provinces.

The difficulty in separating source, propagation and site effects is already emerging, when local magnitudes have to be calculated. One of the purposes of this paper is to deduce the local magnitude of the Liège-1983 event. Fig. 7 depicts the data of some particular stations which have digitally recorded this earthquake at epicentral distances in the range of 329-480 km. This kind of recording allowed the simulation of a seismogram as seen by a Wood-Anderson seismograph system. To assign the local magnitude Richter's amplitude-distance relationship (1935) was used.

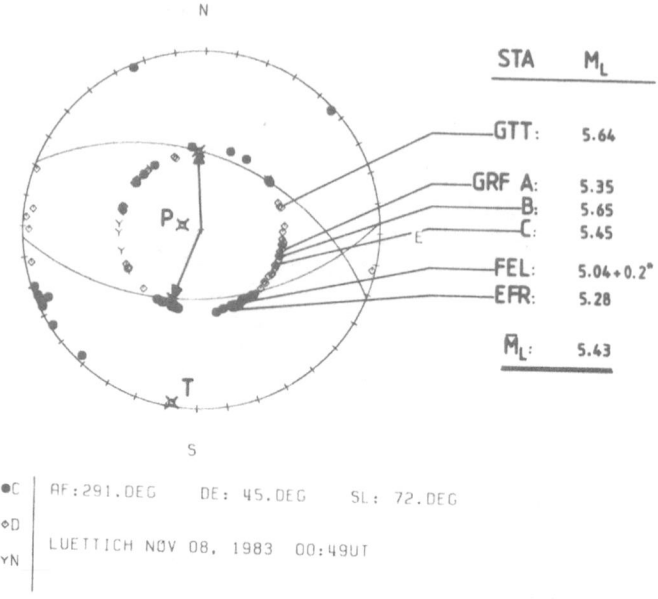

STA M_L

GTT: 5.64

GRF A: 5.35
     B: 5.65
     C: 5.45

FEL: 5.04+0.2°
EFR: 5.28

$\bar{M}_L$:   5.43

● C | AF:291.DEG   DE: 45.DEG   SL: 72.DEG
◇ D |
◇ D | LUETTICH NOV 08, 1983   00:49UT
▽ N |

Fig. 7: Uncorrected local magnitudes of the Liège event calcul-
        ated on the basis of Richter's amplitude-distance
        terms. The focal mechanism is the preferred solution
        KA1 of Faber and Bonjer (1985).

Table 1

STA	Log A	R(km)	$\log A_o$	M_L	ΔM_Q	M_L^Q
GTT	1.44	329.2	4.2	5.64	-0.71	4.93
FEL	0.74	358.3	4.3	5.24*)	-0.36	4.88
EFR	0.98	363.8	4.3	5.28	-0.35	4.93

M_L^Q = Local magnitude, corrected for the difference in anelas-
      tic attenuation between California and Central Europe.

ΔM_Q = Correction term accounting for difference in attentua-
      tion

*) ΔM = +0.2 is added (Lippert, 1979)

        Table 1 demonstrates the dominating role of the distance-term (log A_o) which at our distances obviously controls the local magnitude. The magnitudes at the stations shown in Fig. 7, derived from the recordings of the Liège-event have an average value of M_l = 5.43. This value is considerably greater than those presented at the workshop by Camelbeeck and de Becker (M_l = 4.9), and Ahorner (M_l = 5.1). Listing the magnitudes of the Liège-earthquake with respect to the epicentral distances (Table 2) it turns out that apparently the magnitude increases monotoniously with distance (the data of GTT and GRF-B are omitted and will be discussed separately).

Table 2

Station	Δ (km)	M_L
B*	90.	4.9
BNS	124.1	5.1
FEL	358.3	5.24
EFR	363.8	5.28
GRF-A	422.6	5.35
GRF-C	471.3	5.45

B* = Belgian stations

There are now numerous examples (also with different radiation pattern) available in Central Europe showing this increase in magnitude with distance as discussed, e.g. by Gruber (1983), and Kradolfer (1984). We attribute this effect to an inappropriate use of Richter's amplitude-distance function (derived from earthquakes with low dominating frequencies) for a region with less attenuation compared to California (and therefore in general with higher frequency content). The application of regional specific distance terms would avoid the apparent increase in magnitude with distance, and account for a consistent magnitude determination. However, this magnitude would differ from the California magnitudes especially for small quakes. This can be proved best by comparing the particular seismic moments (Gruber, 1983). We, therefore, prefer a different approach to determine a local magnitude which comes closer to the original definition. When computing the synthetic Wood-Anderson seismogram the difference in anelastic attentuation between Central Europe (excluding areas with sediment coverage) and California (ΔQ_s = 300, Durst, 1981) is taken into account. The procedure is equivalent to deriving the seismic moment in the frequency domain. It has been applied to the digital recordings of the stations GTT, FEL and FEL (Table 1). An average local magnitude of M_L^Q = 4.91 for the Liège earthquake is achieved in this way. Introducing this value into equation (1) the seismic moment can be estimated to be:

$$M_o = 0.71 \ 10^{23} \ \text{dyne cm.}$$

Concerning the digital records of GTT the interpretation of the large, short period (S-wave) amplitudes, which are responsible for the high, uncorrected magnitude value of M_L = 5.64 (Table 1), remains ambiguous. Whether or not we are faced with local site effects - as seems to be the case at the GRF B-subarray (see Fig. 7), where numerous examples are observed (Faber, personal communication) - or with effects related to the rupture process, remains still open. In addition to the extraordinary high frequency content of the P- and S-wave groups in the GTT record, the comparison of the displacement seismograms with those of the station EFR (Fig. 8) shows a striking difference in the ratios of the radial and transverse components. Both stations are close to different nodal planes (Fig. 7), and roughly at the same epicentral distance (Table 1). We expect that waveform-matching of the S-wavelets will contribute to the understanding of the rupture process of the Liège event.

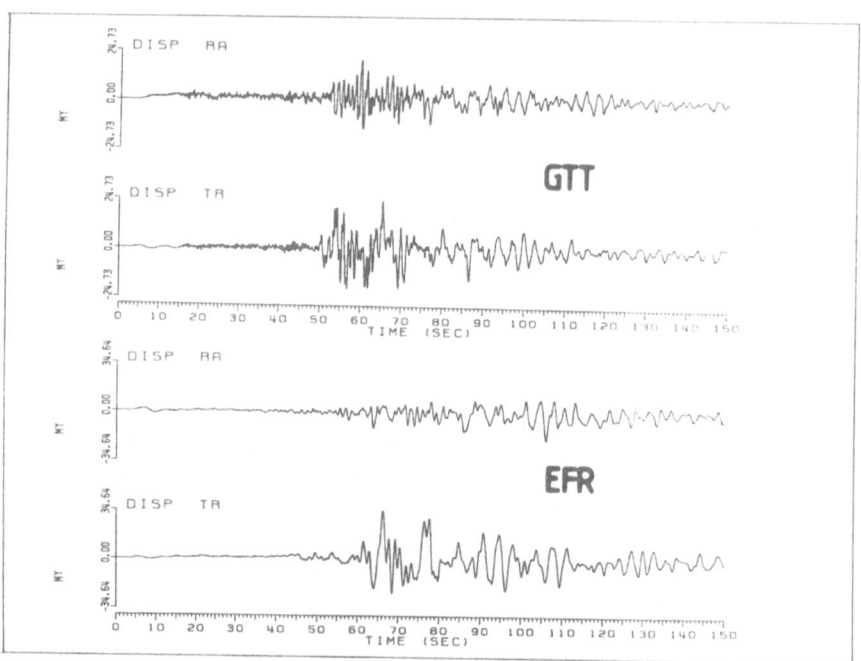

Fig. 8: Liège 1983 mainshock: Radial and transverse components of the displacement restitutions (bandpass filtered between 0.05-6.25 Hz) of the digital velocity-recordings at GTT and EFR (for azimuth see Fig. 7).

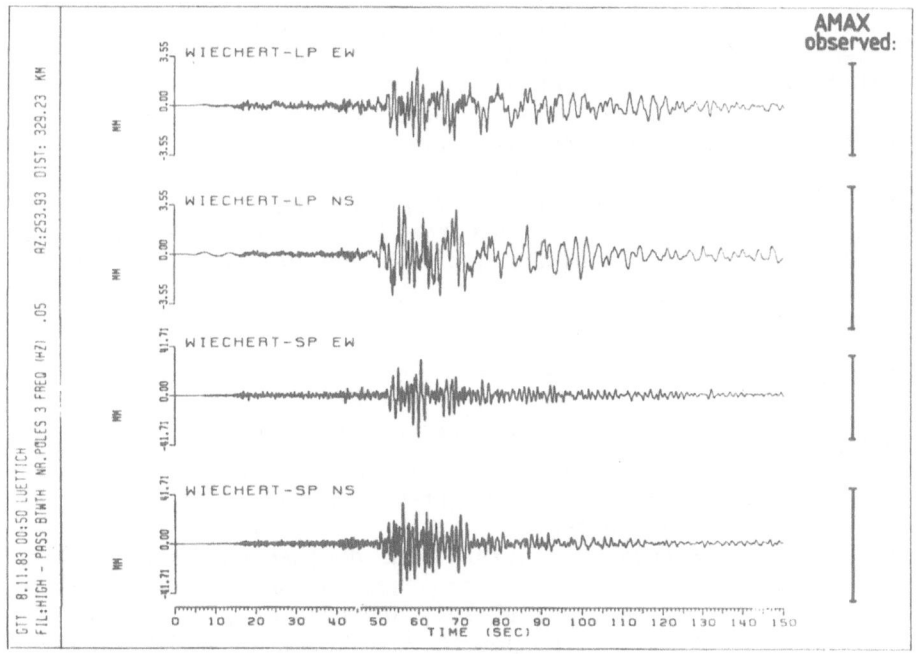

Fig. 9:   Göttingen (GTT) LP- and SP-Wiechert seismograph simula-
          tion of the Liège-1983 earthquake. Due to the difficul-
          ties in reprocuding the original Wiechert-seismograms
          only maximum peak to peak amplitudes are shown for
          comparison.

Making advantage of the data recorded by the temporary
installed digital system at GTT first simulations of the
recordings of the short and long period Wiechert seismographs,
working since decades, were performed. Fig. 9 shows the
simulated seismograms of the Liège event. The bars at the right
side indicate the maximum peak to peak amplitudes of the
original Wiechert-seismograms. The conformity of recorded and
simulated data is at least satisfactory and provides a
quantitative basis for the interpretation of historical recordings at this station.

ACKNOWLEDGMENTS

I am grateful to the staff of the Geophysical Institute,
University of Göttingen, for maintenance of our digital station

and to S. Faber for providing the GRF-data. I appreciated very much the discussions especially with I. Apopei, A. Gruber, K. Koch and D. Seidl concerning the problems related to digital filtering, restitution and simulation of different seismograph-systems. I thank I. Apopei for providing her programs for restitution and simulation of seismic signals and K. Koch for his plot-routines. The computations were carried out on the UNIVAC 1100 at the computer center of the University of Karlsruhe, and on the Raytheon 500 and HP 1000 of the Geophysical Institute. J. Mechie read the manuscript. G. Bartman typed the manuscript. This research was sponsored partly by the SFB 108 (Sonderforschungsbereich) at the University of Karlsruhe.

REFERENCES

Ahorner, L. (1970). Seismo-tectonic relations between the Graben zones of the Upper and Lower Rhine valley. In: Graben Problems, Illies, H., Mueller, St. eds.: pp. 155-166, Schweizerbart Stuttgart.

Ahorner, L. & Schneider, G. (1974). Herdmechanismen von Erdbeben im Oberrheingraben, In: Approaches to Taphrogenesis, Illies H., Fuchs, K. eds: pp. 99-104, Schweizerbart Stuttgart.

Ahorner, L., Baier, B. & Bonjer, K.P. (1983). General pattern of seismotectonic dislocations and the earthquake generating stress field in Central Europe between the Alps and the North Sea. In: Plateau Uplift - The Rhenish Massif, A Case History, Fuchs, K., K. v. Gehlen, Mälzer, H., Murawski, H., Semmel, A. eds: pp. 187-197, Springer Verlag.

Aki, K. (1967). Scaling law of seismic spectrum. J. Geophys. Res. 73, pp. 1217-1231.

Apopei, I. & Bonjer, K.-P. (1984). Moment-magnitude scaling laws for shallow and intermediate depth Vrancea earthquakes, submitted to Tectonophysics.

Archuleta, R.J., Cranswick, E., Mueller, C. & Spudich, P. (1982). Source parameters of the 1980 Mammoth Lakes, California, earthquake sequences. J. Geophys. Res. 87, 4595-4607.

Bonjer, K.-P. (1979). The seismicity of the Upper Rhinegraben. A continental rift system. In: Proc. Int. Research Conf. on Intra-Continental Earthquakes. Petrovski, J., Allen, C. eds: pp. 107-115, Ohrid, Yugoslavia.

Bonjer, K.-P. & Apopei, I. (1984). Seismic dislocation pattern in the southern Rhinegraben / Black Forest area, in preparation.

Bonjer, K.-P. & Gruber, A. (1983). Die Sierentz-Erdbebenserie 1980/81. Berichtsband des Sonderforschungsbereich 108, pp. 91-108, Karlsruhe.

Bonjer, K.-P., Gruber, A., Jung, P., Durst H. & Kaminski, W (1982). On magnitudes and scaling laws of small Rhinegraben earthquakes (abstract). EOS Trans. AGU 63, 1266.

Bonjer, K.-P., Gelbke, C., Gilg, B., Rouland, D., Mayer-Rosa, D., Massinon, B. (1984). Seismicity and dynamics of the Upper Rhinegraben. J. Geophys., in press.

Brace, W.F. & Kohlstedt, D.L. (1980). Limits on lithospheric stress imposed by laboratory experiment. J. Geophys. Res. 85, 6248-6252.

Chen, W.-P., Molnar, P. (1983). Focal depths of intracontinental and intraplate earthquakes and their implications for the thermal and mechanical properties of the lithosphere. J. Geophys. Res. 88, 4183-4214.

Durst, H. (1981). Digitale Erfassung und Analyse der physikalischen Prozesse in den Erdbebenherden aus dem Bereich des Oberrheingrabens. Diplomarbeit, Geophysikalisches Institut, Universität Karlsruhe.

Faber, S. and Bonjer, K.-P. (1985). Phase recognition and interpretation at regional distances from the Liège event of November 8, 1983, this volume.

Gruber, A. (1983). Die Sierentz-Erdbebenserie von 1980/81. Unterschung der Raumzeit-Verteilung von Nachbeben und deren spektrale Herdparameter. Diplomarbeit, Geophysikalisches Institut, Universität Karlsruhe.

Hanks, T.C. (1982). f_{max}, BSSA 72, 1867-1879.

Hanks, T.C. & Thatcher, W. (1972). A graphical representation of seismic source parameters. J. Geophys. Res. 77, 4393-4405.

Hiller, W., Rothé, J.-P. & Schneider, G. (1967). La seismicité du Fossé Rhénan. In: The Rhinegraben Progress Report 1967, Rothé, J.-P., Sauer, K. eds: pp. 98-100, Abh. geol. Landesamt Baden-Württemberg Freiburg/Straßburg.

Illies, J.H. (1981). Mechanism of Graben Formation. Tectonophysics, 73, 249-266.

Jung, P. (1983). Herdparameter und Ausbreitungseffekte krustaler und mitteltiefer Erdbeben in den rumänischen Karpaten unter besonderer Berücksichtigung der Vrancea-Region. Diplomarbeit, Geophysikalisches Institut, Universität Karlsruhe.

Kradolfer, U. (1984). Magnitudenkalibrierung von Erdbebenstationen in der Schweiz. Diplomarbeit, ETH-Zürich.

Lippert, W. (1979). Erstellung von Formeln für die Magnitudenbestimmung von Nahbeben der Jahre 1971-1978 im Bereich des Oberrheingrabens. Diplomarbeit, Geophysikalisches Institut, Universität Karlsruhe.

Richter, C.F. (1935). An instrumental earthquake magnitude scale. Bull. Seismol. Soc. Am. 25, 1-32.

Schneider, G., Schick, R. & Berckhemer, H. (1966). Fault-plane solutions of earthquakes in Baden-Württemberg. Z. Geophys. 32, 383-393.

INTRAPLATE SEISMIC ACTIVITY IN THE BRABANT MASSIF AND THE GRABEN SYSTEMS OF LOWER RHINE-ROER AND NORTH SEA

A. Reinier Ritsema

Royal Netherlands Meteorological Institute
De Bilt, The Netherlands

ABSTRACT

This is a review of the seismicity and potential seismic hazard in the intraplate region of Northwestern Europe. Earthquakes in the region take place on rejuvenated structural lines of Paleozoic, Mesozoic and Cenozoic age. The maximum credible earthquake has a magnitude of about $6\frac{1}{2}$. The recurrency rate for magnitude 6 earthquakes in the total region is of the order of 60-100 years. Locally this time of recurrence is 10-100 times longer. Co-operative measures in seismic data acquisition and exchange, and in scientific investigations in the field are advocated.

INTRODUCTION

The seismicity of the continental intraplate Europe is at least two orders of magnitude lower than that of the interplate Hellenic Arc between the plates of Europe and Africa (fig. 1). Yet, the interest in such earthquake activity however small, is not only academic. The Liège earthquake of November 8, 1983 in which an MSK Intensity of VII was reached (M. de Becker, T. Camelbeeck, this volume) and considerable damage occurred locally to houses of Class B, triggered extra attention of governments and industry in the region for the possible hazards of such events. Justification for this increased interest is found in the vulnerability of existing structures and new structures to be built in the region, such as the raising of dikes and dams, nuclear power plants, multistore buildings, and offshore drilling-, exploration and production-platforms. The greatly increased investments in such structures and the growing opportunity for

P. Melchior (ed.), Seismic Activity in Western Europe, 85–98.
© *1985 by D. Reidel Publishing Company.*

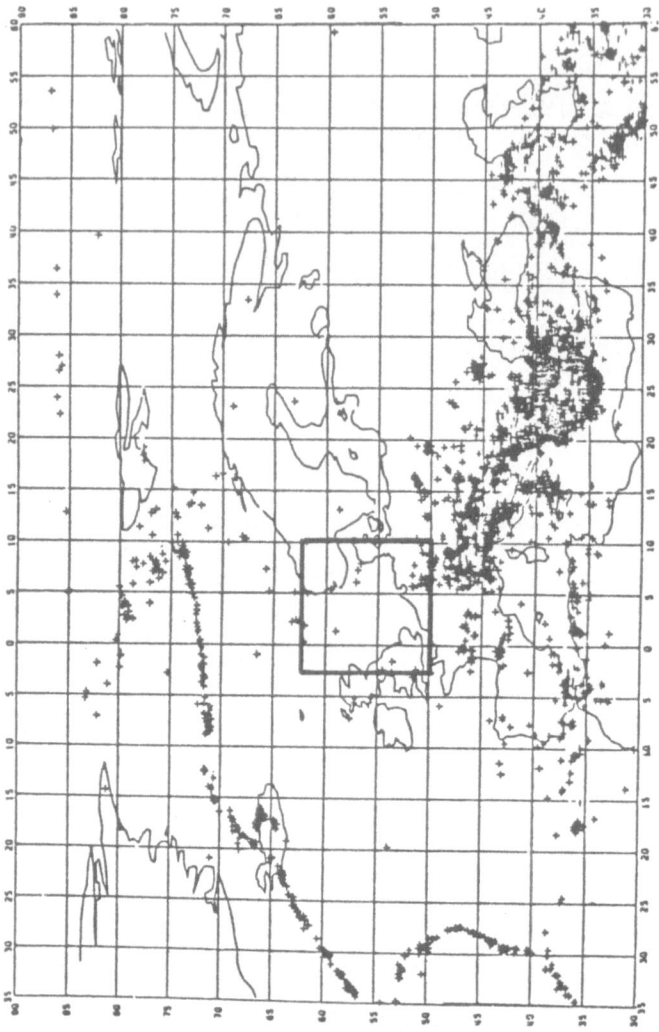

figure 1. Intraplate seismicity of NW Europe in comparison with interplate activity in the South and West. Selection of events: period 1970–1980.

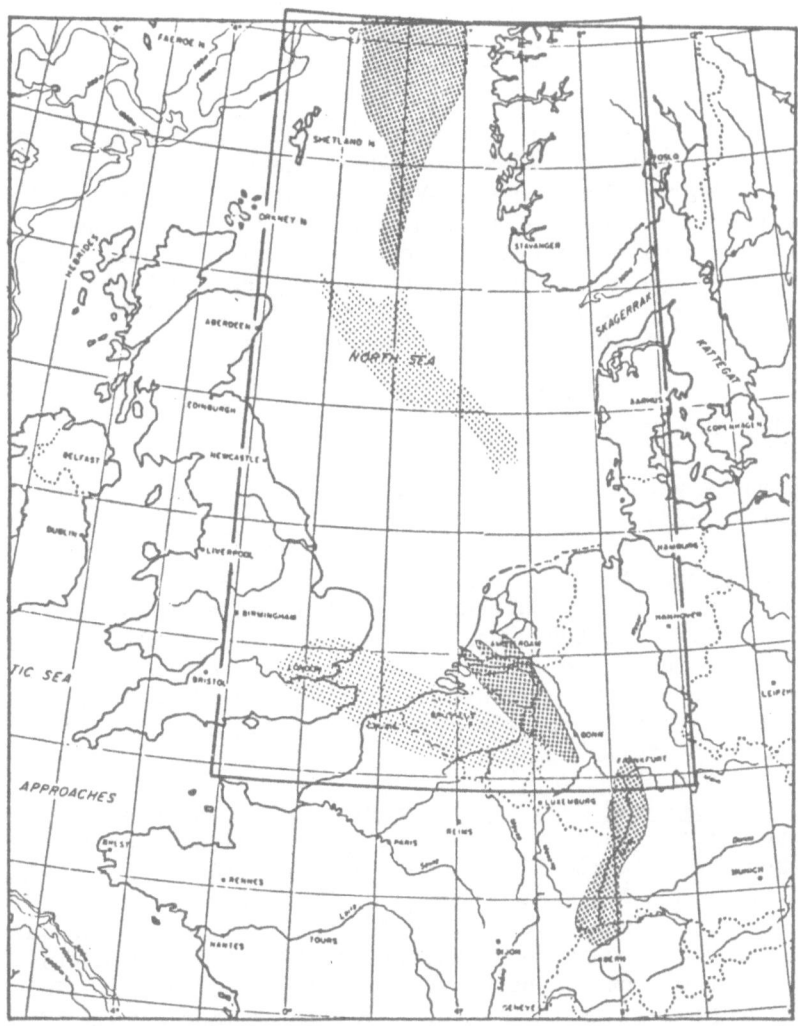

Figure 2: Location of the zones of intraplate seismic activity, from North to South: Viking Graben, Central
Graben, Lower Rhine-Roer Graben, Brabant Massif
and Upper Rhine Graben.

important consequences of an earthquake calamity on the population
which is more dense than ever, and the subsequent environmental
hazards, are additional reasons for extra attention.

DELINEATION OF SEISMIC ZONES

In the intraplate region North of the Alpine plate boundary
between Africa and Europe a clear decrease of seismic activity
is observed from the Hercynian-Variscan structures in the South
to the Cambrian and Pre-Cambrian platform regions in the North.
Main concentrations of activity are found in the Upper Rhine
Graben, not considered further in this paper, the Lower Rhine-
Roer *) Graben, the Brabant Massif, and the Central Graben and
Viking Graben situated in the North Sea (fig. 2).

AGE RELATIONS

These separate zones in the region do not have a common age
(Ziegler, 1982):

The Brabant Massif is of Paleozoic age. Mesozoic cover is
either absent, or very thin and mostly eroded away. Also in the
Cenozoic period the zone was situated above sealevel as a great
anticline. Large-scale differential movements of post-Paleozoic
age are unknown in the Brabant Massif. The present-day seismic
activity most likely takes place on rejuvenated faults of Hercynian
or Caledonian age.

The Viking and Central Graben systems in the present North Sea
generated in the Mesozoic, starting in the Triassic period, at the
time when the Atlantic Ocean between Africa and N. America was
initiated around 185 MY ago (fig. 3). The main development of the
North Sea rift systems took place during the Triassic and Jurassic.
In the Upper Cretaceous and Cenozoic differential movements stopped
in the Grabens, but a general subsidence of the complete North Sea
basin took place at the time. No topographic expressions of the
two underlying Mesozoic Graben structures are to be found at
present at the bottom of the sea. Also for these zones rejuvenation
of old faultlines -this time of Mesozoic age- is suggested as seat
of present seismic activity.

The Roer Graben has its main period of differential movements
during the Cenozoic, the period in which the Alpine orogenesis
took place at the plate boundary of Africa and Europe (fig. 3).

*)In literature the name of this Graben is often misspelled as
"Ruhr" Graben. It is, however, not the river Ruhr from which the
Graben derives its name, but the Rur in Germany or Roer in the
Netherlands.

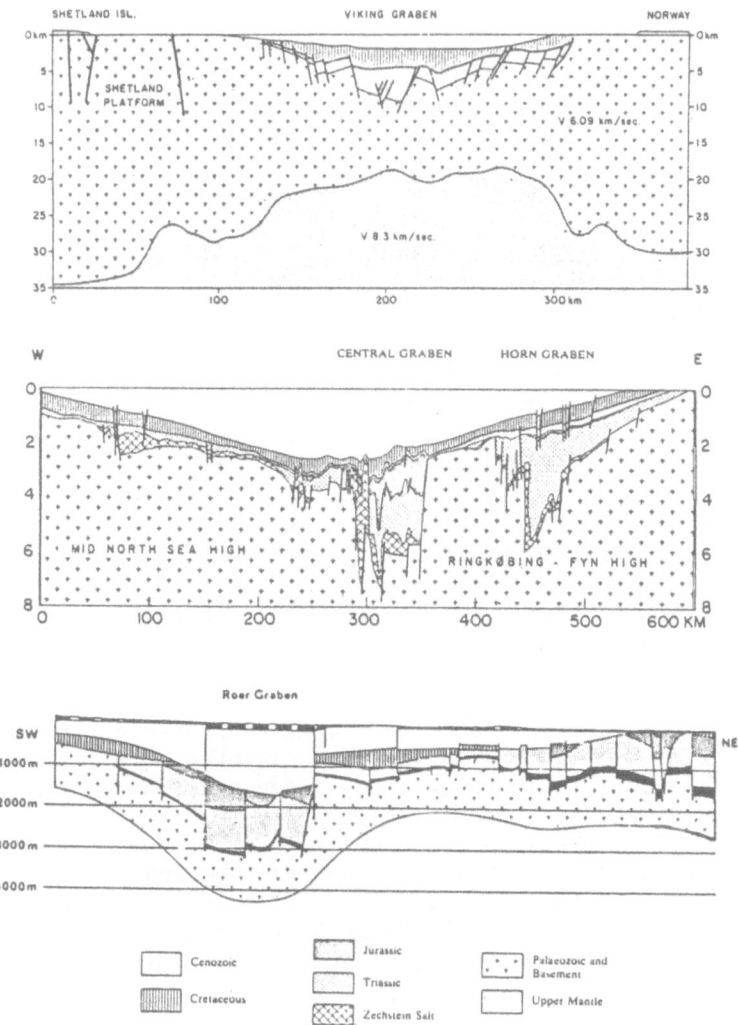

Figure 3: Cross Sections through Viking and Central
          Graben (after Ziegler, 1982) and Roer
          Graben (after Schalke, 1984).

The start of the Graben formation lies in late Eocene about 60 MY ago at the time when Greenland separated from Scandinavia as the beginning of the North Atlantic Ocean. Differential motions continue up till now. Recent and sub-recently generated fault expressions at the surface can be found in the form of steps in the landscape (fig. 4). In NW direction the Graben dies out before reaching the present coast line of the Netherlands.

CAUSE OF PRESENT ACTIVITY

The four seismic zones in the region that presently are active, thus, have originated in different geological periods and under completely different boundary conditions of the plates.

The present activity in the region could be triggered by a reorientation of the stress field starting in the Miocene following the change in direction of convergence between the African and European plates. This change of the stress direction from NNE to about NW, resulted in sinistral motions in the Upper Rhine Graben system as opposed to the former tensional motions and in purely tensional SW-NE motions in the Roer Graben (Illies, 1978; Illies, Baumann, 1982). Such a splitting apart of the fore-land could also be seen as the purely mechanical effect of conti-nent collision. The periods of active subsidence in the Roer Graben, however, do not exactly coincide with the main compres-sional events and orogenic phases in the Alpine fold belt which shows that the relation is not as simple as suggested in this model (Ziegler, 1982). Still, the role of interplate forces on the re-activation of old weak spots, lines and fractures in the Paleozoic and Mesozoic basement definitely is of influence in the localization of intraplate activity, Evidently, concentration of stress occurs at places of anomalous structure and composition, such as hot spots, intrusions, thin crust, fractured lithosphere a.o. Certain is that the Lower Rhine-Roer Graben of Cenozoic age and the Mesozoic North Sea Grabens are not part of one mega-fracture system (Ziegler, 1982).

MAXIMUM CREDIBLE EARTHQUAKE

For seismic risk studies it is of importance to dispose of a magnitude value as the maximal one to be expected during the life-time of a structure for which the study is meant. Notwithstanding the age-difference of the separate seismic zones in the region the maximum credible earthquake is of the same order of magnitude of $6\frac{1}{4}$-$6\frac{3}{4}$ for the whole region. Values mentioned in Ritsema & Gürpinar (1983) diverge between 6.5 for the Roer Graben (Rosen-hauer, Ritsema, Ahorner), $6\frac{3}{4}$ for the Brabant Massif s.l. (v. Gils, Zaczek), 6.7 for the North Sea Grabens (Burton), to 7-$7\frac{1}{2}$ for the whole region by Ringdal.

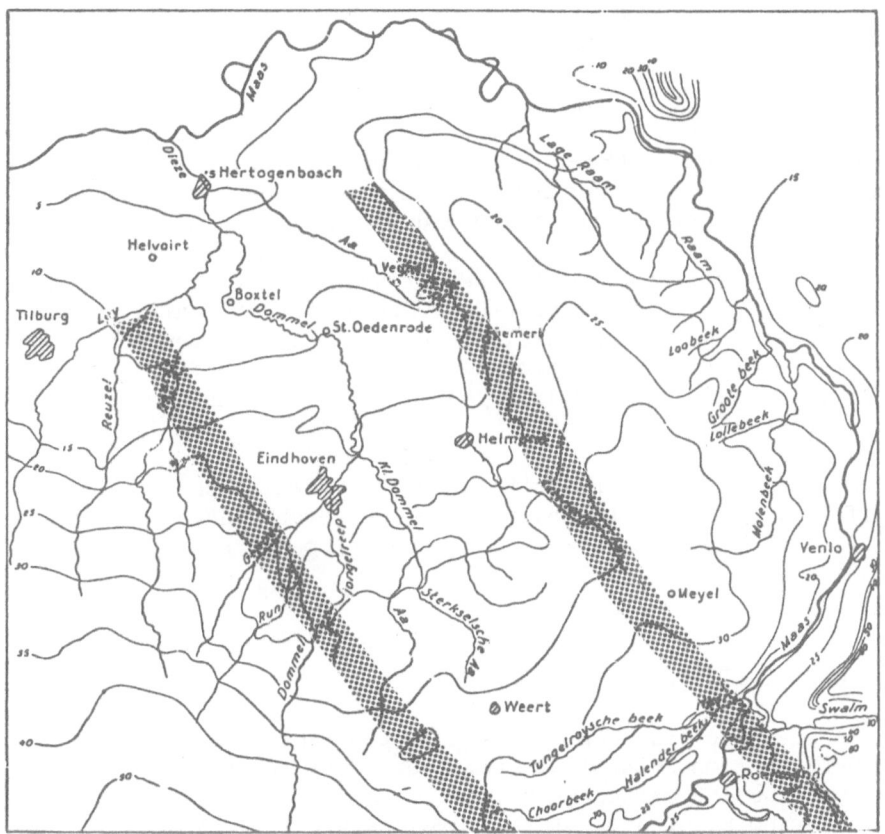

Figure 4: Relief map of part of the Roer Graben in the
provinces of N. Brabant and S. Limburg (after
Zonneveld, 1947) with location of main fault
lines: that of Valkenswaard or the Feldbiss
in the West, and the Peel boundary fault in
the East.

The maximal recorded historic shock has a magnitude of about
$6\frac{1}{4}$ in the Lower Rhine-Roer Graben, about $6\frac{1}{2}$ in the Brabant Massif
s.l., and about $6-6\frac{1}{4}$ in the realm of the North Sea. These values
are considerably lower than observed in historical earthquakes in
the continental USA and Canada, where in the past 3 centuries at
least 5 earthquakes reached magnitudes of 7 or higher (fig. 5).
If a comparison with the other side of the Atlantic is sought
it is probably more correct to compare seismic activity in the

region under consideration with that of Greenland which once
formed the American counterpart for this part of the European
continent. Earthquakes of magnitude ≥ 6 are unknown from the
Greenland region (Gregersen, 1982).

An estimation on the basis of the maximal possible fault
plane or earthquake source in the region using global data of
dip-slip earthquakes (Purcaru & Berckhemer, 1982) reaches at least
magnitude 7, but a great margin of uncertainty exists about the
actual and effective size of individual faults in the region (fig. 6)

A rough estimation of the recurrence rate for magnitude 6
earthquakes in the Lower Rhine-Roer Graben is 150 years, in the
Brabant Massif s.l. 200 years, in the North Sea Grabens 200 years
where the larger concentration clearly lies at the Viking Graben,
in the Netherlands proper 1200 years, and in the northern part
of the Netherlands about 6000 years (see also Ritsema 1966).

Maximal recorded intensities are VII-VIII in the Brabant
Massif s.l., VIII in the Lower Rhine-Roer Graben and probably
the same for the North Sea Grabens (see also Ahorner et al. 1976).

SEISMIC ACTIVITY IN THE NETHERLANDS

A data base has been built up at the KNMI of all earthquakes
with epicenter on Netherlands territory. For the period 1500-1980
this concerns 150 earthquakes, not taking into account numerous
felt aftershocks of some of the more heavy events.

Numbers of Netherlands earthquakes in the KNMI-file
(+ means aftershocks)

	province of South Limburg	province of North Brabant	Nij-megen	province Zeeland	North of the Neth.	Total
1500-1599	5	8++			2	15++
1600-1699	10	1		1	1	13
1700-1799	22++	3			4	29++
1800-1899	14	8+	3		12	37+
1900-1965	13	16+			8	37+
1965-1983	16	2	4		1	23
1500-1983	80^{2+}	38^{4+}	7	1	28	154^{6+}

Figure 5: Location of N. American intraplate earthquakes of
          M > 7 in the past 300 years relative to the NW
          European intraplate earthquake zones. Continent
          configuration of 200 MY B.P.

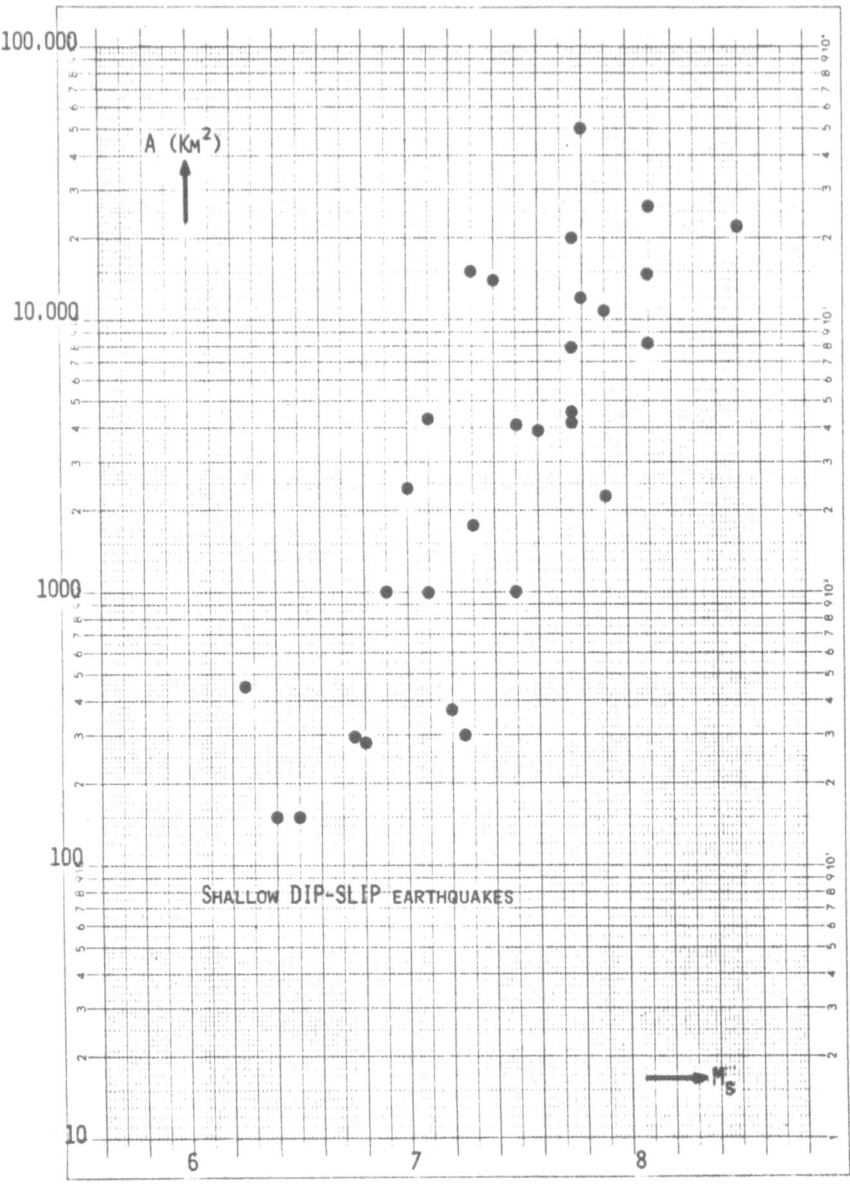

Figure 6: Relation fault plane area versus magnitude for
dip-slip earthquakes, on the basis of data
compiled by Purcaru and Berckhemer (1982).

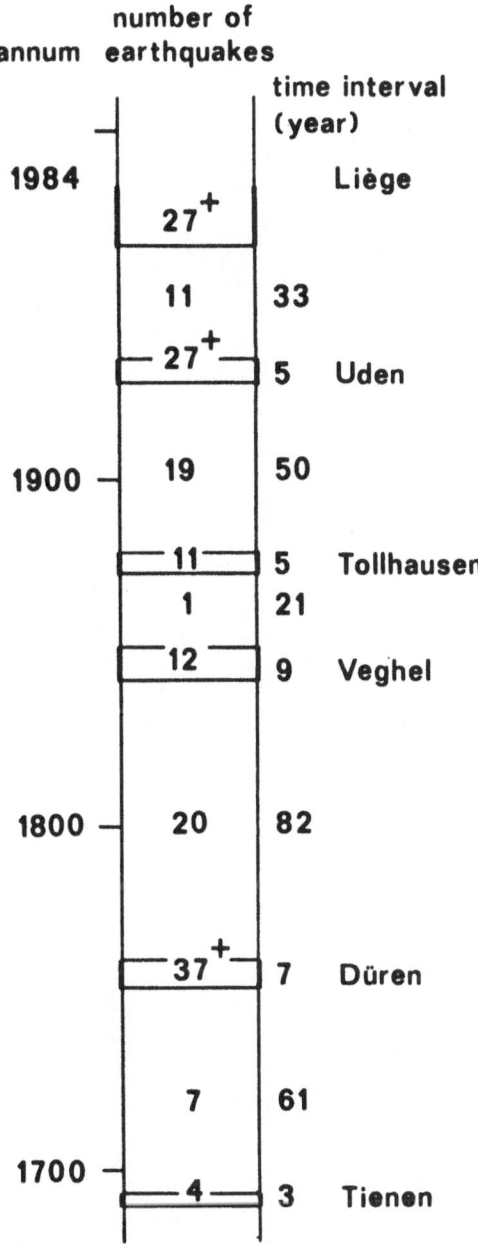

Figure 7: Earthquake activity as observed in
The Netherlands since 1700. Relative short periods of increased activity alternate with longer periods
of low seismic activity.

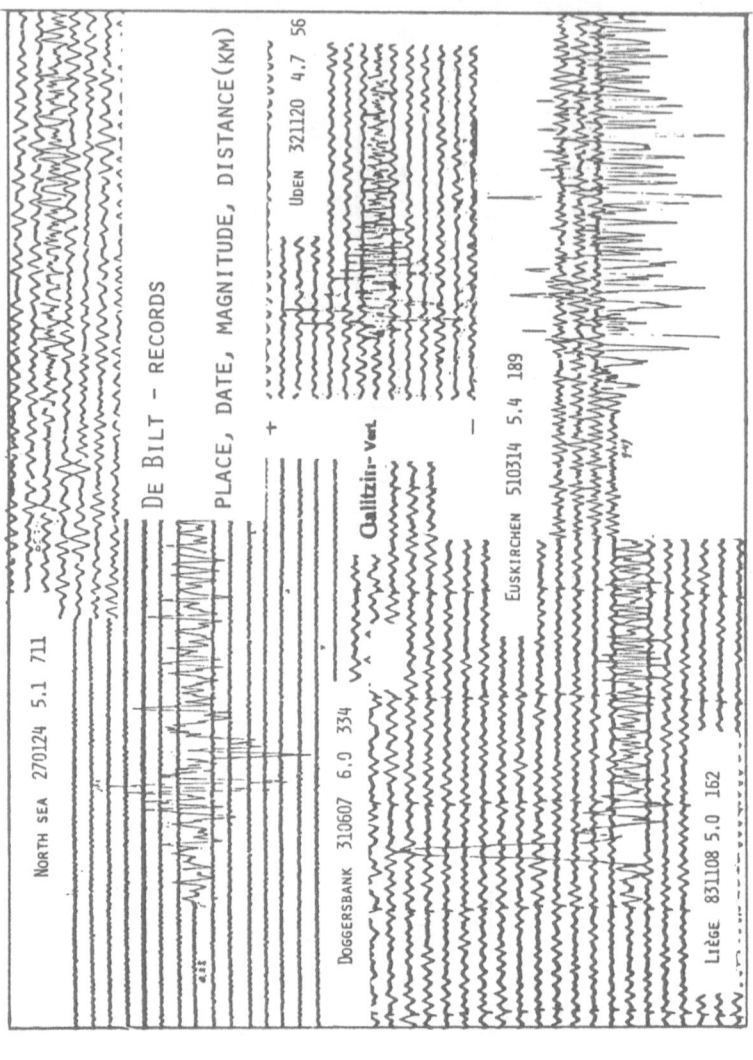

figure 8. Vertical Galitzin seismograms of De Bilt station of some of the more heavy intraplate earthquakes in the region.

In time the activity is not regularly distributed (fig. 7). A concentration in a limited number of bursts is observed. Periods of great activity in the Netherlands often coincide with activity further SE in the German part of the Roer Graben system. Activity in the Brabant massif, on the other hand, mostly does not coincide with the periods of acitivity in the Netherlands. This is a clear indication that earthquake activity within the Roer Graben is linked in a natural way and that such a link does not exist with the earthquakes of the Brabant massif nor with those of the North Sea Grabens.

In figure 8 De Bilt seismograms of some of the more heavy earthquakes in the region are shown.

CONCLUSIONS AND RECOMMENDATIONS

1.   A magnitude 5 earthquake can bring about much damage and
     loss of property and even lives, as is illustrated by the
recent Liège earthquake. Shocks of this magnitude do occur in the
region of the Brabant Massif, Lower Rhine-Roer Graben and Rhenish
plateau roughly once in 25 years. Investments in monitoring
instrumentation and manpower, thus, are justified. This should
preferably take the form of a regional co-operation between the
countries involved, i.e. The Netherlands, Germany, Belgium,
Luxemburg and France. In the case of incorporation of North Sea
shocks also Denmark, Norway and the U.K. should be included. What
is needed than is an instant exchange of data and communication
on computer level between participating parties. Obligatory are
digital seismic records at key stations in the active zones, or
telerecording of analog data by telephone or radio link transmission.

2.   The extra gain of such an intensive co-operation will be
     found in the field of precise delineation of microseismicity
of the region, which together with common study projects, for
example in lithosphere structure to detect zones of weakness,
in satellite geodesy, in geodetic surveying, and in actuo geology
to measure present crustal motions, could initiate a better understanding of the underlying processes of seismicity in general and
of the seismic hazard in the region in particular.

REFERENCES

Ahorner, L., J.M. van Gils, J. Flick, G. Houtgast, A.R. Ritsema,
     1976 - First draft of an earthquake zoning map of NW Germany
     Belgium, Luxemburg and The Netherlands. Publ. 153, K.N.M.I.
     De Bilt: 167-170.

Becker, M. de, T. Camelbeeck, **1985** - This Volume.

Gregersen, S., 1982 - Seismicity and observations of Lg wave
    attenuation in Greenland. Tectonophysics 89, pp.77-93.

Illies, J.H., 1978 - Two stages of Rhine Graben Rifting. In:
    L.B. Ramberg & E.R. Neumann (eds.), Tectonics and Geophysics
    of Continental Rifts, D. Reidel Publ. Co.

Illies, J.H., H. Baumann, 1982 - Crustal dynamics and morpho-
    dynamics of the Western European Rift System, Z. Geomorph.
    N.F. Suppl. Bd. 42, pp. 135-165.

Purcaru, G. & H. Berckhemer, 1982 - Quantitative relations of
    seismic source parameters and a classification of earthquakes,
    Tectonophysics 84, pp. 57-128.

Ritsema, A.R., 1966 - Note on the seismicity of The Netherlands,
    Proc. Kon. Ned. Akad. Wetens., Serie B, 69, no. 2, pp. 235-239.

Ritsema, A.R. & A. Gürpinar (eds.), 1983 - Seismicity and Seismic
    Risk in the Offshore North Sea Area, NATO Advanced Study
    Institutes Series, Serie C, D. Reidel Publ. Co., pp. 420;
    with special reference to papers of L. Ahorner, N.N.
    Ambraseys, P.W. Burton, F. Ringdal, A.R. Ritsema, W. Rosen-
    hauer, J.M. van Gils, Y. Zaczek.

Schalke, H.J.W.G., 1984 - Geological Sections through the onshore
    and adjacent part of the continental shelf, Investerings
    Maatschappij Nederland Energie Services BV, 's-Gravenhage.

Ziegler, P.A., 1982 - Geological Atlas of Western and Central
    Europe, 2 parts, Shell International Petroleum Maatschappij
    & Elsevier SPC, pp. 130 (+ 40 enclosures).

Zonneveld, J.I.S., 1947 - Het Kwartair van het Peelgebied en de
    naaste omgeving, Meded. Geol. Sticht., Serie C-VI-no. 3,
    pp. 223.

SOME NOTES CONCERNING THE SEISMICITY IN BELGIUM - MAGNITUDE
SCALE - DETECTION CAPABILITY OF THE BELGIAN SEISMOLOGICAL
STATIONS

T Camelbeeck

Centre de Géophysique Interne
Observatoire Royal de Belgique (3 Av. Circulaire
- 1180 Bruxelles)

ABSTRACT

A magnitude scale using Lg-wave at a period of 1 second was
defined for Northwestern Europe and used to determine the local
magnitude of recent earthquakes in Belgium. The method gives
coherent evaluation when applied to earlier data as shown by the
study of the Havré earthquake of April 3, 1949.
In addition, the cumulative frequency distribution of earthquakes felt in Belgium and the detection capability of the belgian
seismological stations are discussed.

1. INTRODUCTION

Studies concerning the seismicity and the seismic hazard in
a given region require an uniformisation of the magnitude information.
For the countries with a low level of seismicity, local magnitude is an adequate measure of the size of an earthquake. Even
if heterogeneity exists between different networks, a well calibrated one can give accurate estimation of source parameters and
homogeneity in his data.
Recently, Herrmann and Kijko [1] have shown the uniformity for
different short-period instruments of the magnitude evaluation by
the measurement of the ground displacement produced by the vertical component of the Lg-wave. This method was developped for the
belgian stations and applied to recent and earlier data.

99

P. Melchior (ed.), Seismic Activity in Western Europe, 99–108.
© 1985 by D. Reidel Publishing Company.

2. ATTENUATION OF THE Lg-WAVE

The Lg-wave is the largest arrival in the short-period records of continental near-earthquakes. It travels with a group velocity of about 3,5 km/s.

The observed decrease of the Lg-wave amplitude A(Δ) with distance can be fit by the spatial decay of an Airy-phase of a surface wave propagating in an anelastic medium [2]. For epicentral distances lower than 1000km, this decrease is described [3] by:

$$A(\Delta) = a \ \Delta^{-5/6} \ exp(-\gamma\Delta) \tag{1}$$

where γ is the anelastic attenuation factor.
Using the seismograms from a single station for several earthquakes, a γ-determination is possible. This method is described by Nuttli [4].
Two data sets were used: all the near-earthquakes of Central and Southern parts of Europe recorded on the vertical short-period seismometer at the station Dourbes (DOU) during the year 1980 and at the station Warmifontaine (WRM) during the period 1965-1971. The magnification curves of the instruments (Benioff seismometer at Dourbes and Spengnether series D-H at Warmifontaine) are given in the figure 1.
All the amplitude data (at a period comprised between 0.5 and 1 sec) were equalized respectively to those of a M_L = 4.0 earthquake assuming $\Delta(log\ A) = \Delta M_L$, where ΔM_L is M_L minus 3.0 (4.0 respectively).
The magnitude values for this equalization are the M_L-values given by the LDG network for the first data set and the M-values given by the B.C.I.S. for the second one.
Rewriting equation (1) as:

$$A^*(\Delta) = C.A(\Delta) \ \Delta^{5/6} = a \ exp(-\gamma\Delta) \tag{2}$$

where C is a normalization factor such as $A^*(\Delta) = A(\Delta)$
                 for Δ = 10km.
and plotting the data on a semi-logarithmic paper (figures 2 and 3), a least-square analysis provides respectively the γ-values of 0.0026 km^{-1} and 0.0023 km^{-1}, the ground displacement being 2.42 μm and 3.16 μm for a M_L = 3 earthquake at an epicentral distance of 10 km.

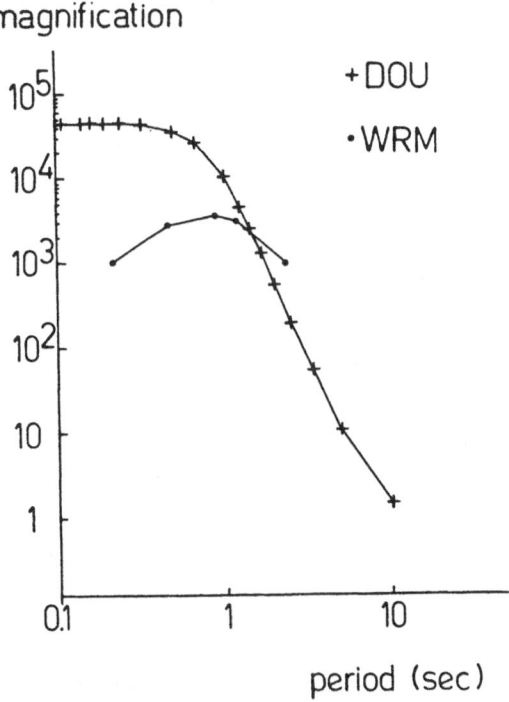

Figure 1. Magnification curves of the vertical short-period seismometers at the stations Dourbes (Benioff seismometer) and Warmifontaine (Spengnether).

3. MAGNITUDE SCALE

The local magnitude of an earthquake is defined [5] by

$$M_L = \log A(\Delta) - \log A_o(\Delta) \qquad (3)$$

where $A(\Delta)$ is the maximum trace amplitude (in mm) at an epicentral distance Δ measured on the seismogram of a standard Wood-Anderson seismometer.

$A_o(\Delta)$ is the distance variation of the maximum trace amplitude of a magnitude zero earthquake.

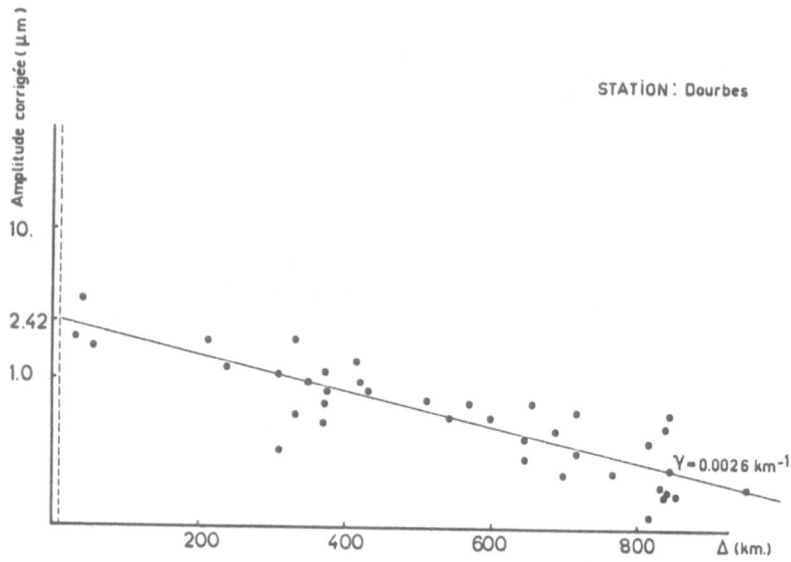

Figure 2. Lg-wave anelastic attenuation at a period of
1 second.
The amplitudes are corrected to a M_L =3 earthquake.

A general formula, similar to (3) can be obtained using ground
displacement measurement of the vertical component of the Lg-wave
(equation 1):

$$m_{bLg} = - \log a + \left(\frac{5}{6}\right) \log \Delta + \gamma\Delta \log e + \log A(\Delta) \qquad (4)$$

where the expression $- \log a + (5/6) \log \Delta + \gamma\Delta \log e$ is equivalent to the term $- \log A_o$ in relation (3).

The values of $- \log A_0$ are calculated (table 1) by choosing a in
such a way that a m_{bLg} = 3.0 earthquake is an event with an amplitude of 2.42 μm at an epicentral distance of 10 km.

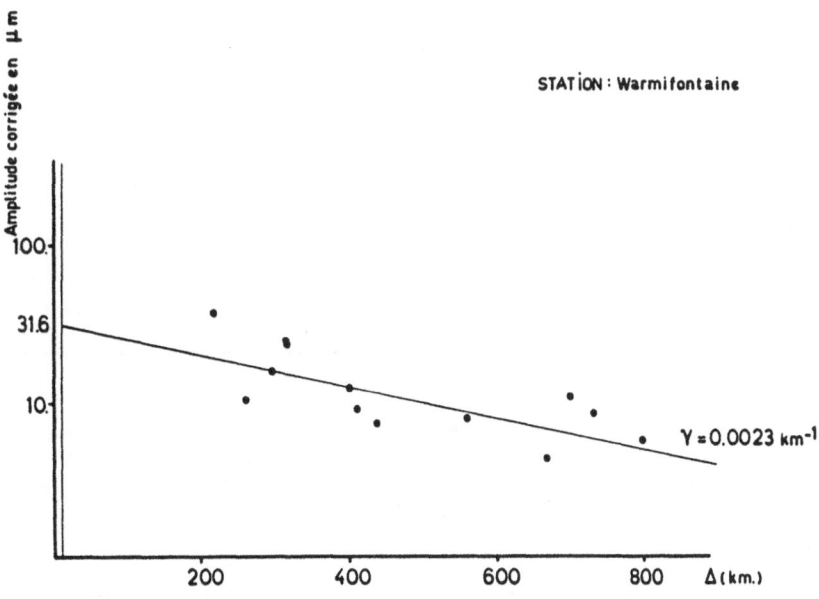

Figure 3. Lg-wave anelastic attenuation at a period of
1 second.
The amplitudes are corrected to a M=4 earthquake.

TABLE 1. m_{bLg} magnitude scale (belgian stations):
$m_{bLg} = \log A(\Delta) - \log A_o$ A is in μm

Δ (km)	$-\log A_o$	Δ (km)	$-\log A_o$
10	2.61	260	4.07
20	2.88	280	4.12
30	3.03	300	4.17
40	3.15	320	4.22
50	3.24	340	4.26
60	3.32	360	4.31
70	3.39	380	4.35
80	3.45	400	4.39
90	3.50	430	4.45
100	3.55	460	4.51
110	3.59	490	4.56
120	3.64	520	4.62
130	3.68	550	4.67
140	3.71	580	4.73
150	3.75	610	4.78

Δ (km)	$-\log A_o$	Δ (km)	$-\log A_o$
160	3.79	640	4.83
170	3.82	670	4.88
180	3.85	700	4.93
190	3.88	750	5.01
200	3.91	800	5.09
220	3.97	850	5.17
240	4.02	900	5.25

Since march 1983, the m_{bLg} values determined by the belgian stations are sent to the International Centers. The agreement obtained is good compared with the magnitudes given by other networks even for belgian earthquakes (table 2).

TABLE 2. Comparison of the magnitude determined by the belgian stations; the LDG network and the station BNS for recent belgian earthquakes.

date	origin time	Lat °N	Lon °E	Belgian m_{bLg}	LDG M_L	BNS M_L
1982 SEP 14	19h24m	50.40	4.19	3.4	3.4	
1982 SEP 14	19h29m	50.40	4.19	2.6	2.6	
1983 AUG 4	7h 8m	50.43	4.48	3.3	3.5	3.3
1983 AUG 9	1h32m	50.43	4.48	3.3	3.4	3.2
1983 SEP 24	9h49m	51.10	5.76	3.1	3.2	3.2
1983 NOV 8	1h25m	50.65	5.55	2.9	2.9	
1983 NOV 8	2h13m	50.65	5.48	3.5	3.4	3.3

4. EVALUATION OF MAGNITUDE FOR EARLIER DATA

The attenuation curve obtained in this paper can be used to standardize earlier data (before 1960). To verify the accuracy of the method, an evaluation of m_{bLg} for the Havré earthquake of april 3, 1949 was calculated using the data from stations at regional distances (without applying station corrections). The magnitude is comprised in the range of values 4.3-4.4 which were obtained in comparing the records obtained with the same instrument at the station Heerlen for the earthquakes of Havré and Carnières (march 28, 1967 - M_L = 4.5).

The m_{bLg} values obtained in studying the data from the short-
period instruments with a natural period of 1 second at the
station Paris, Zürich and Stuttgart figure in Table 3.
The average value is $m_b(Lg)$ = 4.36 which is very close to the
proposed value.

TABLE 3. Ground displacement of the Lg-wave (vertical component)
and m_{bLg}, determined at stations Paris, Stuttgart and Zürich for
the Havré earthquake of April 3, 1949 at 12h33m.

Station	Δ(km)	Seismometer	A(Lg) µm	m_{bLg}
PARIS	217	Grenet T_0=1s	1.74	4.2
STUTTGART	423	T_0=1s	0.91	4.4
ZURICH	485	T_0=1s	0.97	4.5

5. DATA FILE CONCERNING THE SEISMICITY OF BELGIUM

        A revision of the data concerning the earthquakes felt in
Belgium was recently undertaken by the "Centre de Géophysique
Interne" at the Royal Observatory of Belgium. The figure 4 is
the map which contains all the data actually available. It can
be noted that no earthquake having occured before 1900 figures in
the map with a magnitude or a depth information.
This map will be discussed elsewhere.
The cumulative frequency magnitude statistic calculated with the
data figuring in this preliminary file is given for the periods
1900-1960 and 1960-1983 in figure 5. Those curves give valuable
information about the annual probability of occurence of earthquakes with a magnitude greater than a given value M (for M
greater than 4) (table 4).

TABLE 4. Number of earthquakes felt in Belgium per year (N/a)
having a magnitude greater than M and period T between successive
earthquakes for a given level of magnitude.

M	N/a	T(year)
5.5	0.02	50
5.0	0.08	12.5
4.5	0.13	7.7
4.0	0.40	2.5

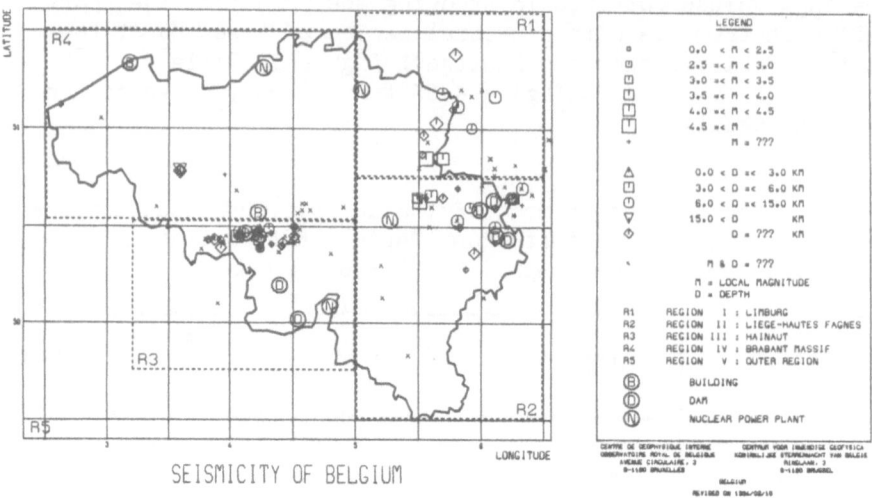

Figure 4. Seismicity of Belgium.

The evident lack of data below magnitude 4 which is apparent for
the period 1900-1960 has two main reasons :
- a poor detection capability (only the station Uccle
  was in operation)
- bad or no magnitude determination of the events which
  have been recorded.

6. DETECTION CAPABILITY OF THE BELGIAN STATIONS

The detection of earthquakes with magnitude lower than 4
was improved since 1960 by the installation of new stations.
It is presently possible to detect an earthquake of M_L = 2.5
with an epicenter anywhere in Belgium. The detection capability
of the existing belgian stations is shown in figure 6. Even if
Membach and Dourbes permits the detection of micro-earthquakes,
the data from those stations are not sufficient to locate them
precisely. It can be concluded that installation of portable
stations in the active zones of Belgium is a need for two reasons :
- the detection of very small events ($M_L \sim 1$)
- the precise location of these events

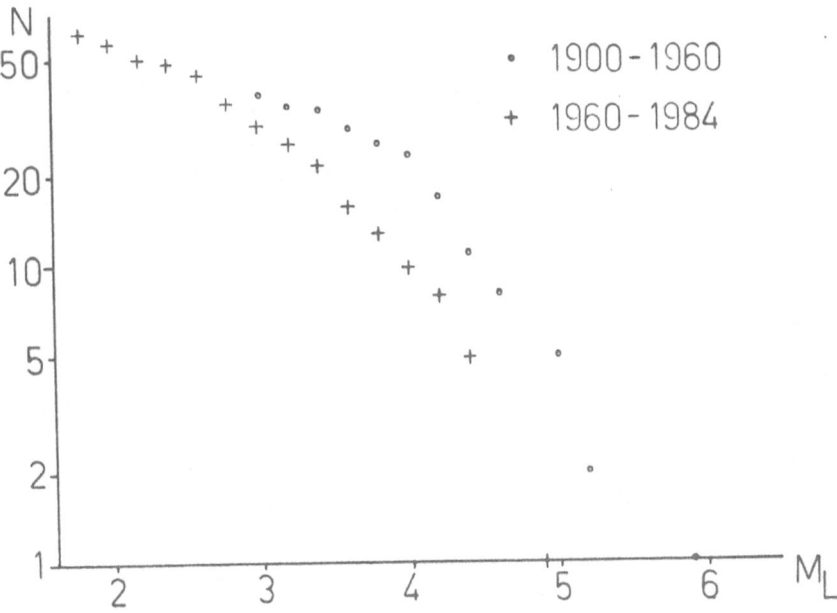

Figure 5. Cumulative frequency-magnitude distribution of
the earthquakes felt or having their epicenter in Belgium
for two periods: 1900-1960 and 1960-1984.

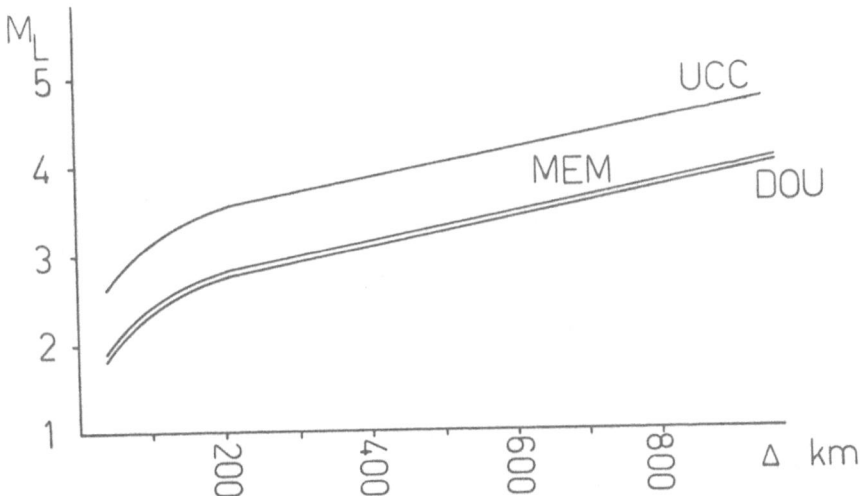

Figure 6. Actual detection capability of the belgian
seismological stations-MEM(Membach)-DOU.(Dourbes and UCC(Uccle).

7. CONCLUSION

The value of the γ- attenuation factor of the Lg-waves at a period
of 1 second in continental Western Europe is 0.0025 km^{-1}. This
corresponds to an apparent Q of 350. The m_{bLg} values obtained
using this attenuation curve are coherent and similar to that
determined by other european networks. Application of the method
to earlier earthquakes permits a reasonable evaluation of their
magnitude. This will provide useful results to the study of the
seismicity in Belgium.

REFERENCES

1. Herrmann R.B. and Kijko A. (1983)
   Bulletin of the Seismol. Soc.of America, vol. 73 pp 1835-1850

2. Nuttli O.W. (1973)
   Journal of Geophysical Res. 78, pp 876-885

3. Herrmann R.B. and Nuttli O.W. (1982)
   Bulletin of Seismol. Soc.of America, vol.72, pp 389-397.

4. Nuttli O.W. (1980)
   Bulletin of the Seismol. Soc.of America,vol.70,pp 469-486.

5. Richter (1958)
   Elementary Seismology
   WH Freman and Compagny.

RECENT SEISMICITY IN HAINAUT - SCALING LAWS FROM THE SEISMOLO-
GICAL STATIONS IN BELGIUM AND LUXEMBURG

T. Camelbeeck

Centre de Géophysique Interne
Observatoire Royal de Belgique

ABSTRACT

A relationship is established between the local magnitude M_L
determined with the data from seismological stations in Belgium
and Grand-Duchy of Luxemburg and the seismic moment M_0 derived
from the study of the coda waves recorded at the station of
Dourbes. This was done in studying 17 earthquakes having occurred
in Hainaut during the period 1965-1971. Assuming circular sources,
relationships between source parameters and local magnitude were
obtained. This provides us with means for estimating source parameters using M_L determinations and fault size evaluations. Two
examples are discussed: the earthquakes of 15 december 1965 (12h
7 m) and 28 march 1967 (15h 49m).

1. INTRODUCTION

In the far-field approximation, seismic sources can be described with several physical quantities relatively independant
of the details of the kinematic models:

   - the seismic moment M_0
   - the seismic energy E_S
   - the apparent stress $\eta\bar{\sigma} = \mu\, E_S/M_0$
            μ is the rigidity of the medium
            (we suppose $\mu = 3\ 10^{11}$ dynes/cm^2)
            η is the efficiency of seismic radiation
            $\bar{\sigma}$ is the average of initial stress σ_0
            and final stress σ_1 along the fault
            plane.

109

P. Melchior (ed.), Seismic Activity in Western Europe, 109–126.
© *1985 by D. Reidel Publishing Company.*

$$\text{- the stress drop } \Delta\sigma = C\, M_0/S^{3/2} = \sigma_0 - \sigma_1$$

S   is the fault area
C   depends on the fault shape and
    slip direction for a circular
    crack
    $C = 7/16\ (\pi)^{3/2}$

These parameters are difficult to estimate in the case of small
events, since the seismic energy pattern is of the short period
type which is strongly affected by heterogeneities of the Earth's
crust. A neat description of the dimension of the source can then
be given by the local magnitude calculated with a well calibrated
network.
Considering the theoretical displacement source spectrum and the
definition of local magnitude, a relationship between local magnitude, stress drop $\Delta\sigma$ and fault size R was established for circular cracks by Randall (1973). For small sources (radius R smaller
than one kilometer), M_L varies as log ($\Delta\ \sigma\ R^3$) whereas for greater
source radii (greater than twenty kilometers), the variation is
following log ($\Delta\ \sigma\ R$).
     Relationships between source parameters and local magnitude
can be established for small events in finding a link between
seismic moments determined by the study of the coda of local
events and local magnitudes. This can be performed if the quality
factor of the S-waves and the backward turbidity coefficient $g(\pi)$
in the crust are known. This paper is meant to look for such a
relationship.

2. RECENT SEISMICITY IN HAINAUT

     The Hainaut seismic zone is the region of Belgium between
the cities of Mons and Charleroi. It is depicted as region III
on the first map of figure 1 where all the data actually available
about the seismicity of Belgium are presented {1}.The geographical
extent of the zone is more apparent on the second map where the
numbers are refered to the earthquakes studied in this paper
(Table 1).
     Since 1900, some earthquakes of relatively low magnitude
have caused any damages. These principal events are:

1911 june	1	22h52m	Gosselies	M ∿ 4.2	I_0 = VII
june	3	14h36m	Gosselies	M ∿ 4	I_0 = VI
1949 april	3	12h34m	Havré	M = 4.3	I_0 = VII
1954 jul	10	17h18m	Quaregnon	M ∿ 4.2	I_0 = VI
1965 dec	15	12h 7m	Strépy	M = 4.4	I_0 = VII
1966 jan	16	12h33m	Chapelle-lez-Herlaimont	M = 4.4	I_0 = VII
1967 march	28	15h49m	Carnières	M = 4.5	I_0 = VII
1968 aug	13	16h57m	Haine-Saint-Pierre	M = 4.1	I_0 = VII
1976 oct	24	20h33m	Maubeuge	M = 4.2	I_0 = VI

I_0 being given in the M.S.K. scale.

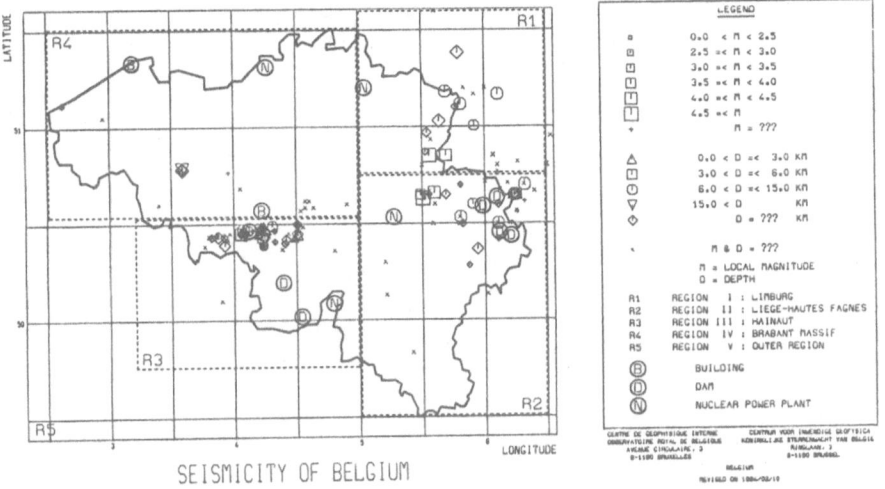

SEISMICITY OF BELGIUM

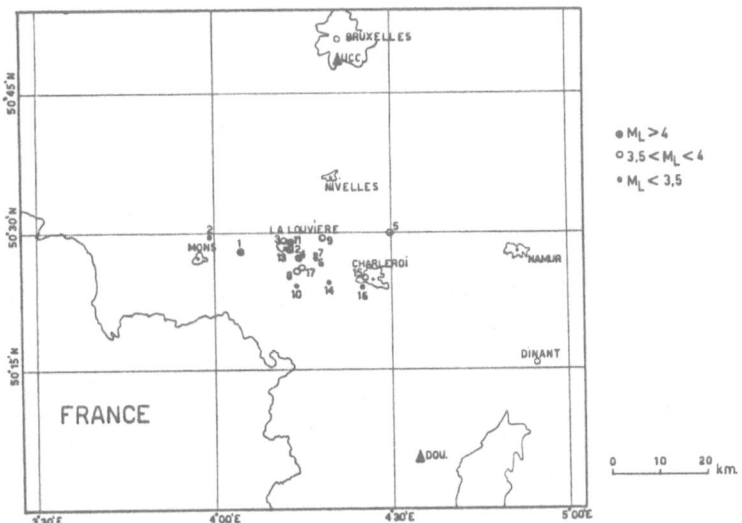

Figure 1. Location of the Hainaut seismic zone in the
general context of the seismicity in Belgium.

This simple enumeration gives an idea of the seismicity in Hainaut.
As in 1964, new stations were installed in Belgium, it is possible
to study the microseismicity (restricted to events with $M_L > 2$).
Figure 2 shows the cumulative frequency magnitude statistic of

earthquakes from the Hainaut zone since 1964. For $M_L \geq 2.5$ the
relation between log N and M_L appears linear (log M_L = a-b M_L)
with b = 0.6.

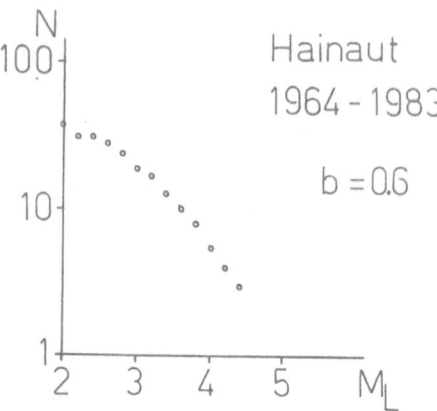

Figure 2. Cumulative frequency magnitude statistic
between 1964-1983.

Some characteristics are common to these events:

- shallowness of the hypocenters (less than 8 km)
  which explains the principal macroseismic characteristics (small macroseismic radii with high epicentral
  intensity).

- occurrence under the form of swarms with fore- and aftershocks sequences. This appears on the figure 3 where are
  shown the number of events associated to the four
  principal events of the period 1965-1970.

3. ANALYSED DATA

The events studied in this paper are those having occurred between
1965 and 1971 for which at least two belgian stations are available to determine a local magnitude. These events are listed
in Table 1 and their localisation is featured in figure 1.
    The local magnitudes were determined from m_{bLg} for station
WRM, from the duration of the coda for station DOU and from the
maximum amplitude measurements on the three components for the
stations LUX and UCC.
    The adopted value for the local magnitudes is the average
of those obtained at the different stations.

Table 1. List of the earthquakes studied

Event N°	Date	Time	Latitude (°N)	Longitude (°E)	Depth km	DOU	LUX	UCC	WRM	Local magnitude adopted
1 (a)	15/12/65	12 07 15.6	50.47	4.07	9±2	4.5		4.3	4.5	4.4
2 (a)	15/12/65	14 02 14.0	50.50	3.98	0→3	3.5		3.5	3.2	3.4
3 (a)	16/01/66	06 51 34.7	50.48	4.20	10±5	3.7		3.9		3.8
4 (c)	16/01/66	12 32 50.5	50.45	4.23	5	4.4		4.5		4.4
5 (b)	20/03/66	00 08 15.1	50.50	4.50			3.8	3.9	3.7	3.8
6 (b)	04/04/67	18 04 48.2	50.45	4.28			3.4	3.6	3.1	3.3
7 (c)	09/04/67	04 53 48.6	50.46	4.28			2.6			2.6
8 (a)	12/08/68	07 26 41.9	50.43	4.22	5±5	3.9	3.5	3.8		3.7
9 (a)	13/08/68	16 17 29.0	50.49	4.30	6±4	3.7	3.4	3.8		3.6
10 (b)	13/08/68	16 40 41.0	50.40	4.20			2.7	2.8	3.1	2.8
11 (a)	13/08/68	16 57 15.1	50.48	4.21	1→4	4.4	3.9	4.2		4.1
12 (a)	23/09/68	04 08 11.9	50.47	4.20		2.7	3.0	3.2	3.1	3.0
13 (a)	23/09/68	05 47 15.5	50.47	4.20		2.9		3.0	2.8	2.9
14 (c)	16/01/70	23 34 57.0	50.41	4.32		3.0			2.6	2.8
15 (a)	03/11/70	08 45 58.8	50.42	4.41	7±2	3.9	3.0			3.7*
16 (c)	03/11/70	12 07 35.0	50.40	4.40		3.0				3.0
17 (a)	20/12/70	13 48 34.0	50.43	4.24	6±3	3.6		3.5		3.5

* For this event, the adopted local magnitude is obtained in comparing the macroseismic effects with other ones (3.0 M_L determined at LUX seems too small)
(a) Localisation by the author.
(b) Localisation by ISC
(c) macroseismic localisation (Van Gils)

Table 2. Seismological stations in Belgium and Luxemburg during the period 1965-1971.

Station	Latitude (°N)	Longitude (°E)	Elevation (m)	Period of operation	Type	Comp	Vs	Vs	Vmax
DOU.	50.0960	4.5942	224	permanent	Benioff	V	1.0	0.8	59.000
GIP	50.5922	5.9742		oct.1968→apr.1970	Grenet	V	0.7	0.76	11.100
LUX	49.6000	6.1333	270	may1967→oct.1972	Stuttgart	V / EW / NS	12.0 / 8.2 / 7.5	11.5 / 8.1 / 7.5	
UCC	50.7983	4.3594	105	permanent	Wiechert	V / EW / NS	4.1 / 4.4 / 4.4		700 / 977 / 895
WRM	49.8333	5.3806	242	sep.1964→sep.1970	Sprengnether	V	1.5	1.5	3800

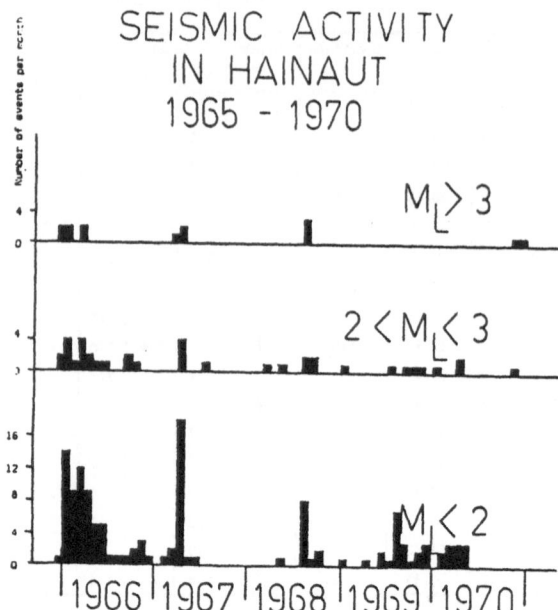

Figure 3. Number of events per month recorded during
the period 1965-1970 at the station Dourbes with a
given level of magnitude.

4. BELGIAN SEISMOLOGICAL NETWORK

    During this period, five stations were in operation in
Belgium and Grand-Duchy of Luxemburg. Dourbes (DOU) and Uccle
(UCC) are permanent whereas La Gileppe (GIP), Luxemburg (LUX)
and Warmifontaine (WRM) were temporary stations. In Table 2 are
given all the information concerning these stations.

5. SEISMIC CODA OF LOCAL EARTHQUAKES - QUALITY FACTOR Q -
   BACKWARD TURBIDITY COEFFICIENT

    Observed agreement between Q of S-waves and Q of coda waves
supports the hypothesis that the coda of local earthquakes are
S-waves backscattered by the heterogeneities in the Earth's crust
(Aki [2]- Tsujiura [3]). Assuming single scattering, Aki and
Chouet [4] obtained a simple relationship linking the spectrum
of primary waves and the power spectrum of scattered waves at
lapse time t (time measured from the earthquake origin time) and
at frequency ω when the station is close to the epicentre:

$$P(\omega,t) = \frac{V}{2} g(\pi) \mid \phi_o(\omega,vt/2) \mid^2 \tag{1}$$

where   V is the S-wave velocity in the crust
                 (v=3.45 km/s)
             g(π) is the backward turbidity coefficient
         $\phi_0(\omega,vt/2)$ is the displacement spectrum of primary
                 wave at distance vt/2 from the source.

Supposing that the attenuation of the S-waves is caused only by
the scattering due to heterogeneity (Aki [2]) the backward turbidity coefficient is simply related to the Quality factor of
S-waves by:

$$g(\pi) = \frac{\omega}{Q \, V} \tag{2}$$

An average Q-value estimate, in the frequency band 1 to 2 Hz, has
been determined using the estimation of the γ attenuation factor
of the Lg-waves by the method of Nuttli {5}.
    The average value of γ obtained with the data from the sta-
tions Dourbes and Warmifontaine is γ = 0.0025 km^{-1} [1]. This
gives a Q value of 700 at 2Hz and 350 at 1 Hz. Q = 700 is the
value obtained from the coda shape study.
    Hence, the values adopted in the following sections will be

            Q = 700
            g(π) = 0.0050 km^{-1}

6. SEISMIC MOMENT DETERMINATION

In the far-field approximation, the displacement spectrum $\bar{u}_s(\omega,r)$
of primary S-waves at a distance r from the source is:

$$\bar{u}_s(\omega,r) = \frac{M(\omega)}{4\pi\rho V^3} \frac{F^S}{r} C_{fs} \tag{3}$$

        where M(ω) is the displacement far-field source spectrum
              ρ is the density (= 2.7 g/cm^3)
              F^S is the radiation-pattern coefficient
              C_{fs} represents the effects of the free surface
                 at the station site.

The source spectrum can be written in the form (Masuda - Suzuki
[6])

$$M(\omega) = M_o \frac{1}{(1 + (\frac{\omega}{\omega_C})^\alpha)^\beta} \tag{4}$$

with $\alpha.\beta = 2$

where M_0 is the seismic moment

ω_c is the corner frequency simply related to the
source radius R by a relationship of the kind

$\omega_c = k \, V/R$

k being a constant depending on
the source model

Noting that $\phi_0(\omega,r)$ is $\bar{u}_c(\omega,r)$ averaged over the focal sphere and
considering the attenuation of primary waves, a relationship
between $P(\omega,t)$ and $M(\omega)$ is obtained:

$$P(\omega,t) = 2 \, v \, g(\pi) \frac{M^2(\omega) < F^S . C_{fs} >^2_{FS}}{t^2 \, (4\pi\rho v^4)^2} e^{-\omega t/Q(\omega)} \qquad (5)$$

where $<F^S . C_{fs}>_{FS}$ is the average value of $F^S C_{F.S}$ over
the focal sphere. This term is not
easy to evaluate. Here it is taken
as unity. $(<F^S> \sim 1/\sqrt{2})$.

For seismometers of the short-period WWSSN type, the decreasing of
the predominant frequency ω_p in the coda (due to anelasticity) is
easily observed on a seismogram of high quality. This phenomenon
has been calculated for the vertical Benioff instrument installed
at Dourbes, in determining the maximal amplitude at each lapse
time t of the resultant filter (Herrmann [7]) composed of the
source spectrum $M(\omega)$, the instrumental response $I(\omega)$ and the
attenuation $\exp(-\omega t \, /2Q)$. In table 3 are figured the values of
the predominant frequency in function of the reduced time $t^* = t/Q$
for a source having an infinite corner frequency (this assumption
is valid as long as the corner frequency of the source spectrum
is higher than the natural frequency of the seismometer). This
allows to evaluate the power spectrum of the coda by measuring
the average peak-to-peak amplitude $\sqrt{<y^2(t)>}$ at lapse time t using
the formula of Aki [8] :

$$<y^2(t)> = \frac{1}{(2\pi)^{12}} \left| \frac{Q(\omega_p)}{-dt/d\omega_p} \right|^{1/2} P(\omega_p,t) \qquad (6)$$

Table 3 shows also the calculated values of the coda shape function defined by :

$$F_c(\omega_p,t^*) = \left| \frac{Q(\omega_p)}{-dt/d\omega_p} \right|^{1/4} \frac{I(\omega_p)}{t^*} e^{-\omega_p t^*/2} \qquad (7)$$

Table 3. Predominant frequency and coda shape function for the
vertical Benioff seismometer at the station Dourbes

$t^* = t/Q$	$f_p(Hz)$	$Fc(f_p, t^*)$
0.01	1.942	.330+07
0.02	1.873	.150+07
0.03	1.812	.928+06
0.04	1.757	.635+06
0.05	1.709	.469+06
0.06	1.664	.362+06
0.07	1.624	.288+06
0.08	1.586	.236+06
0.09	1.551	.196+06
0.10	1.519	.165+06
0.11	1.488	.141+06
0.12	1.460	.121+06
0.13	1.433	.106+06
0.14	1.407	.926+05
0.15	1.383	.814+05
0.16	1.360	.724+05
0.17	1.338	.642+05
0.18	1.318	.573+05
0.19	1.298	.521+05
0.20	1.278	.471+05
0.21	1.260	.421+05
0.22	1.243	.383+05
0.23	1.226	.353+05
0.24	1.209	.321+05
0.25	1.194	.294+05
0.26	1.178	.271+05
0.27	1.164	.246+05
0.28	1.150	.229+05
0.29	1.136	.212+05
0.30	1.123	.195+05
0.35	1.062	.137+05
0.40	1.008	.983+04
0.45	.961	.729+04
0.50	.919	.549+04

Hence, using (6), (7) and (5), the average peak-to-peak amplitude
taken from the seismogram $\sqrt{\langle A(t)^2 \rangle}$ can be written as :

$$\langle A(t)^2 \rangle^{1/2} = \frac{1}{(2\pi)^{1/4}} F_c(\omega_p, t) \frac{[2g(\pi)]^{1/2} M(\omega_p)}{Q(\omega_p) 4\pi V^{3.5} \rho} \tag{8}$$

The coda source spectrum at frequency ω_p is then determined by the
relation

$$M(\omega_p) = 26 \times 10^{26} \frac{\sqrt{A(t)^2}}{F_c(\omega_p, t)} \quad \text{where } \sqrt{A(t)^2} \text{ is in cm} \\ \text{and } M(\omega_p) \text{ in dyne-cm.}$$

The average peak to peak amplitude from the seismogram $\sqrt{A(t)^2}$ is
calculated in taking into account that the enveloppe curve through
the maximal amplitude points A_{max} (figure 4) represents more
radiated energy than in reality.
It has been calculated that $\sqrt{A(t)^2} \sim A_{max}/3$ when the average value
is taken for a time interval of 5 seconds around the considered
lapse time.

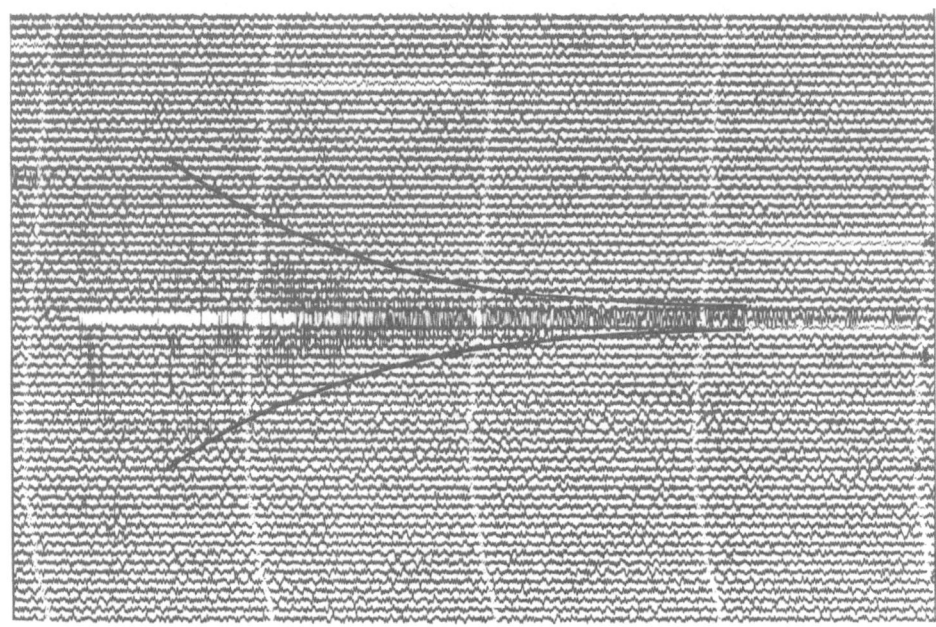

Figure 4. Seismogram (reduced by a factor of 0.57) at the station
   Dourbes (Benioff vertical seismometer) of the Haine-Saint-Pierre
   earthquake (august 13, 1968, 16h57m - M_L = 4.1). The decreasing
   of the enveloppe through the maximal amplitude points $A_{max}(t)$ is
   shown (1 minute corresponds to 6 cm on the original seismogram).

The coda source spectrum M(ωp) of the 17 events studied are represented in function of lapse time t in figure 5. It can be seen
that the coda shape is well corrected indicating that the adopted
Q-value is good.

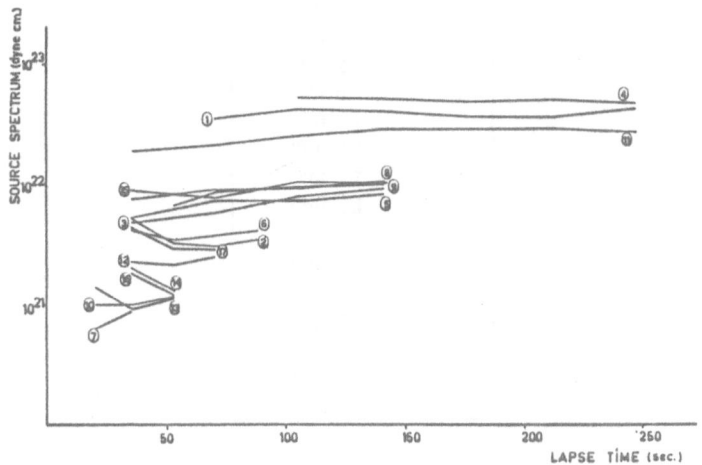

Figure 5. Coda source spectrum for the events studied (in dyne.
          cm).

According to Street and al |9| for earthquakes in the Central
United States having a seismic moment of 6.8 10^{21} dyne.cm, the
corner frequency ($f_C = \omega_C/2\pi$) is 1.67 Hz. In this case ω_p becomes
smaller than ω_C when the lapse time is greater than 35 s.So far
for smaller events (M_L < 4) the coda source spectrum is the seismic moment, hence the adopted value for M_0 is the average value
of M(ωp). For larger events, it is more indicated to take the value of M_0 as being the largest value of lapse time M(ωp).

Following remarks should be considered :
- using a single scattering theory in the coda study leads to an
  overestimation of the seismic moment and the quality factor
  (Gao et al |10|).
- free surface effects are neglected.
- Q and g(π) are assumed to be constant in the frequency range of
  interest.
- attenuation is due to scattering only.

The M_O values adopted for the events studied in this paper figure
in table 4

Table 4. Seismic moment of the 17 events studied.

EVENT Nº	SEISMIC MOMENT dyne cm.
1	$3.6 \ 10^{22}$
2	$3.2 \ 10^{21}$
3	$6.9 \ 10^{21}$
4	$4.8 \ 10^{22}$
5	$7.2 \ 10^{21}$
6	$3.8 \ 10^{21}$
7	$7.2 \ 10^{20}$
8	$8.3 \ 10^{21}$
9	$7.7 \ 10^{21}$
10	$1.1 \ 10^{21}$
11	$2.2 \ 10^{22}$
12	$2.2 \ 10^{21}$
13	$1.1 \ 10^{21}$
14	$1.5 \ 10^{21}$
15	$8.5 \ 10^{21}$
16	$1.4 \ 10^{21}$
17	$2.9 \ 10^{21}$

7. RELATION BETWEEN SOURCE PARAMETERS AND LOCAL MAGNITUDE

A least-square analysis provides the following linear relationship
(figure 6) :

$$\log (M_O) = 0.99 \ M_L + 18.22 \tag{10}$$
$$M_O \text{ is in dyne.cm}$$

For a circular crack, a simple relationship exists between seismic
moment, fault radius R and stress drop $\Delta\sigma$ (Keilis-Borok [11]) :

$$M_O = \frac{16}{7} \Delta\sigma R^3 \tag{11}$$

Combining relations (10) and (11), local magnitude and sources
parameters R and $\Delta\sigma$ are linked by :

$$\log (\Delta\sigma) + 3 \log (R) = .99 \ M_L + 11.86 \tag{12}$$

$\Delta\sigma$ is in bar
R is in cm

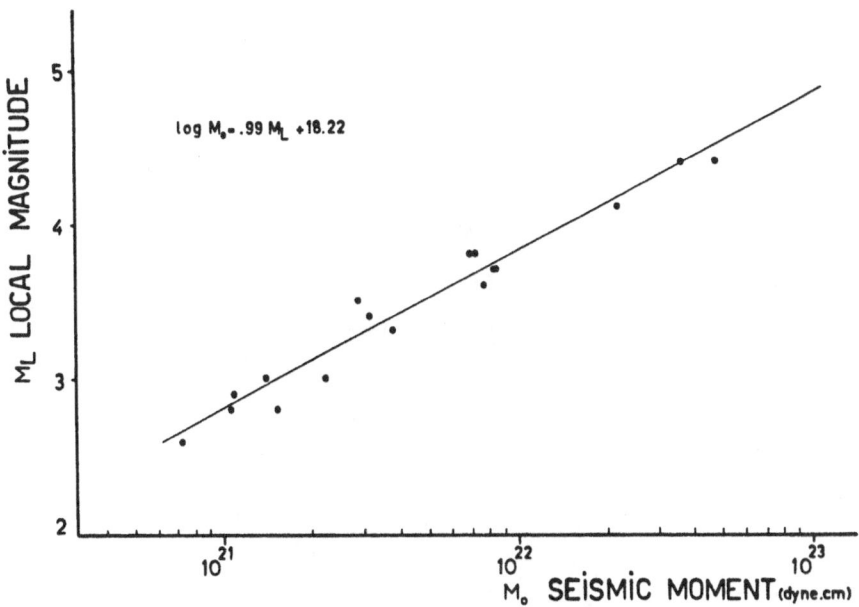

Figure 6. Relation between local magnitude and seismic moment.

Figure 7 represents relation (12) for the case of a stress drop of 1 bar. For other values of $\Delta\sigma$ it is sufficient to translate verti- cally the logarithmic scale. The theoretical results of Randall [12] also are figured out. For the magnitude range concerned in this study, the fitting of the two curves is satisfactory to a scale factor near what can be explained as follows :
- the overestimation of the seismic moment.
- the differences between the local magnitude scales in Belgium and in California.

The curve has been extrapolated for greater magnitudes by compa- ring with the theoretical one. Seismic energy may be related to the seismic moment and the corner frequency according to Randall [12]

$$E_S \cong 3.0 \, \frac{M_o^2 \, \omega_c^3}{(2\pi)^3 \, \rho \, V^5} \qquad\qquad (13)$$

Using the expression $\omega_c = 2.31 \, v/R$ (Brune [13]), for the events considered here, following expression is obtained :

$$\log (E_S) = .99 \, M_L + \log (\Delta\sigma) + 12.26 \qquad\qquad (14)$$

$\Delta\sigma$ is in bar
E_S is in erg

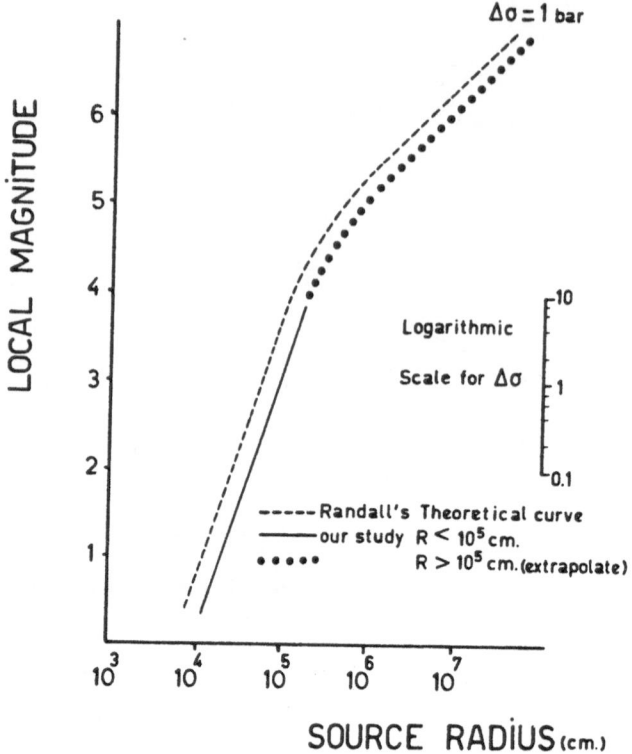

Figure 7. Relation between source parameters and local magnitude.

With formula (13), the apparent stress can be related with M_0 and R :

$$\eta \, \overline{\sigma} = .157 \times 10^{-6} \, \frac{M_0}{R^3} \qquad (15)$$

$\eta \overline{\sigma}$ is in bar
M_0 is in dyne.cm
R is in cm

All these relationships are useful tools to determine source parameters. In the next section, two examples are discussed wherein source parameters are estimated using M_L determination and fault size evaluation.

8. APPLICATION : Earthquakes of dec 15, 1965 and march 28, 1967 in Hainaut.

For these two events, respectively of magnitude 4.4 and 4.5, the source radius has been calculated in estimating the pulse rise time $\tau^{1/2}$ with the aid of the seismograms (figure 8) obtained by the LDG network. Boatwright [14] gives a relationship between pulse

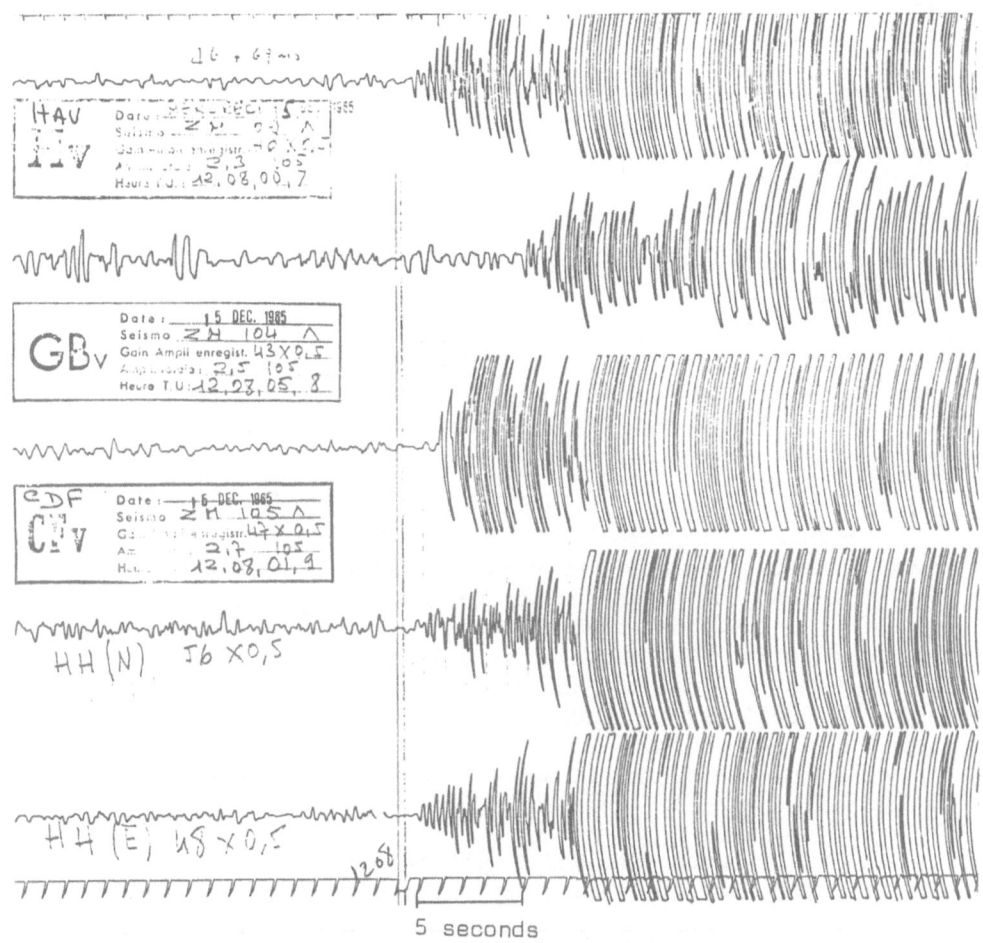

5 seconds

Figure 8.Seismograms from the stations HAU,GBF and CDF (L.D.G. network)for the earthquake of dec 15, 1965-12h 7m (M_L = 4.4).

rise time, source radius, rupture velocity v_r and takeoff angle of the ray θ :

$$\tau_{1/2} = \frac{(13-12\ \xi)}{16}\ \frac{R}{v_r} \qquad\qquad \xi = \frac{V_r}{V}\ \sin(\theta) \qquad\qquad (16)$$

Averaging on all the possible θ values and supposing $v_r = 0.8\ v$, the following relation is obtained

$$R = 5.38\ \ \tau_{1/2} \qquad\qquad (17)$$

R in km

$\tau_{1/2}$ in sec

Considering seismometers being ideal velocity-meters and taking into account the wave attenuation, the time τ_{sis} from the first break to the first zero crossing is related to $\tau_{1/2}$ by :

$$\tau_{sis} = \tau_{1/2} + 0.5 \int_0^\Delta \frac{dX}{Q_\alpha\ V_\alpha}$$

where Q_α is the Q-factor for Pn-waves (in this case). If Q = 500 and for a distance of ca 300 km (v_α = 8.1 km/s), $\tau_{sis} = \tau_{1/2} + .046$. For several LDG stations, the τ_{sis} values are indicated in Table 5. The average values of $\tau_{1/2}$ obtained are .142 sec and .158 sec for the dec 15, 1965 and march 28, 1967 events respectively. Introducing these values in (17), the radii R are 760 m and 850 m.
The values of the source parameters for these two events figure in Table 6.

Table 5. τ_{sis} values determined with the data from the LDG. Stations for the earthquakes of dec 15, 1965 and march 28, 1967

PULSE WIDTH (SEC) (P_n-WAVE)									
Station / Event	LBF	LOR	TCF	HAU	GBF	CDF	FLN	SSC	SSF	LFF
15/12/65	.206	.142	.165	.192	.300	.118	.163	.220		
28/03/67		.234	.265	.233		.213			.079	.202

Table 6. Source parameters of the earthquakes of dec 15, 1965 (12h7m) and march 28, 1967 (15h49m).

Date	Time	Latitude (°N)	Longitude (°E)	M_L	Mo dyne.cm.	Es erg	R cm.	$\Delta\sigma$ bar	$\eta\sigma$ bar
15/12/65	12 07 15.0	50.47	4.07	4.4	$4.17\ 10^{22}$	$1.9\ 10^{18}$	$7.6\ 10^4$	40.8	14.7
28/03/67	15 49 25.4	50.48	4.23	4.5	$5.25\ 10^{22}$	$2.7\ 10^{18}$	$8.5\ 10^4$	37.3	13.0

9. CONCLUSION

The method presented here seems very useful when studying earthquakes for which only traditional paper seismograms are available.
Local magnitude determinations combined with source radius estimates allow the evaluation of the principal source parameters.
The method, presented by Randall (1973) using theoretical relationships, can be calibrated in establishing a relationship between local magnitudes and seismic moments obtained from a given
network. This method was applied using the data from the belgian
seismological stations.
Seismic moments were determined using coda waves recorded at
Dourbes.
As a conclusion, the method applied appears very satisfactory for
what concerns these determinations, principally for two reasons :

- by averaging over the focal sphere, it operates independantly
  of the focal mechanism.
- it is less dependant of the station-site conditions than those
  based on body waves studies.

If the values obtained for the seismic moments seem overestimated,
this probably is ascribable to the fact that multiple scattering
has been neglected and that the backward turbidity coefficient is
not sufficiently known. Notwithstanding this, the procedure is a
promising one.

ACKNOWLEDGEMENTS

I thank Dr. B. Massinon for providing the seismograms from the
LDG network for the earthquakes studied in this paper.

REFERENCES

[1] T. CAMELBEECK - this volume.

[2] AKI, K., 1980 - Journ. of Geoph. Res. vol. 85, pp. 6496-6504.

[3] TSUJIURA, M., 1978 - Bull. of the Earthq. Res. Inst., vol 53,
       pp. 1-48.

[4] AKI, K., CHOUET, B., 1975 - Journ. of Geoph. Res. Vol. 80,
       n° 23, 3322-3342.

[5] NUTTLI, O.W., 1980 - Bull. of Seism. Soc. America, 70, pp.
       469-485.

[6] MASUDA, T., SUZUKI, Z., 1982 - Phys. of Earth and Plan. Int. 30, pp. 197-208.

[7] HERRMANN, 1980 - Bull. of Seism. Soc. America, vol. 70, pp. 447-468.

[8] AKI, K., 1969 - Journ. of Geoph. Res., vol. 74, pp. 615-631.

[9] STREET and alter, 1975 - Geophys. Journal 41, pp. 51-63.

[10] GAO and alter, 1983 - Bull. of Seism. Soc. America, vol. 73, pp. 377-390.

[11] KEILIS BOROK, V.I., 1959 - Ann. Geof. 12, pp. 205-214.

[12] RANDALL, J.M., 1973 - Bull. of Seism. Soc. America, vol. 63, pp. 1133-1144.

[13] BRUNE, J.N., 1970 - Journ. of Geoph. res. 75, pp. 4997-5009.

[14] BOATWRIGHT, 1980 - Bull. of Seism. Soc. America, Vol. 70, pp. 1-28.

TECTONIC APPROACH OF THE SEISMICITY IN THE SOUTHERN BELGIUM AND THE NORTHERN FRANCE.

Colbeaux Jean-Pierre, Dupuis Christian.

Univ. Sc. et Techn. Lille. France - Fac. Polyt. Mons. Belgium.

The basement of the southern Belgium and the northern France appears divided into a tectonic mosaic by a faulted network. Two major Blocks (B. Ardennais and B. Brabançon) separated by the Zone de Cisaillement Nord Artois - ZCNA - (North Artois Shear Zone) are composed of about ten Sub-Blocks.
The tectonic activity, directed by this structures, decreases from the Paleozoic to the Mesozoic and then increases during the Cainozoic. At the present time, the earthquakes are located on the boundaries of the Blocks and Sub-Blocks, especially on the ZCNA.

I. OUTLINES OF THE STRUCTURE OF THE SOUTHERN BELGIUM AND THE NORTHERN FRANCE.

Since the end of the variscan orogenesis, the structural evolution of the southern Belgium and the northern France, is characterised by the permanence of a faulted network in the Basement (Block pattern) active through successive crisis until today (Polyphase tectonic).

1) A Block pattern (fig. 1).

Several multifield studies (7), (8), (9), conforted by detailed works (13), (4), (21), have shown the interference between the faulted network of the basement and the sedimentation/erosion phenomenons. They lead to define different Blocks and Sub-Blocks

127

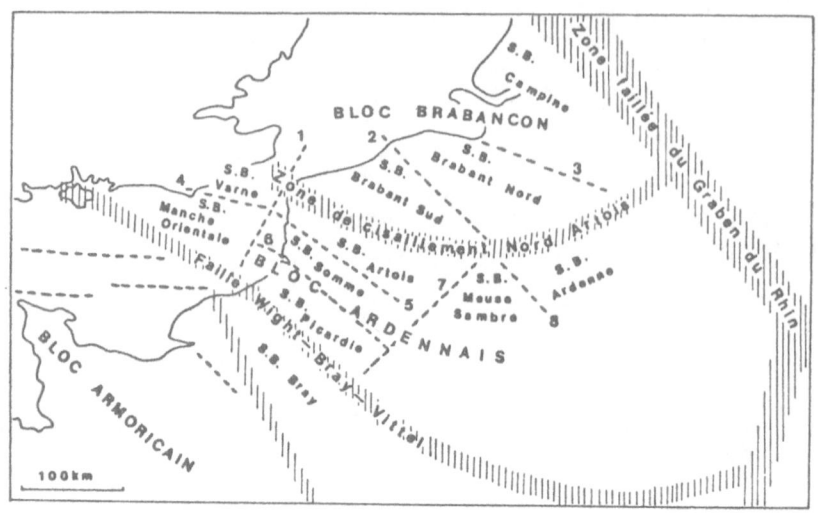

Fig. 1. - Block pattern of the basement
        (Blocks and Sub-Blocks) in the
        Belgium and northern France.
        Faults or faulted zones :
        1 - Zone Faillée du Pas-de-Calais.
        2 - Zone Faillée Sud-Brabant.
        3 - Zone Faillée Nord-Brabant.
        4 - Faille de Bassurelle.
        5 - Zone Faillée Montreuil -
            Bassurelle.
        6 - Zone Faillée de la Somme.
        7 - Faille du Vermandois.
        8 - Faille ? after (7), (8) and
            (9).

with a homogeneous behaviour during a certain time.
Blocks and Sub-Blocks are delimited by faults or
faulted zones. The two main Blocks (B. Ardennais
and B. Brabançon) are separated by an important
structural line : the Zone de Cisaillement Nord
Artois - ZCNA (North Artois Shear Zone). It coincides with the variscan front, the arched form of
which can be interpreted as a pre-variscan paleogeo-
graphic inheritance (7).
In the considered area ten Sub-Blocks can be distinguished. They are limited by faults of lesser im-

portance than the ZCNA. Their influence on the sedimentation is not so frequent and not so long as
those of the Blocks.

2) A Polyphase tectonic.

Mesostructural studies (begun in 1975) of many palaeo-,
meso- and cainozoic ontcrops reveal a sequence of
deformations, each characterised by a peculiar ori-
entation of the maximum compression δ 1.
This can be summarized as follows :
A - Wesphalian C - D : asturian phasis of the varis-
    can orogenesis, folding before "longitudinal"
    overthrust, $\simeq$ E-W oriented structures, δ 1 N-S ;
B - post-Westphalian - ante stephanian : high angle
    conjugated fractures with a N-S oriented bisec-
    trix, δ 1 N-S ;
C - Stephanian - Permian : dextral strike-slip move-
    ments along the ZCNA with "transverse" folds and
    wedges development (N-S) in the longitudinal
    overthrust (A), δ 1 E-W ;
D - upper Permian - Trias : movements along the longi-
    tudinal and transverse overthrust, δ 1 N-S ;
E - upper Jurassic - lower Cretaceous : dextral stri-
    ke slip movements along the ZCNA ;
F - middle Eocene : high angle conjugated fractures
    with a N-S orientated bisectrix in the Cretaceous
    and movements along the preexisting joints into
    Paleozoic rocks ?, reaching of Artois Sub-Block
    separating the Paris basin from the Brussel basin,
    stylolites development in Paris basin (5), (12),
    (14) ;
G - Miocene (?) : high angle conjugated fractures
    with a E-W oriented bisectrix in the Palaeo- and
    Mesozoic formations, δ 1 E-W ;
H - end of the lower Pleistocene, opening of the Pas-
    de-Calais(Strait of Dover) on the Zone Faillée
    du Pas-de-Calais like a graben structure, δ 1(?) ;
I - middle and recent Pleistocene, fracturing of
    Quaternary formations NNE-SSW, WNW-ESE, (10),
    δ 1 NW-SE

II. RELATIONSHIPS BETWEEN THE EARTHQUAKES AND THE
    BLOCK TECTONIC.

    It is clear that the Block pattern remains active
    since a long time. So it is well-founded to search
    the continuation of such a tectonic in the today

seismicity.   Figure 2 emphasizes the coincidence
between the distribution of the earthquakes and
the faulted network.   In spite of the dissimila-
rity of the used documents (19), (24), (18), several features may be hold :
1 - the alignment of the most part of the earthqua-
    kes in the ZCNA proximity, with a magnitude
    gradient increasing from west to east ;
2 - the earthquakes concentration is some regions
    (Lille, Mons, Liège) ;
    Such areas which belong to urban zones, can be
    suspected of human amplification, nevertheless
    their distribution is in good agreement with
    the division in three different parts of the
    variscan front in Belgium (Dupuis. in 22).
    It is noteworthy that the Mons region, in parti-
    cular, coincides with a fault converging
    (fig. 2).
3 - the hypocentres swarm on the Rhine graben margin ;
4 - the good alignment of some earthquakes on the
    Zone Faillée Sud-Brabant (2, fig. 2).

The today compressive stress δ 1 deduced from in-
situ measurements (15), (20) and from the earthquakes
focal mechanism (1) have a average orientation situated in the N 130° - 140° interval.
As a assumption, we can propound the followed mechanism. The Rhine-graben, parallel to δ 1, is in extension. On the other hand the Blocks and the Sub-
Blocks limits form an angle with the δ 1 stress and
are affected by relative movements.
Concerning the ZCNA, this movement is dextral,
without friction for the N 110° western branch, with
friction for the eastern branch, the orientation of
which makes an approximate right angle with δ 1.
This can explicate the stronger intensity of the
eastern earthquakes.

CONCLUSIONS.

The seismicity herits from the Block pattern and seems
to be the continuance of the tectonic activity.
The main seismic line is bound with the ZCNA, the
activity of this one appears to be dominated by shear
deformations.
Nevertheless it became obvious that it is necessary
to collect detailed information in view to complete
the tectonic approach : morphologic studies at different scales, levellins anomalies, remote sensing, neotectonics, fracturation, model test.

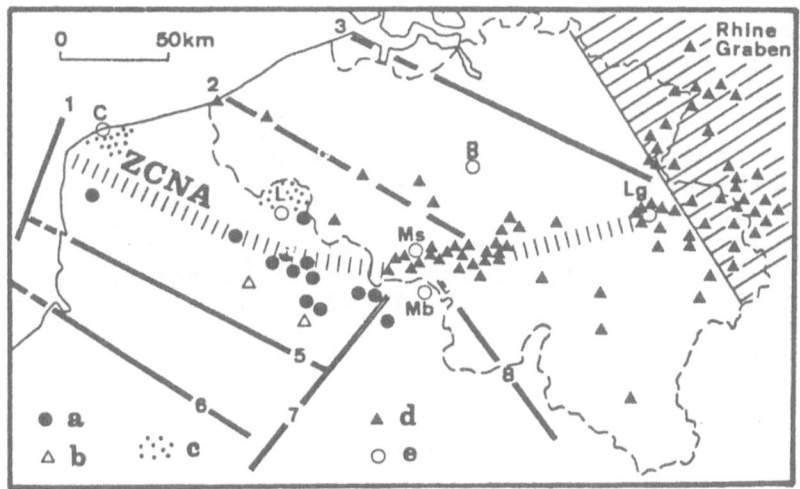

Fig. 2. - Relationships between the sismicity and the Block pattern in the Belgium and the northern France. Approximate
Blocks and Sub-Blocks boundaries after Colbeaux et al.
1977. Number as in fig. 1.
- a : earthquake after (19);
- b : individualised hypocentres;
- c : maximum known intensity (VII, M.S.K.) areas
b and c after (24);
- d : hypocentres simplified after (18);
- e : localities : C, Calais, Li, Lille, Ms, Mons,
Mb, Maubeuge, B, Bruxelles, Lg, Liège.

Such not very expansive works can lead towards a more
precise location of the dangerous zones and can constitute an instrument for interpretation of the radon
migration zones distribution especially in the Mons
basin (6).

BIBLIOGRAPHIE.

(1)   Ahorner, L. 1970, International Upper Mantle
      Project, Scientific Report, 27, pp. 155-166 -
      297-316, 5 fig.
(2)   Ahorner, L. 1975, Tectonophysics, 29, pp. 233-
      249, 9 fig. 3 tabl.
(3)   Autran, A.,Gerard, A., Weber, Ch. 1976, Bull.
      Soc. géol. Fr., 18, pp. 1119-1132.

(4)   Becq-Giraudon, J.F., Colbeaux, J.P., Leplat J.
      1981, Ann. Soc. Géol. Nord., Cl, pp. 117-121,
      3 fig., 1 tabl.
(5)   Bergerat, F. 1977, Rev. Géo. Phys. et Géol. dyn.,
      (2), XIX, pp. 325-338, 6 fig.
(6)   Bouko, Ph., Charlet, J.M., Quinif, Y. 1985, Nato
      Workshop 243/84, Brussels 20/22 March 84. Royal
      Observatory of Belgium - Brussels. A paraître.
(7)   Colbeaux, J.P., Beugnies, A., Dupuis, Ch.,
      Robaszynski F. 1977 a, 5ème R.A.S.T., Rennes.
(8)   Colbeaux, J.P., Beugnies, A., Dupuis, Ch.,
      Robaszynski, F., Somme, J. 1977 b, Ann. Soc.
      géol. Nord, XCVII, pp. 191-222, 27 fig., 1 tabl.
(9)   Colbeaux, J.P., Dupuis, Ch., Robaszynski, F.,
      Auffret, J.P., Haesaerts, P., Somme, J., 1980
      Bull. Inf. Geol. du Bassin de Paris, 17, pp.
      41-54, 9 fig.
(10) Colbeaux, J.P. 1983, Bull. A.F.E.Q., 3-4, p. 183-
      192, 9 fig.
(11) Colbeaux, J.P. 1983, Bull. Soc. belge de Géolo-
      gie, en cours de publication.
(12) Debrand-Passard, S., Gros, Y. 1980, Bull. Soc.
      Géol. France, 22, pp. 647-653, 3 fig.
(13) Dupuis, Ch. 1979, C.R. Acad. Sc., 288, pp. 1587-
      1590, 3 fig.
(14) Illies, J.H. 1977, Geol. en Mijnbouw, 56, 4, pp.
      329-350, 18 fig.
(15) Illies, J.H., Greiner, G. 1978, Geol. Soc. of
      America Bull., 89, pp. 770-782, 12 fig. 1 tabl.
(16) Macar, P. 1976, Géomorphologie de la Belgique.
      Hommage au Professeur P. Macar. Université de
      Liège.
(17) Marlière, R. 1951, Bull. Soc. belge Géol. Paléontol.
      et Hydrol., T. LX, pp. 17-27.
(18) Melchior, P. et al. 1985, Nato Workshop 243/84
      programme.
(19) Montessus de Ballore (de). 1906, la science sis-
      mologique. Les tremblements de terre. 1 vol. in
      8°, 5-79 pp. 222, fig. et cartes. A. Colin, Paris.
(20) Paquin, Ch., Froidevaux, C., Souriau, M. 1978,
      Bull. Soc. Géol. France, 20, pp. 727-731, 4 fig.
      1 tabl.
(21) Robaszynski, F., 1981, Cretaceous Research, 2,
      pp. 197-213, 5 fig.
(22) Robaszynski, F., Dupuis, Ch. 1983, Belgique, Guide
      Géologique régional. Masson éditeur.
(23) Van Gils, J.M., Zaczek Y. 1978, Tijdschrift der
      Openbare Werken van Belgie, 6, pp. 502-539

THE PROBLEM OF MICRO-EARTHQUAKES IN THE SOUTHERN PART OF THE NETHERLANDS

Th. de Crook

Royal Netherlands Meteorological Institute
De Bilt, The Netherlands

ABSTRACT

From 1940 to 1983 no earthquakes in the southern part of The Netherlands with a magnitude smaller than two are detected. This study shows that detection and location of earthquakes with magnitude two or lower with the existing network is mostly impossible and that the capability to detect and locate can be improved only by an extention of the seismic network.

INTRODUCTION

The detection and location of microearthquakes are of interest for The Netherlands, because this country has a low seismicity and the magnitude of the earthquakes generally is low ($M \leq 5$). Microearthquakes are used to identify active faults, which are important to evaluate the seismic risk. Microearthquakes are also interesting for the determination of the structure of the earthcrust and the study of the seismicity induced by mining.

All earthquakes with epicenter within The Netherlands, recorded in the period 1940-1983 are listed in table 1. Figure 1 shows the location of the seismic network in The Netherlands and gives a definition of the areas the Southern part of The Netherlands and the South of Limburg. The frequency distribution in figure 2 is derived from the earthquake list in table 1. The dashed lines give the expected earthquakes for $M_L \leq 2$ according to the formula $\log N = a - bM$ (Karnik, 1971). The earthquakes of table 1 are plotted in figure 3. It illustrates the well-known fact that the Southern part of The Netherlands is seismically the most active region (Ritsema, 1966; Ahorner a.o., 1976).

P. Melchior (ed.), Seismic Activity in Western Europe, 133–143.
© *1985 by D. Reidel Publishing Company.*

Table 1.

```
*************************************************************************************
******************* EARTHQUAKES IN THE NETHERLANDS REGION SINCE 1940 ****************
*************************************************************************************
```

DATE	SOURCE	N	COORDINATES	ORIGINTIME	DEPTH	M	ML	I	RMAX	REC	FELT	LOCATION
420113	DBN :		52.1 N- 5.1 E					2		-	UTRECHT AREA	UTRECHT
600625	BNS :	28	51.19N- 5.68E	14:29:13.4	12.5	4.2		5	135	DBN	DORDRECHT,BREDA	S OF WEERT
	BCIS:		51.5 N- 5.7 E	14:29:13.0						WIT	ARNHEM AND ZWOLLE	
										HEE		
670428	DBN :		50.9 N- 5.8 E							HEE	VALKENBURG	VALKENBURG
	BNS :			03:17:45				2		BNS		
670428	DBN :		50.9 N- 5.8 E							HEE	VALKENBURG	VALKENBURG
	BNS :			05:03:37				2		BNS		
670428	BNS :		50.87N- 5.83E	16:05:07			2.4	2		HEE	VALKENBURG	VALKENBURG
										BNS		
										DOU		
670501	HEE:		50.87N- 5.83E	20:09:39			2.0	2		HEE	VALKENBURG	VALKENBURG
										BNS		
691213	BNS:		51.2 N- 5.9 E	22:54:44			2.6			BNS		W OF ROERMOND
										DOU		
										WRM		
										GIP		
710218	DBN :	70	51.05N- 5.95E	23:41:23	10-20	4.0		5	150	HEE	LIMBURG,	KONINGSBOSCH
	BCIS:		51.03N- 5.93E	23:41:23.0		4.5			120	RSB	EASTERN-NOORDBRABANT,	
	ISC :	45	51.04N- 5.64E	23:41:23	16					DBN	SOUTHERN-GELDERLAND,	
	NEIS:		50.9 N- 6.0 E	23:41:24.2	27	4.2				WIT	DOETINCHEM,ARNHEM AND OSS	
	BNS :		51.05N- 5.95E	23:41:25	22	4.4		4-5	130			
711115	BCIS:	4	50.9 N- 6 E	21:17:10.0			2.1	2		HEE	KLIMMEN	KLIMMEN
										RSB		
										WIT		
721230	DBN :		51.8 N- 5.8 E	05:33:52						HEE		NIJMEGEN
	BCIS:	5	51.8 N- 6.1 E	05:33:51						WIT		
	BNS :			05:33:52			2.6					
721230	DBN :		51.8 N- 5.8 E	05:48:28	5-10		3.2	3-4	40	HEE	NIJMEGEN,DIDAM,	NIJMEGEN
	BCIS:		51.8 N- 6.1 E	05:48:26	0					WIT	WAGENINGEN,TIEL,	
	ISC :	25	51.68N- 5.64E	05:48:27.6	33						UDEN.	
	NEIS:		51.6 N- 5.7 E	05:48:28.4	33							
	BNS :		51.8 N- 5.9 E	05:48:27			3.2					
721231	DBN :			08:00:20		2				HEE		NIJMEGEN
	BCIS:		51.8 N- 6.1 E	08:00:20						WIT		
	ISC :	9	51.78N- 5.7 E	08:00:21.1	33							
	NEIS:		51.8 N- 5.8 E	08:00:21.4	33							
	BNS :			08:00:20			2.7					
750621	DBN :			16:49:52			3.2	4		HEE	VAALS-AREA	VAALS
	BCIS:		50.8 N- 6.1 E	16:49:52				4		WIT		
	ISC :	20	51.1 N- 5.6 E	16:49:51	28							
	NEIS:		50.8 N- 6.0 E	16:49:51.3	28							
	BNS :						3.2					
750727	DBN :						1.5	1		WTS	-	VALKENSWAARD
	BNS :		123KM	01:23:50.6			2.6			HEE		

DATE	SOURCE	N	COORDINATES	ORIGINTIME	DEPTH	M	ML	I	RMAX	REC	FELT	LOCATION
760629	DBN :					4		5		HEE	BETWEEN WEERT	ROERMOND
	CSEM:		51.2 N- 6.1 E	05:00:30						WTS	AND ROERMOND	
	ISC :	15	51.14N- 5.7 E	05:00:29.1	0							
	BNS :		51.21N- 5.81E	05:00:32			3.3					
790717	DBN :	3	51.8 N- 5.8 E	16:48:56.7		3		3	30	WTS	SW OF NIJMEGEN;	NIJMEGEN
	CSEM:		51.74N- 5.9 E	16:48:56.1	7					HEE	GRAVE,MOOK,CUYK,	
	ISC :	11	51.74N- 5.87E	16:48:55.1	5					WIT	OVERASSELT,BEERS,	
	BNS :		51.83N- 5.78E	16:48:55.5			3.0	4			MALDEN,HATERT,	
											MILL	
800605	DBN :	4	51.2 N- 5.8 E	12:11:40		4			100	ENN	STRONG IN AREA	HEIJTHUIZEN
	CSEM:	48	51.20N- 5.88E	12:11:40.7	10		3.8(VKA,STR)			WTS	ROERMOND/WEERT;	
	NEIS:	29	51.22N- 5.82E	12:11:39.5	22					WIT	EINDHOVEN,SITTARD,	
	BNS :		51.25N- 5.81E	12:11:39.0	18		3.8	5		DBN	ARNHEM,WAGENINGEN,EDE,	
											DRIEBERGEN,ULFT,ZEVENAAR	
800606	DBN :	3	51.2 N- 5.8 E	04:24:10		2				ENN	HEIJTHUIZEN	HEIJTHUIZEN
										WTS		
										HEE		
811220	DBN :	6	50.92N- 5.79E	10:38:50.9	10					ENN		KLIMMEN
	CSEM:	13	50.90N- 5.79E	10:38:50.9	10					HEE		
	BNS :	5	50.87N- 5.89E	10:38:52.7	22		2.8			WTS		
820302	DBN :	8	51.05N- 5.84E	01:27:27	10		4	4-5	35	ENN	STRONG IN AREA	BORN
	CSEM:	32	51.01N- 5.76E	01:27:26.3	10	4.0	3.7			WTS	SITTARD/GELEEN;	
	NEIS:	10	50.98N- 5.89E	01:27:27.0	24					HEE	AKEN,MAASMECHELEN,	
	BNS :	9	51.03N- 5.85E	01:27:27.1	14		3.6			WIT	ROERMOND,MAASTRICHT,	
										DBN	HERKENBOSCH,NEER.	
820522	DBN :	3	51.1 N- 5.8 E	04:41:07		3		3	15	ENN	SITTARD,GELEEN,	ECHT
	CSEM:	22	51.06N- 6.03E	04:41:08.0	10					WTS	HERKENBOSCH,	
	BNS :						3.0			HEE	SUSTEREN	
820522	DBN :	5	51.10N- 5.94E	06:00:02.5		4		4-5	40	ENN	ELSLO,ROERMOND,	ECHT
	CSEM:	52	51.25N- 5.75E	06:00:04.1	10					WTS	GELEEN,MAASBRACHT,	
	NEIS:	32	51.12N- 5.80E	06:00:00.8	10		4.4			HEE	SCHINVELD,STEIN,BORN,BAARLO,	
	BNS :		51.08N- 5.96E	06:00:03.4	14		3.8	4-5	70	WIT	KERKRADE,SIMPELVELD,DUSSELDORF,	
										DBN	BUNDE,MAASTRICHT,BEESEL ENZ	
830205	DBN :	5	51.39N- 5.39E	07:51:16.4	10		3	3	10	WTS	WAALRE,EINDHOVEN	WAALRE
	BNS :			07:51:16			2.8			BNS		

MEANS OF ABBREVIATIONS :

DBN	: DE BILT (NETHERLANDS)
CSEM	: CENTRE SEISMOLOGIQUE EUROPEO-MEDITERRANEEN (STRASBOURG)
BCIS	: BUREAU CENTRAL INTERNATIONAL DE SEISMOLOGIE (STRASBOURG)
ISC	: INTERNATIONAL SEISMOLOGICAL CENTRE (NEWBURY, UK)
NEIS	: NATIONAL EARTHQUAKE INFORMATION SERVICE (BOULDER, USA)
BNS	: BENSBERG (W-GERMANY)
N	: TOTAL NUMBER OF STATIONS USED FOR DETERMINATION
M	: MAGNITUDE (RICHTER SCALE)
ML	: MAGNITUDE (LOCAL)
I	: INTENSITY (MERCALLI SCALE)
RMAX	: MAXIMUM RADIUS (KM) OF FELT AREA
ENN	: CODE FOR SEISMIC STATION EPEN (NETHERLANDS)
WTS	: CODE FOR SEISMIC STATION WINTERSWIJK (NETHERLANDS)
HEE	: CODE FOR SEISMIC STATION HEERLEN (NETHERLANDS)
WIT	: CODE FOR SEISMIC STATION WITTEVEEN (NETHERLANDS)
RSB	: CODE FOR SEISMIC STATION RAVENSBOSCH (NETHERLANDS)
VKA	: CODE FOR SEISMIC STATION VIENNA (AUSTRIA)
DOU	: CODE FOR SEISMIC STATION DOURBES (BELGIUM)
GIP	: CODE FOR SEISMIC STATION GILEPPE (BELGIUM)
WRM	: CODE FOR SEISMIC STATION WARMIFONTAINE (BELGIUM)
STR	: CODE FOR SEISMIC STATION STRASBOURG (FRANCE)

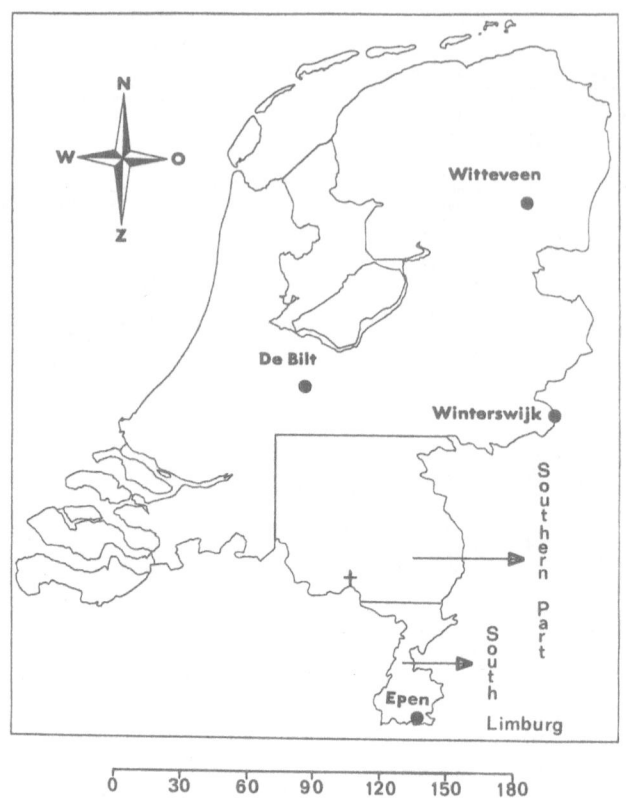

Figure 1. Locations of the seismic stations in The Netherlands.
          Station codes: DBN, De Bilt; WIT, Witteveen; WTS,
          Winterswijk; ENN, Epen.

          It is observed that microearthquakes of $M_L < 2$ are not
detected in The Netherlands. This raises the question whether or
not these earthquakes occur in the region.
In the next sections it is explained that detection and location
of microearthquakes with $M_L < 2$ is mostly impossible in the
Southern part of The Netherlands; so the possibility of the
occurrence of these earthquakes still exists. With an extended
network this question can be decided.
          Theoretically it is likely that microearthquakes with $M_L < 2$
do occur, see figure 2. Actually, a single example has been found
which could be located, namely that near Hulsberg (South of

Limburg) on 27 May 1984, recorded at ENN, MEM and HEE. In addition there are every year in the order of hundred onsets present in ENN records only for which no epicenter determination is possible. It is expected that only an extension of the network will enable us to locate such small events.

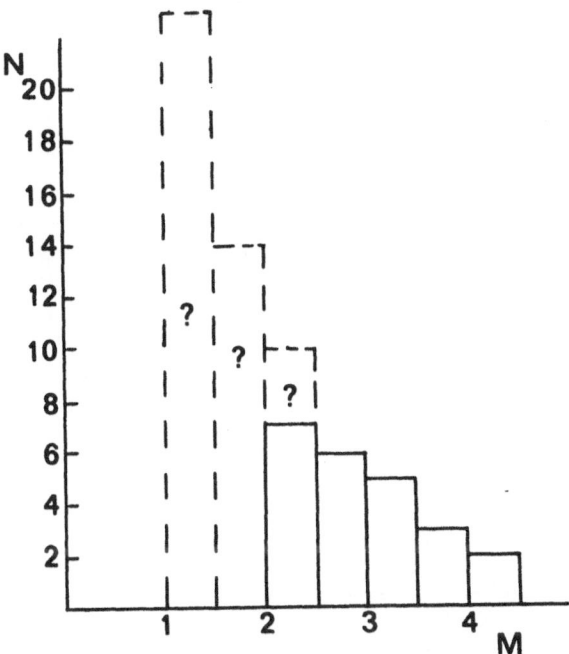

Figure 2. Frequency distribution for earthquakes of The Netherlands for the period 1940–1983. For each class is $M \leq M_L < M + \frac{1}{2}$. Total number of recorded earthquakes 23. N is number of earthquakes, M is local magnitude and ? are extrapolated number of earthquakes of $M_L \leq 2$.

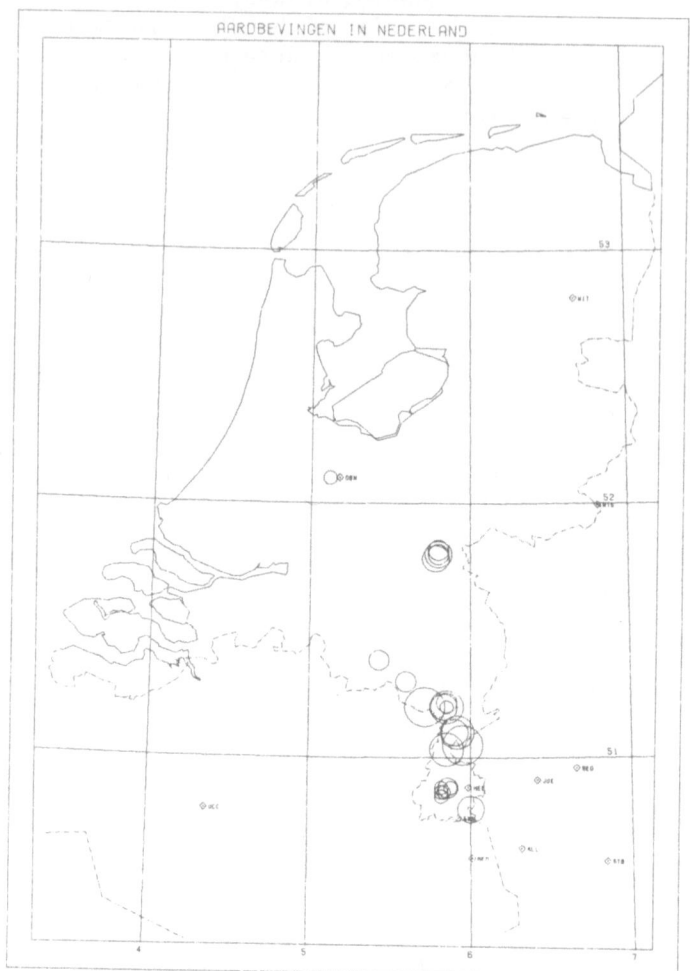

Figure 3. Earthquakes in The Netherlands, 1940–1983. The size of
    the symbols is an indication for the magnitude of the
    shock according to the classes given in the figure 2.

DETECTION CAPABILITY OF THE NETHERLANDS SEISMIC NETWORK

    The detection capability curve, figure 4, has been
constructed from the HEE (prior to August 1979), ENN and WTS
seismograms of about 40 earthquakes in The Netherlands and
adjacent areas (1976–1983) with $1 < M_L < 3.5$. The ENN and WTS

records are of the same quality (seismometer both Willmore MK II, short period, vertical component, magnification both 50.000 at 1 Hz and about the same noise level). The station HEE (Heerlen) has a higher noise level. In the detection area the earthquakes can be detected by P and S onsets which are recognizable. The detection area depends on the noise level at the station, the local geology (attenuation) and the magnification of the seismometer.

The full line in figure 4 could be determined rather well for $1 < M_L < 2.5$ from the seismograms of ENN (HEE) and WTS. The dashed line between M = 0 and 1 has been obtained by extrapolation to the point (M_L = 0, distance = 5 km). This point has been taken from a paper by Bock and Mayer-Rosa, 1981.

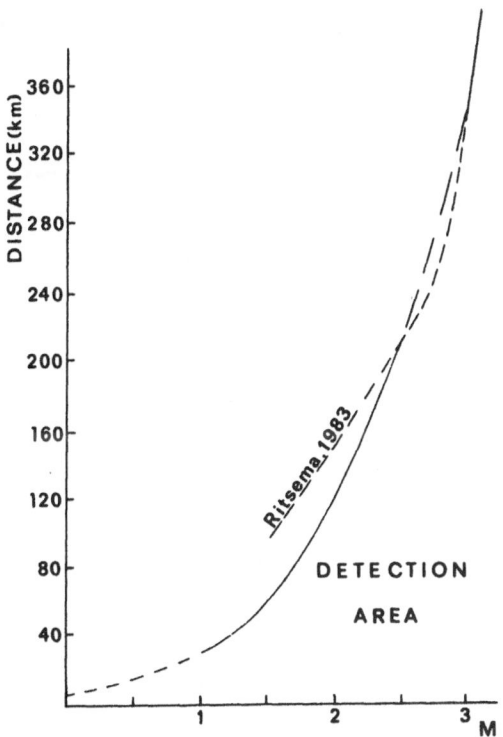

Figure 4. Detection capability of ENN and WTS for micro-earthquakes.

The dashed line for $2.5 < M_L < 3$ could not be determined very well because of sparsity of data. It has been obtained by extrapolation of the magnitude-distance curve to the point (M_L = 3, distance =

345 km), determined by Ritsema, 1983 for the station WTS. All of
the used earthquakes in this magnitude range are in agreement
with this line. The line of Ritsema for m_b > 3 is well supported
by data, but for m_b < 3 (the dashed line) the results were
preliminary.

In figure 5 the concentric circles around ENN and WTS denote
the detection areas for different values of M_L which are derived
directly from the detection capability curve of figure 4. The
figures 6 shows the number of stations available for detection in
different ares in the Southern part of The Netherlands for M =
1.5, 2, 2.5 and 3 respectively. The detection areas for DBN and
WIT are estimated partly from Ritsema, 1983 and partly from
seismograms.
For magnitudes smaller than 1.5 the detection area is limited to a
small area around station ENN. For M $\geq$ 3.5 three or four stations
are available for detection and thus for location of earthquakes
in the Southern part of The Netherlands (Ritsema, 1983).

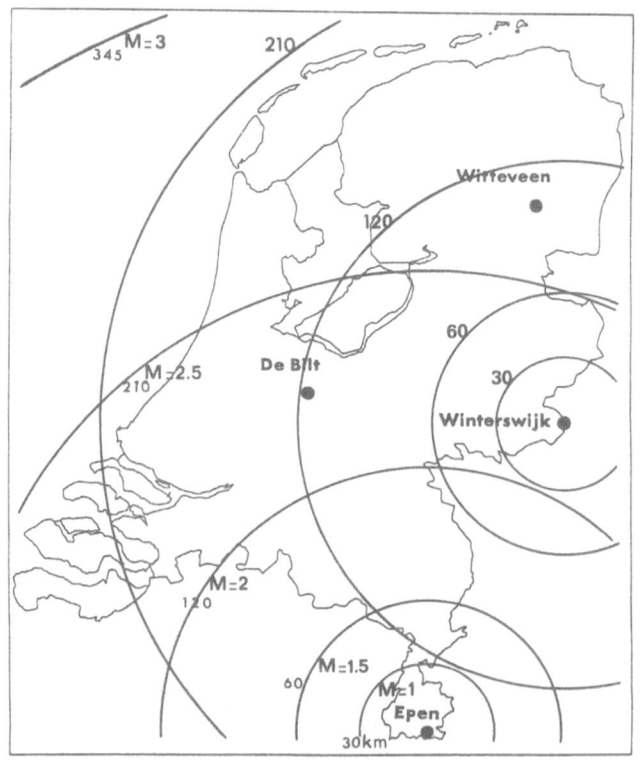

Figure 5. Detection areas for ENN and WTS.

From the figures 6 it follows that the detection of earthquakes of M > 2 is always possible in the Southern part of The Netherlands, because at least one station is available. For M ≤ 2 detection becomes more uncertain. The noise level is higher under bad weather conditions (especially in fall and winter time) and the detection capability may be sometimes of the order of one unit of magnitude lower (Ritsema, 1983). This explains that

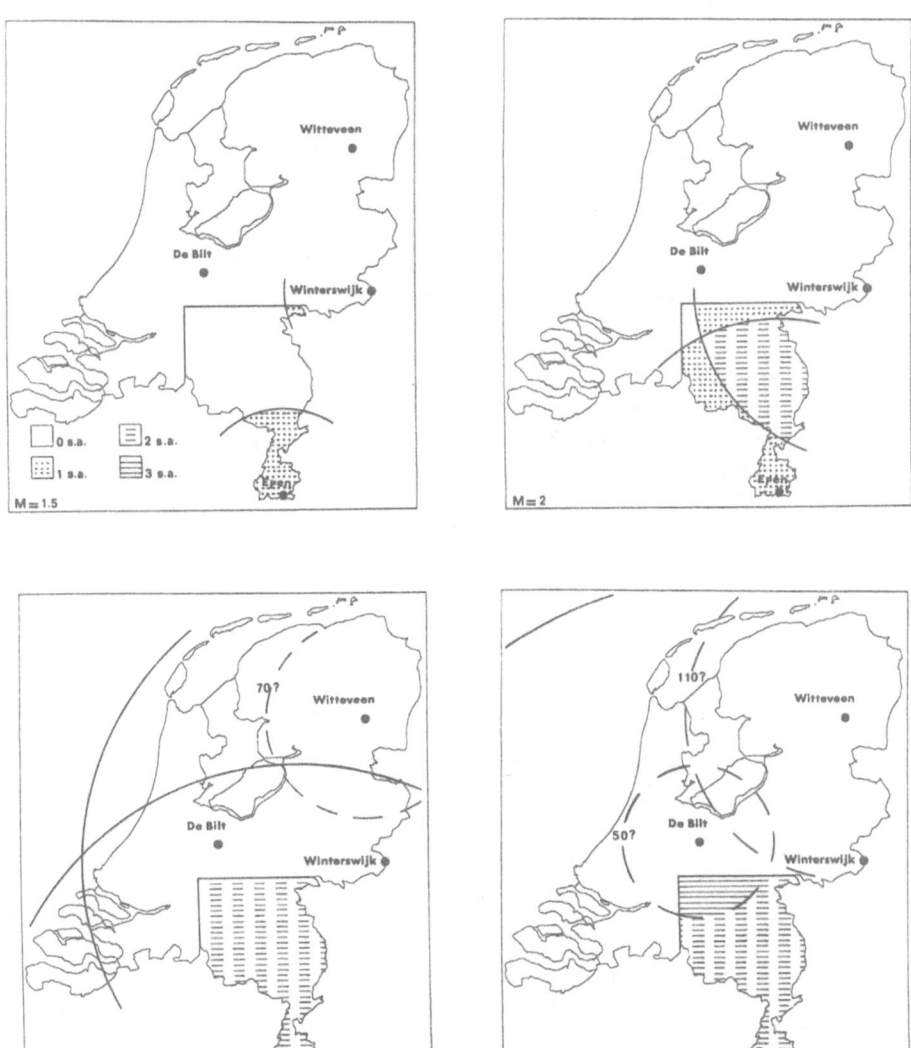

Figure 6. Number of stations available (s.a.) for detection in different areas in the southern part of The Netherlands for M = 1.5, 2, 2.5 and 3.

earthquakes with M about 2 sometimes remain undetected for a great part of the Southern part of The Netherlands. The conclusion that for M < 2 detection is mostly impossible seems to be justified, except for the South of Limburg where the detection capability is higher, but still only one station is available. The station ENN is operational since 1979 and the station WTS since 1974. Prior to 1979 seismometers with a lower magnification and at other locations with a higher noise level were used, and the detection capability accordingly was considerably lower.

LOCATION CAPABILITY

Earthquakes with $M_L \geq 3.5$ can be located by the Netherlands stations for the whole of the Southern part of The Netherlands (see previous section). With the assistance of the stations close to the border in Belgium and Germany, however, loation of earthquakes with $M_L \geq 2$ in the Southern part of The Netherlands – and with $M_L \geq 1.5$ in the South of Limburg – is possible; except under bad weather conditions the location capability is lower.

Only with a local network or array earthquakes with $M_L < 1.5$ can be located. The stations ENN, MEM (Membach) and HEE, see figure 3, can be considered as such a local network for the south of South Limburg. Figures 7 shows the single example of such a microearthquake, location Hulsberg (50° 53.3'N, 5° 51.5'E), depth 3 km, recorded at the three stations on 27 May 1984, $22^h\ 36^m\ 58.65^s$ UT.

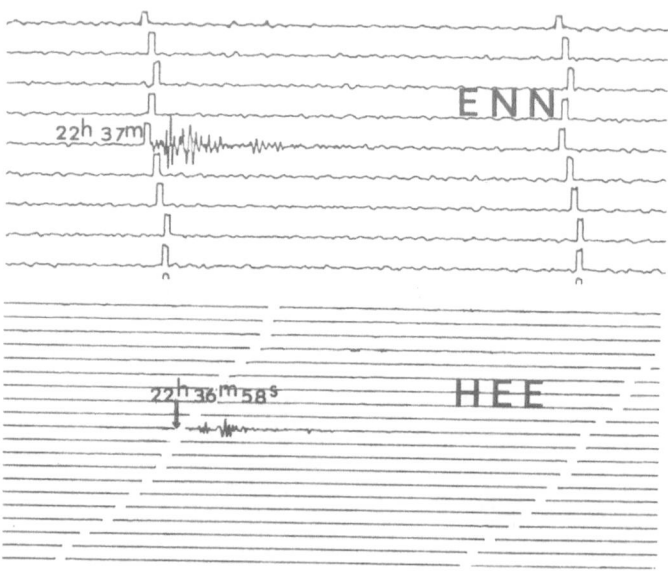

Figure 7. Microearthquake in South Limburg, 1 < M < 1.5. 27 May 1984.

IMPROVEMENTS

The detection and location capability for earthquakes in the Southern part of The Netherlands can be improved by an extension of the seismic network. The optimum location for a new station is in the centre of the Southern part e.g. Valkenswaard (+ in figure 1). For earthquakes around magnitude two, the detection and location capability could be strongly improved. Since the noise level is high in the centre of the Southern part the use of a borehole seismometer seems advisable.
For earthquakes with M_L < 1.5 a local network or array is necessary. A good co-operation and data exchange with stations in Belgium and Germany will help considerably to complete the seismicity map of the Southern part of The Netherlands and adjacent areas.

CONCLUSIONS

For earthquakes with M_L > 2 the detection and location capability in the Southern part of The Netherlands is sufficient, but not for earthquakes with M_L ≤ 2. The detection and location capability can be improved by the extension of the seismic network. Firstly, by a new station in the centre of the Southern part, especially for earthquakes around magnitude two in the region, and secondly by a local network or array near Epen station for earthquakes with M_L ≤ 1.5.

REFERENCES

Ritsema, A.R., 1966, Note on the seismicity of the Netherlands, Proceedings Kon. Ned. Akademie van Wetenschappen, Series B, 69, no. 2, pp. 235-239.
Ahorner, L., a.o., 1976, First draft of an earthquake zoning map of Northwest Germany, Belgium, Luxemburg and the Netherlands, Proceedings of the E.S.C. Symposium in Luxemburg on Earthquake risk for nuclear power plants, pp. 39-41, KNMI, De Bilt.
Bock, G. and D. Mayer-Rosa, 1980/81, Seismic Observations at Lake Emosson, Switzerland, Pageoph. Vol. 119, pp. 185-195, Birkhäuser Verlag, Basel.
Ritsema, A.R., 1983, Note on the detection and location capability of the Netherlands seismic stations for earthquakes in the North Sea Basin, Proceedings of the NATO advanced study workshop on Seismicity and Seismic Risk in the Offshore North Sea Area, pp. 195-197, D. Reidel Publishing Company, Dordrecht.
Karnik, V., 1971, Seismicity of the European Area/2, D. Reidel Publishing Company, Dordrecht.

SEISMICITY AND CRUSTAL STRUCTURE IN ITALY *

Carlo Morelli

Istituto di Miniere e Geofisica applicata
Università, 34123 Trieste, Italia.

ABSTRACT

The extensive geophysical surveys in Italy, especially the extended DSS profiles program (from 1956), have permitted a first
knowledge of the main crustal conditions, which are very varied
and cover most of the normal and extreme features. Of particular
importance for the understanding of the actual seismicity and of
the character of the seismogenetic zones are :
- a long strip of Moho vertical discontinuities, faults and sub-
  ductions along the axis of the Apennines, from the Po Plain to
  Calabria, to which correspond the maxima of seismicity;
- the presence of infra-crustal and crustal doublings all around
  the Adriatic microplate, the borders of which are also characterised by a strong seismicity.

Actual seismicity is an index of the tectonic stresses and mass
unbalances presently acting. They may involve the most recent geological periods (Fig. 1), or be the continuation on very long geo-
dynamic or orogenetic processes; e.g., the Northward push of the
Adriatic (African) plate has formed the Alpine system in the Tertiary, and is causing very severe earthquakes also in the present
(Fig. 2).

A first index of strong lateral mass variations, and consequently of possible unbalanced conditions in the crust's interior and
bottom, can be offered in regional sense by the strong gradients

---

Contribution n° 99, IMGA - Università, Trieste, Italy.

P. Melchior (ed.), Seismic Activity in Western Europe, 145–155.
© *1985 by D. Reidel Publishing Company.*

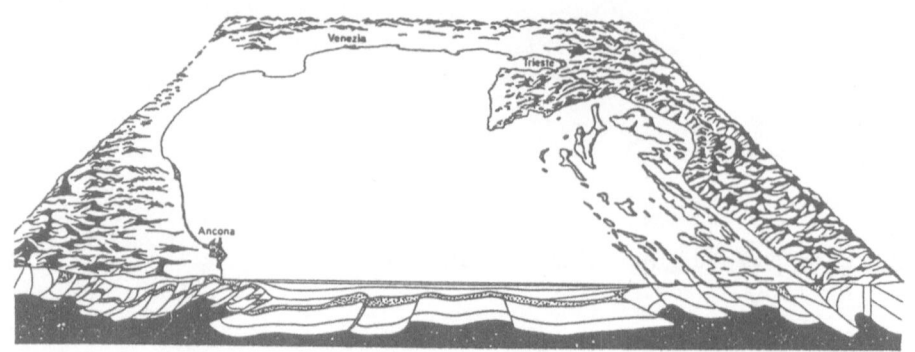

Fig. 1. Reversed centripetal faults in the recent sedimentary
strata are responsible for the strong seismicity on both sides
of the Adriatic sea (courtesy AGIP).

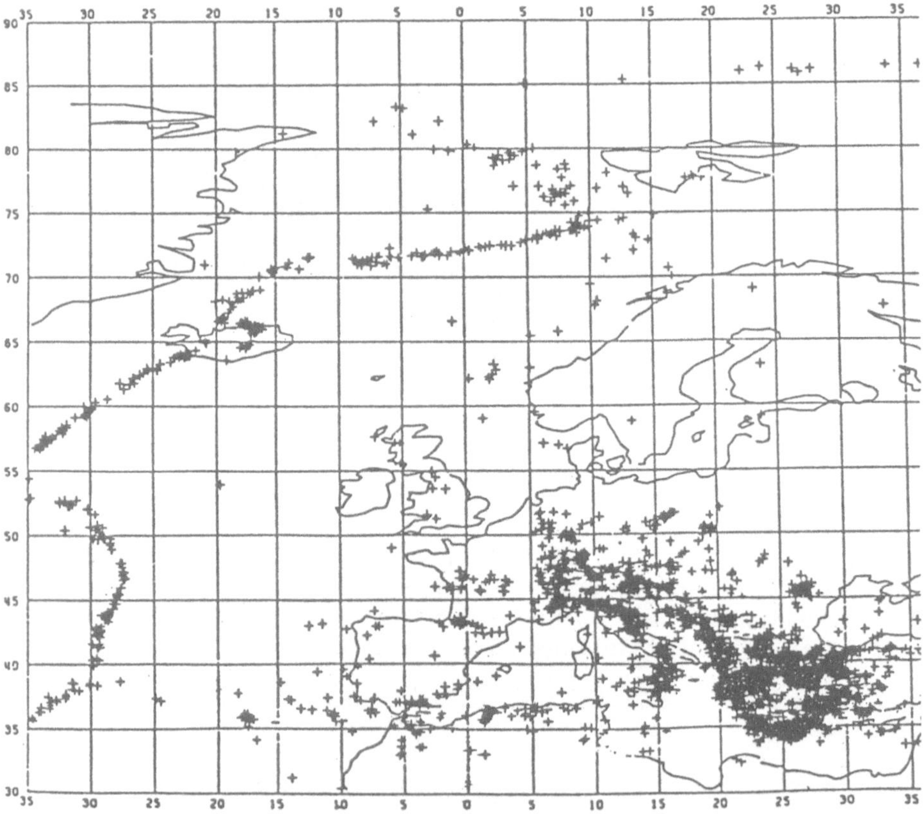

Fig. 2. The Africa - Europe border is characterized by strong seismicity also around the Adriatic microplate (CSEM).

in the Bouguer gravity anomalies (Fig. 3). But the main problem is
always the identification of the anomalous crustal structures :
they represent the geological memory of tectonic paroxysms of the
past, and potential sites for reactivation. Typical consequences
of the continent - continent collisions are the well known over-
thrusts discovered by Geology; and the less known infracrustal and
crustal doublings discovered recently by Geophysics (Fig. 4).

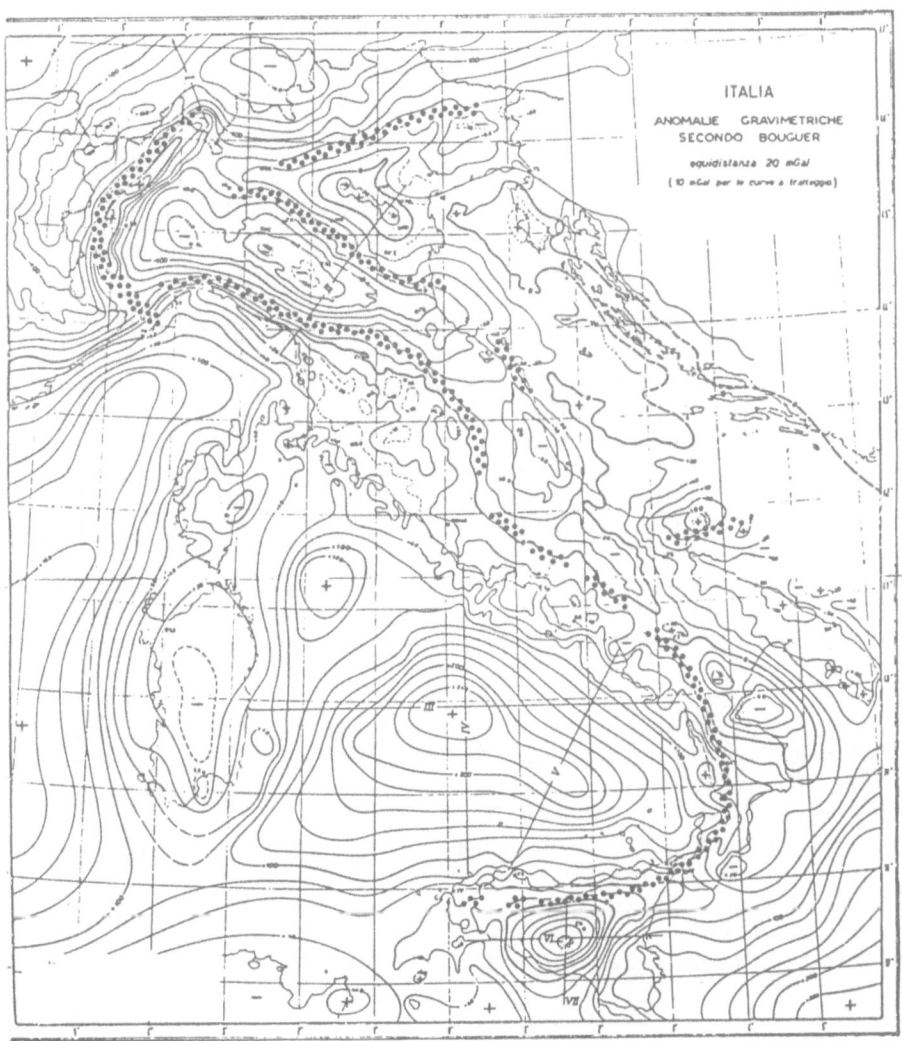

Fig. 3. Strong gradients in the Bouguer gravity anomaly {1} indi-
        cative of mass differences, can also be preferential areas
        for earthquakes.

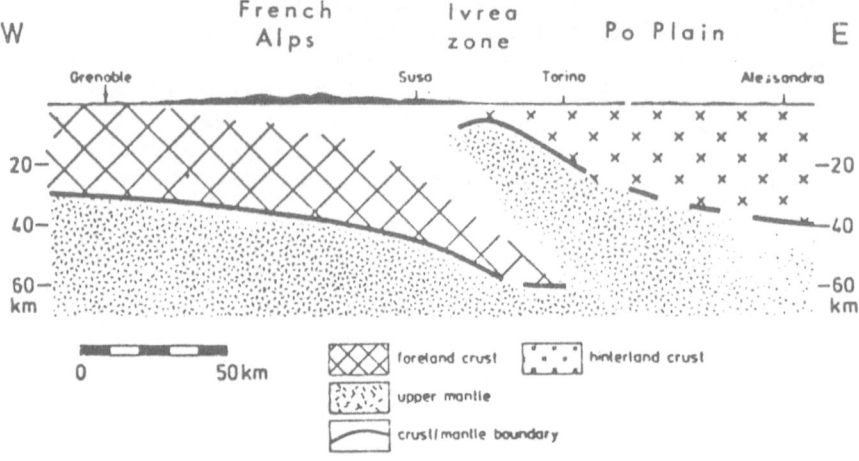

Fig. 4. The Ivrea zone is the first example of crustal doubling
        {2}.

The very extensive program of Deep Seismic (Refraction) Soun-
ding (DSS) performed in Italy (Fig. 5, 6), mostly in european
cooperation (European Seismological Commission) and with financial
grants mainly by the Consiglio Nazionale delle Ricerche, has permitted to discover - inter allia - that around the Adriatic (mi-
cro-) plate infracrustal and crustal doublings appear to be a
general features (Fig. 7) : although, obviously, not all these
mass shifts are necessarily yet active.

On the contrary, the mass deformations induced in the crust by
big vertical Moho faults, or by strong gradients in the Moho iso-
baths (Fig. 8), seem to represent very active seismogenetic zones
(Fig. 9).

Also the long term isostatic adjustments, with vertical strata
displacements of the order of many kms, can give origine (expe-
cially in the so caused flexure zones) to fractures, i.e. to
earthquakes. Examples are in Italy :
a) the continuous subsidence of the Po Plain (Miocene at - 6 kms
   in the Po Delta region : mean - 1 mm/y;
b) Mesozoic at - 14 km at the Apennine border (Fig. 10) : mean
   - 0.2 mm/y;
c) the continuous raising of the Alps (+ 1 mm/y).

The velocity inversions revealed by DSS in the Upper and Lower
Crust, when connected with partial melting (Fig. 11), can also
give origin to earthquakes. But normally these are attributed to
stresses of tectonic origin in the corresponding layers (Fig. 12).

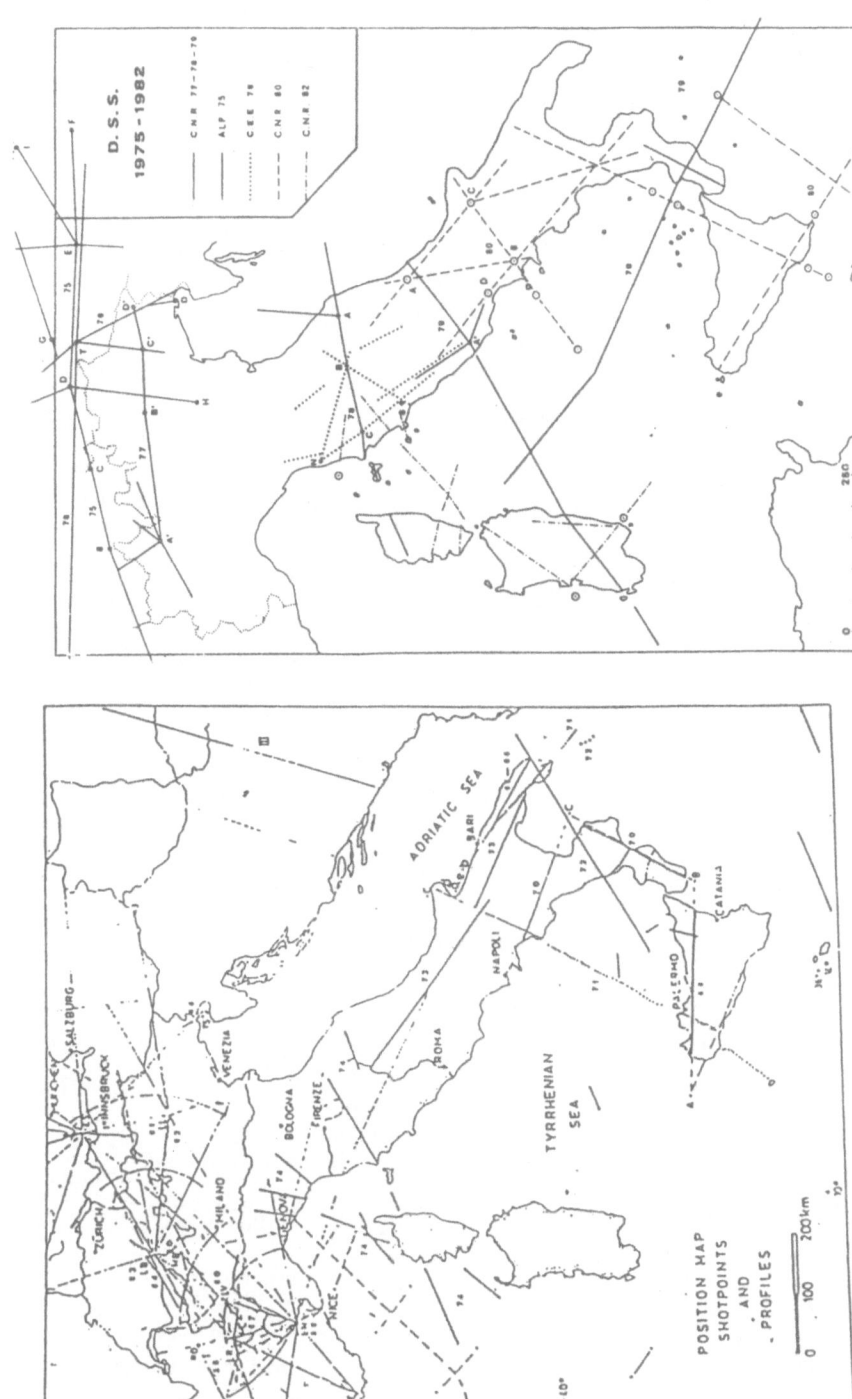

Fig. 5. The DSS profiles in Italy 1956-74.

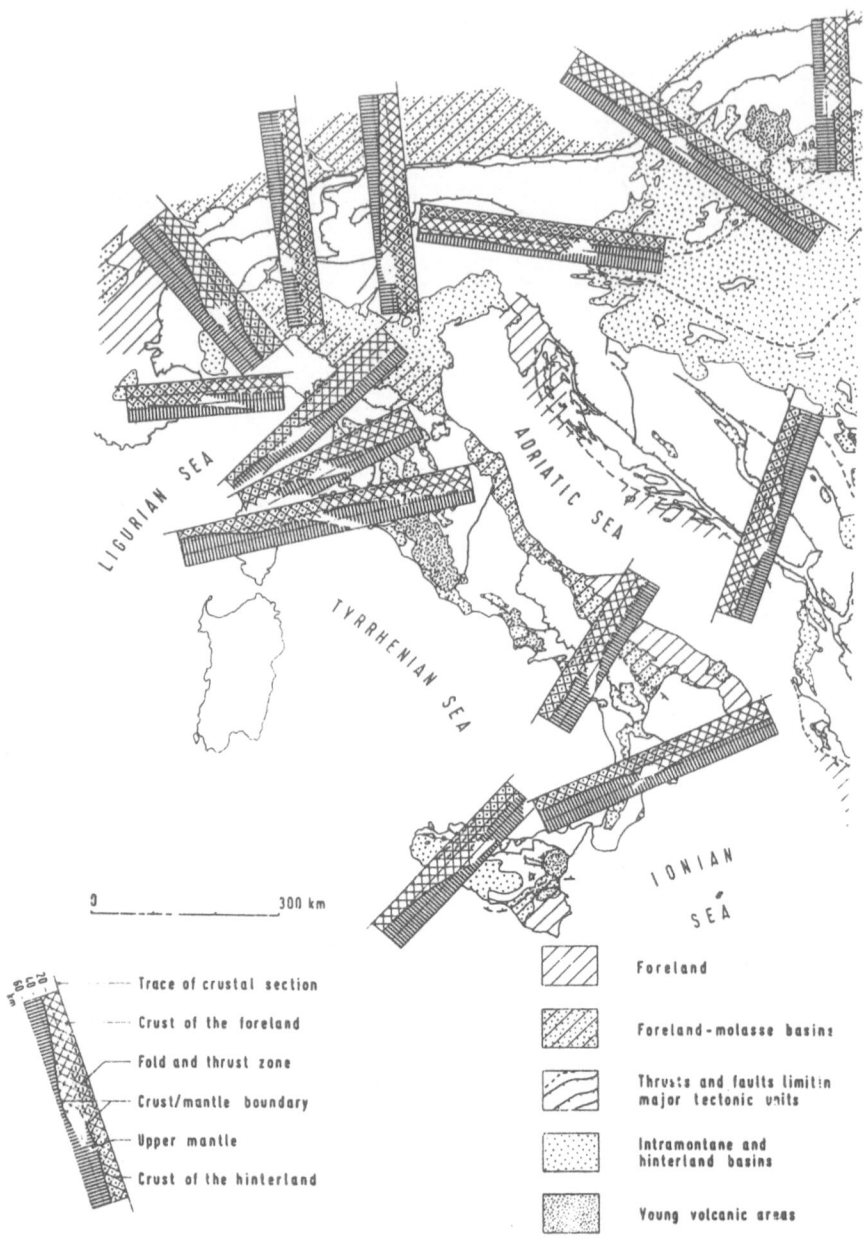

Fig. 7. Crustal doublings around the Adriatic microplate {3}.

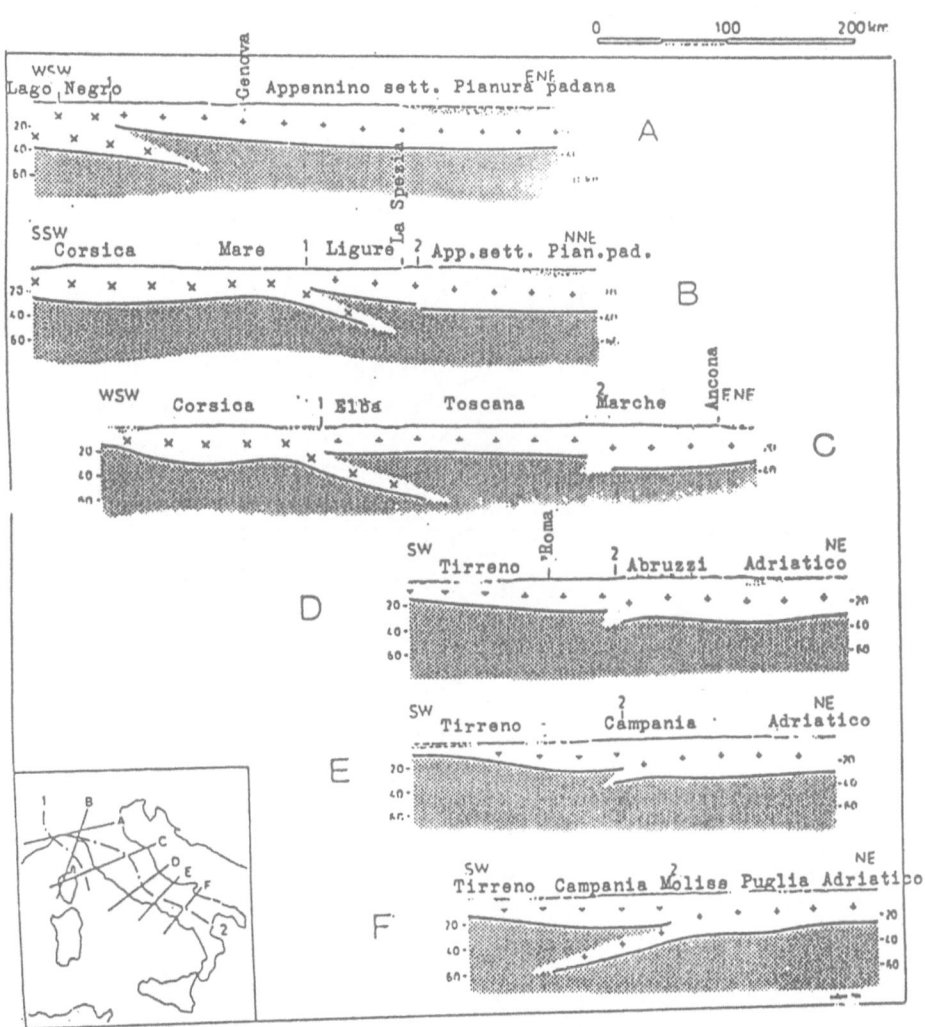

Fig. 8. Main Moho discontinuities, correlable in the Italian
peninsula {4}.

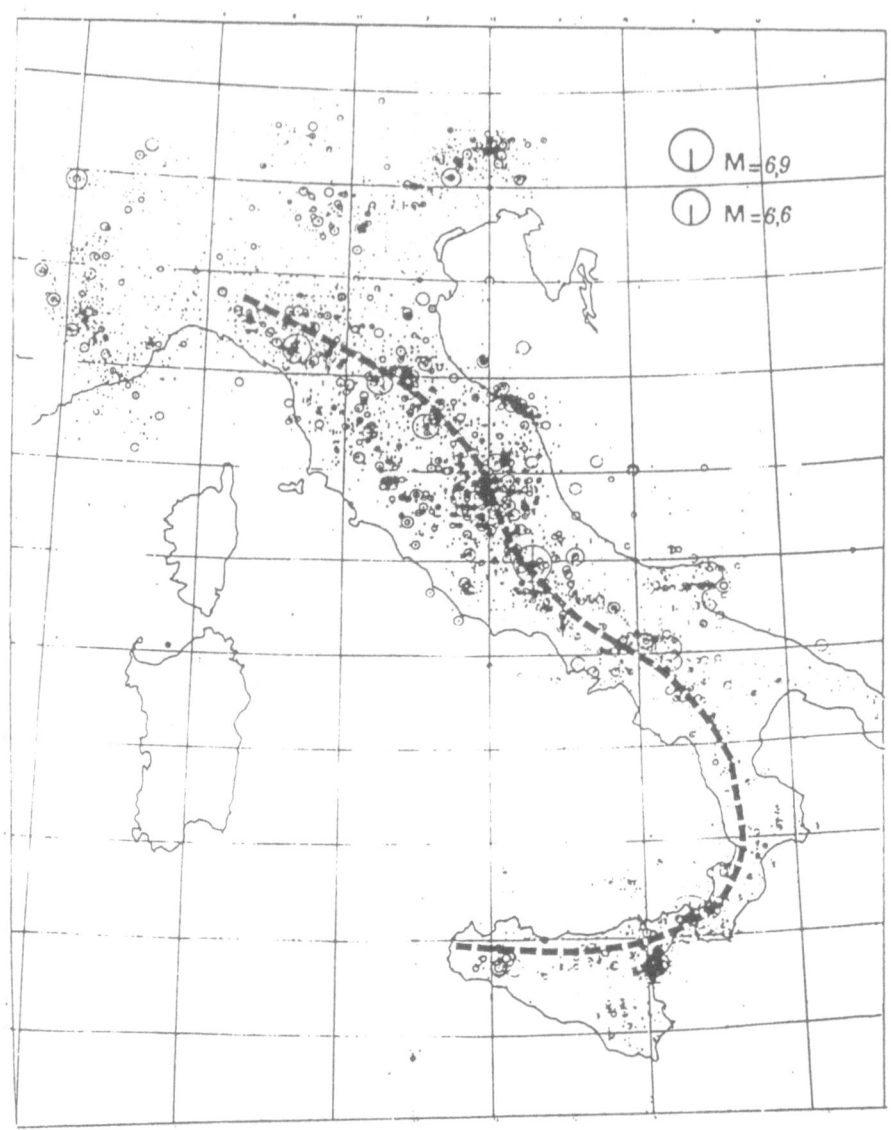

Fig. 9. Seismicity in Italy 1884 - 1971 {5} and main discontinuity
        of Fig. 8.

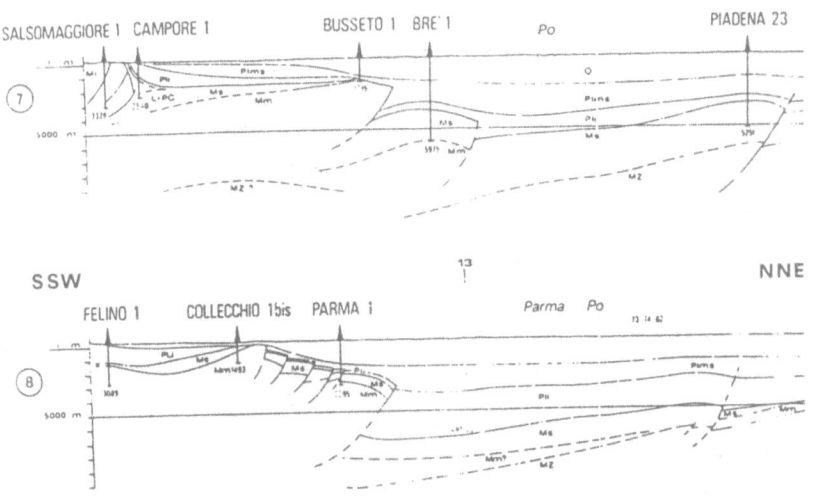

Fig. 10. Geological SSW - NNE sections in the Po Plain (from {6}).

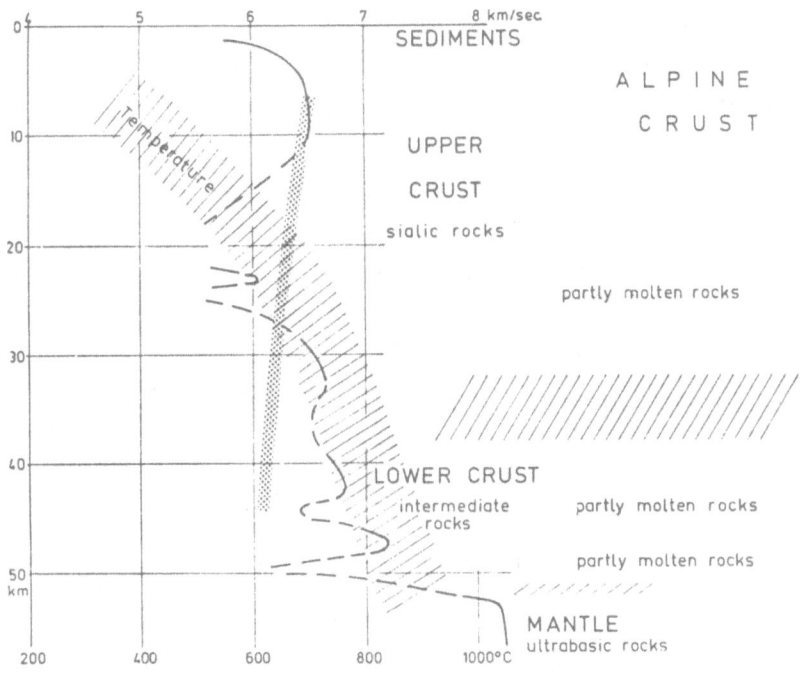

Fig. 11. Velocity inversions in the Central Alps {7}.

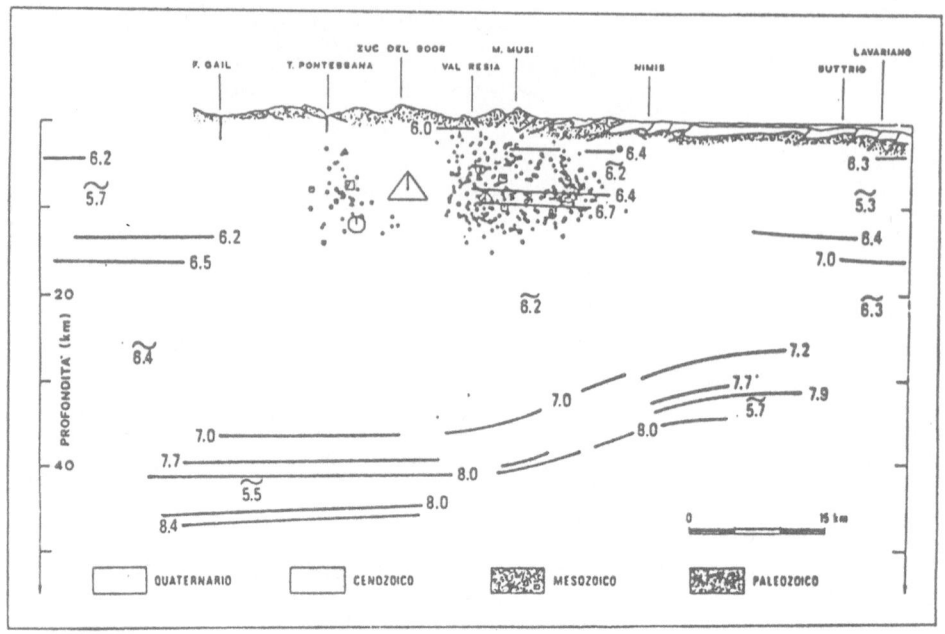

Fig. 12. DSS profile Trieste - Tirol, with foci of the Friuli
         seismogenetic zone {7}.

In conclusion, the knowledge of the crustal structure is a necessary premise for the understanding of the physical conditions acting at the earthquakes foci area, and for the physical characterisation of the seismogenetic zones.

REFERENCES

1. Morelli, C. 1975, Quaderni de "La Ricerca Scientifica" n. 90, CNR, Roma, 427-447.

2. Giese, P. 1984, Schweiz. Mineral. Petrol. Mitt., 48, n. 1, 261-284.

3. Giese, P. and Reutter K.-J. 1978, In : H. Closs, D. Roeder, Schmidt K., (Eds). Alps, Apennines, Hellenides, Schweizerbart, Stuttgart, 565-588.

4. Wigger, P.J. 1984, Berliner Geowiss. Abh., Reihe B/Heft 9, 110 Pg.

5. Carrozzo, M.T., de Visintini, G. and Giorgetti, F. 1974, Boll. Geof. teor. appl. XVI, 61, 3-21.

6. Pieri, M. and Groppi, G. 1981, P.F. Geodinamica CNR, publ. n. 414.

7. Giese, P. and Morelli, C. 1975, Quaderni de "La Ricerca Scientifica" n. 90, CNR, Roma, 453-489.

8. Carulli, G.B., Giorgetti, F., Nicolich, R. and Slejko, D. 1982, In : A. Castellarin & G.B. Vai (a cura di) : Guida alla geologia del Sudalpino centro-orientale. Guide geol. reg. S.G.I., 361-370, Bologna.

DEVELOPMENT OF A PROGRAM FOR RADON MEASURE IN SEISMIC ACTIVE ZONE, THE CASE OF MONS BASIN (BELGIUM).

Bouko Ph., Charlet J.-M., Quinif Y.

Faculté Polytechnique de Mons (Belgium).

In this paper, a complete bibliography about the radon methods and their application in earthquake prediction has been discussed.
Afterwards a review of the techniques for the radon measure is presented with special considerations about the different methods developed at the "Faculté Polytechnique de Mons".
Finally, some applications of these techniques on several sites of the Mons basin (a sismic active zone) are shown and a new full program of radon measures in this basin is presented. In these first applications, we particularly develop the problem of the meteorological influences and the discrimination between a superficial and deep radon flux.

1. RADON AND EARTHQUAKE PREDICTION (1 to 47).

Radon 222 is the only inert and water-soluble gas produced in the radioactive decay of uranium 238. It is continually generated from the radium, this one exists everywhere in the crustal materials. After its generation the gas diffuses to some extend into the pore spaces within rock and is transported on long distance into soil, porous and fracturated zones by soils gases or carrier fluids. This explains that radon can be considered as a "tracer" for many studies : uranium prospection, geothermal engineering, faults location, volcanology and earthquake prediction.

P. Melchior (ed.), Seismic Activity in Western Europe, 157–174.
© *1985 by D. Reidel Publishing Company.*

        Since the Tashkent earthquake(in 1965) (46), nu-
merous reports of radon anomalies have been published.
These anomalies precede or follow moderate or large
earthquakes, immediately or after a long time.   Such
some anomalies have been noticed in USSR, China, Japan
and USA.   Furthermore, in situ experiments using explo-
sive charges to generate seismic waves have been made.
In Europe, some recent studies have been developed in
Austria, France and Iceland.   These reports have been
based on the observation of modifications in radon level in groundwater (wells or springs) or in soil gases.
Some of the radon anomalies were recorded at a long
distance (several hundred kilometers from the epicen-
trum of the reportly related earthquakes).   The recorded radon anomalies are positive or negative, at short
(several hours or days) or long time (several weeks,
years).   As a general rule, the short time anomalies
seems to be related to small magnitude eartquakes and
the long time to high magnitude' eartquakes .
        In the point of view of the interpretation, some
authors have developed theories and models to explain
the apparition and evolution of the radon level variations preceding earthquakes.   So, they try to answer
to the three fondamental questions : where, when and
how will occur the earthquake.   The origin of these
anomalies is often explained by the modification of
the physical characteristics of rocks and fluids in
relation with the variation of the "stress field"
bounded to sismotectonic movements.
        The apparition and evolution of the premonitory
signs is described in two models : the "dilatance-
diffusion" model and the "dilatance-instability" model.
The Fleischer's model (8) and the Scholz's works (38,
43) would allow to determine the magnitude, the epi-
centrum and the date of the hypothetical earthquake
from the duration and the extend of the detected anomalies in radon concentration.   But, up-to-now, nobody
has been able to predict an earthquake particularly
on account of the great influence of the meteorological and seasonal conditions on the kinetic of the
radon escape.
        This research must be managed to a better know-
ledge of the "behaviour" of the radon in relation
with the sismicity and on the study and the construction of a very performing radon detector which can
work in-situ and continuously.   This ideal system
would be a self governing apparatus.   On the other
side, a perfect knowledge of the influence of meteorological parameters on the radon is absolutely necessary.

A statistical study of the correlations between radon countings and climatic parameters (air temperature, soil temperature, humidity, barometric pressure, soil moisture, water table, precipitations, wind, freeze-thaw sequences, solid earth tide,...) on various sites should permit to determine the "meteorological" noise and to correct the measured radon anomalies. It is sure that the control of the radon emanation is not the "absolute tool" but in addition with others techniques into "pluridisciplinary" programs, it can probably help to predict earthquakes.

2. RADON DETECTION TECHNIQUES.

The application of the radon methods in uranium exploration has lead to a great development of the field equipment and techniques of measurement (48, 51). We can distinguish several types of techniques : measures on the spot, integrating methods with a short period (several days and weeks), integrating methods with a long period (about some hundred years to some millions years).

The measures on the spot (sometimes called active methods) use a "sniffer" which collects gas by pumping from the soil ; the radon concentration in the gas is determined from its alpha emission detected by an ionization chamber or better by a scintillometer (ZnS detectors). We are using an emanometer Saphymo-Stel type EPP 10 which utilizes a ZnS (Ag) scintillometer operating at constant flow. The features of this technique are : its very short integrating period (a few minutes) and thus its very great sensitivity for the fast variations of the radon flux.

The integrating methods with a short period (sometimes called passive methods) collect radon over several days or weeks. One always uses a radon detector or collector covered with a plastic cup at the bottom of the holes of about 0.5 meter deep (fig. 1). We have developed three techniques : charcoal detectors, thermoluminescence detectors (TLD), alpha track plastic detectors.

A charcoal detector uses an activated charcoal which absorbs the radon. The deposit of its solid daughters emitting gamma radiation are counted by gamma-ray spectrometry. In principle, the gamma activity corrected from the half-life of the radon (3.8 days) is proportional to the absorbed radon. This relation used by various searchers is not quite true. Indeed, the quantity of radon fixed on the activated charcoal also depends on the absorption-desorption conditions into the detectors. We have studied the influence of all these

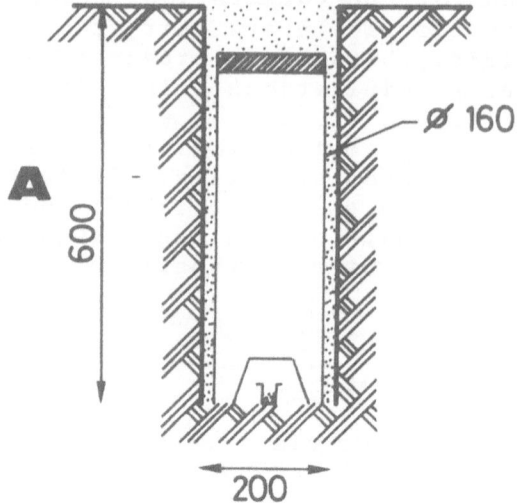

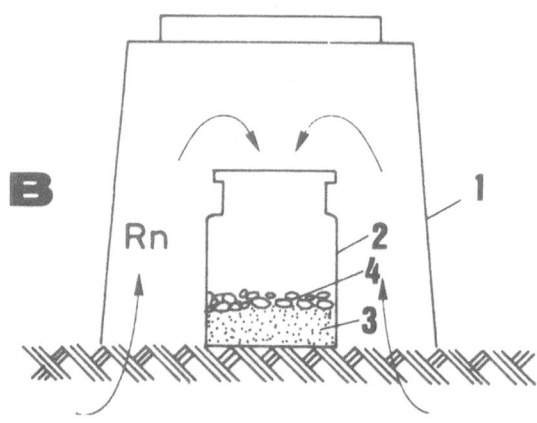

Fig. 1. - A : radon measurement station.
         B : "Boukoal" detector :
             1 : plastic cup.
             2 : glass flask.
             3 : activated charcoal.
             4 : silicagel.

parameters. This absorption aptitude can be increased
in dry and cold conditions. Besides, we have to take
into account the diffusion of radon through the wall
of the detectors in the corrections applied to the
experimental measures.

So, we have developed a new version of the charcoal
method (fig. 1) with a glass flask containing 5 gr of
selected activated charcoal and 5 gr of silicagel
covered by a plastic cup. We name it "Boukoal" because
it has been developed by Mr. BOUKO, a searcher of our
laboratory. In another side, we have calculated by
many experiences the desorption law against temperature
for correcting the measures.

In regard to others charcoal systems (¥) (49,50),
the detectors which we have developed improve the sensitivity and cancel the losses of radon previously noticed. Nevertheless whatever the charcoal detectors
may be, the measured radon concentration reaches a
"stationary level" in relation with the physical laws
of the radioactive decay and also with the absorption-
desorption mechanisms (table 1).

Table 1.

Schema for utilization of charcoal on the field.

A. First field operation : implantation of detectors
   in the ground for a period of two weeks. Two physical phenomenons occur :
   1. An absorption of radon on activated charcoal in
      agreement with the radioactive decay law

      $$N = 1 - N_0 e^{-\lambda t}$$ (λ : radon decay constant).
   2. An absorption-desorption of radon in relation
      with :
      - wet or dry conditions (competition between the
        absorption of radon and water on activated
        charcoal) ;
      - temperature (absorption increases if tempera-
        ture decreases).

(¥) ROAC (Radon on activated charcoal) of the South
    Africa (50)
    CARP (Charcoal absorbed radon prospecting) of
    Dr. Mc. Laughlin (Ireland, 49).

B. Second field operation and measurement : recovery
of the detectors sealed to avoid the escape of radon ;
measurement of the gamma activity from the radon
daughters using a NaI (Tl) scintillator.
The use of a gas-tight flask allows to consider only
the radioactive decay law of the radon :

$$N = N_o e^{-\lambda t}.$$

Others systems based on the radioactive dammages in
various materials allow an increase of the integrating
time. At the "Faculté Polytechnique de Mons", we have
developed two techniques : various TLD powders used in
dosimetry and alpha track plastic detectors (49) in collaboration with Dr. Mc. Laughlin (University College,
Dublin). However, some disadvantages also appear with
these systems. The TLD powders are more sensitive for
the gamma-radiation of the environment than for the
alpha-particles of the radon itself. In the alpha-
track plastic detectors the direct relation between the
track densities and the radon concentration is not fully
understood. Furthermore, its relation can varied with
the environment conditions. Finally, the counting of
the alpha-tracks is only possible in laboratory and so
a great delay is necessary between the recovery of the
detector and the reading. As a set-off, the TLD detectors can be read on the field with a portable apparatus
like this developed at the "Faculté Polytechnique de
Mons".
Others techniques have been developed and more particularly the alpha-card performed by Alpha Nuclear.
It consists of a Rutherford plate on which solid daugh-
ter products of radon emitting alpha particles are deposited (51). All these techniques introduce an additional problem : the presence of thorium and its daugh-
ter "Radon 220" (¤) in the soil. In the charcoal technique, it is possible to eliminate very easily the
thoron effect by a mathematical correction based on its
shorter half-life.
These remarks show the direct or integrating measures with a short period introduce difficulties. So
we think as a very useful advancement any comparaison
between these methods and any development of detectors
using several techniques at the same time. We are actually developing an original system with an autonomous
detector based on a continuous measure of the radon flux.

(¤) Radon 220, sometimes called thoron, is a gas decay
    product of the thorium 232 radioactive serie. Its
    half-life is 51,5 sec against 3,82 days for Radon 222.

In addition to these techniques, we have also proposed and developed integrating techniques with a long
period (52). We are using a natural detector : the TL
minerals of the superficial formations (quartz and
feldspars). We think that this method allows the localization of fossil zone for the radon migration and
thus the permanent zones of seismic activity.

3. SIMULTANEOUS APPLICATION OF SOME RADON DETECTION
   TECHNIQUES ON SEVERAL SITES OF THE MONS BASIN.

In view to compare the different methods of radon
detection, the effect of the different sources of radon
(superficial or deep) and the factors modifying its
emission, we selected several sites in a seismic active
zone, the Mons basin.
Both sites have been the subject of a preliminary
but rather complete study. These sites have been selected for the following characteristics (see table 2) :

Table 2.

Geological features of the test sites.

Blaton site.	Ciply site.
Radon source.	Radon source.
uraniferous beds (thick-ness : a few meters, ura-nium grade : $\pm$ 70 ppm) in folded paleozoic for-mations.	uraniferous phosphatic chalks (thickness : about five meters, uranium grade : 30-40 ppm) in subhorizontal Cretaceous formations.
Fracturation.	Fracturation.
a fault system in the paleozoic shales, silici-tes and limestones.	a Cretaceous fault zone which affects the chalks.
Overburden	Overburden.
a few meters of Tertiary argileous sand.	five meters of very pure calcarenite and a few meters of Quaternary loess with lithological varia-tions.

- a local source of radon which allows a clear localization of a radon migration zone ;
- fractures or faults zones from which a migration of
  radon can easily occur ;
- overburden formations to examine how the radon can
  migrate through the superficial formations and what
  are the optimum conditions of migration.

On the Blaton site, we have performed a profile
with nearly all the techniques developed in radon methods
(49). On the Ciply site, we have applied some techniques
on a surface covering about thousand square meters (53).
Besides, with a view to interpret the results, we have
always carried out a lithological and radiometric study
of the superficial formations.

We shall discuss in this paper only on the application of the direct and integrating short time techniques.
We shall examine the effect of the climatological conditions, of the radiometry and lithology of the overburden
and of the superficial or deep migration zones of radon.

Each survey by measure of radon on the spot shows
great fluctuations. One generally interprets these fluctuations in relation with climatological conditions.
A good understanding of this effect is necessary to interpret others periodic fluctuations (in relation with
seismic activity e.g.). Thus, many surveys or a conti-
nous registration are suitable.

Emanometry field measurements carried out on Blaton
site over a period of one year show great variations although the general shape of the profiles is always the
same (fig. 2). However, it is interesting to note that
the relative intensity of the radon peaks varies from
a period to another. So, the maximum between points 16-
19 is the most affected by the periodic fluctuations
(probably in relation with the climatological conditions),
the stronger intensities appear after a frozen period
(17-12-80). We can think that the radon source is probably not very deep (a few meters of tertiary sands over-
lying the radioactive beds located in the Paleozoic formations) and thus is strongly influenced by the meteorological fluctuations. As a set-off, the anomaly situated
between points 34-36 is more constant, less affected by
time fluctuations. In fact, this anomaly is not in relation with some radioactive beds in the Paleozoic formations ; it is located over the Visean limestones affected by a fault zone and, because its deeper origin, it
is less influenced by the climatological fluctuations.

The same conclusions can be stated from the study of
the Ciply site (53). A statistical multivariable analysis for data and parameters shows that the results of the

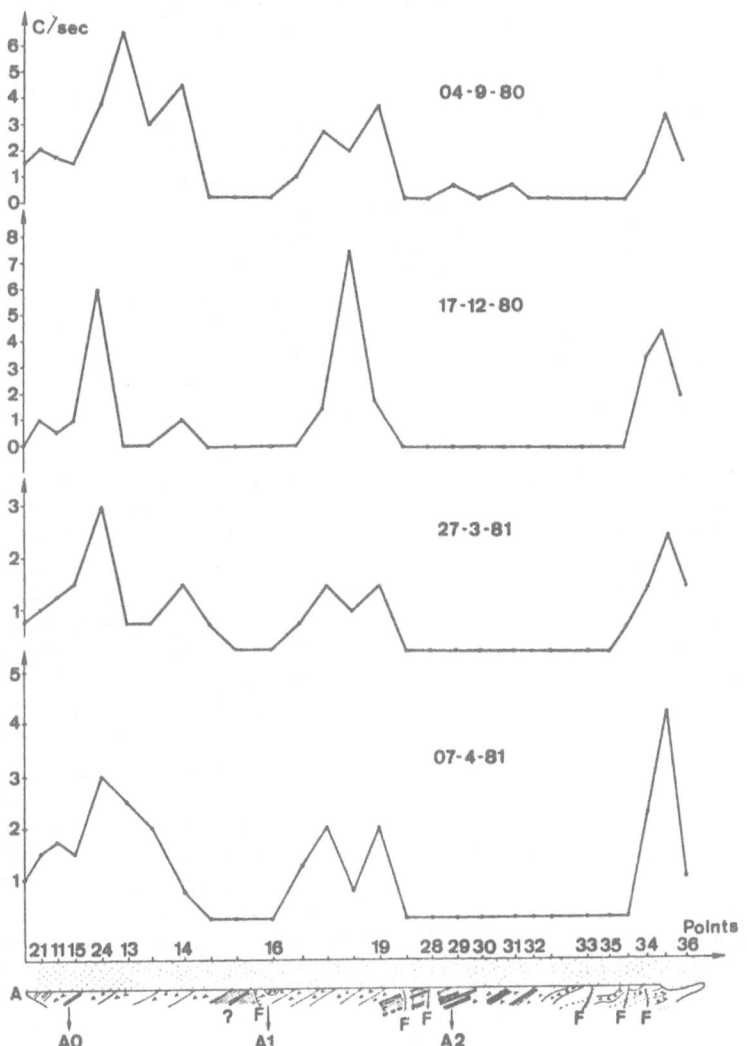

Fig. 2. - Four emanometries at different dates,
         on the Blaton profile.
         F : fault.
         A_o, A_1, A_2 : anomalous concentrations
                        of uranium.
         13, 14, 16,... : measurement points.

emanometry can be very variable (fig. 3). Sometimes
they are situated near the charcoal measures and the
lithological features of the superficial formations
and sometimes near the radiometric pole. One can think
that emanometry is influenced, following the climatological conditions by the lithological features of the
superficial formations (e.g. relation lithology - per-
meability) or by its internal radioactivity. In the
first case, the most important radon source is probably
deep, in the second case it is superficial. This important conclusion shows that by a surveillance of a
site, it would be possible to distinguish a superficial
or a deep radon source.

        Other conclusions concern the comparaison between
the different techniques of radon detection in overburden gases. There is a general good agreement between
the different integrating techniques and the amanometry
approach (49). However, there are also differences.
For example, on Blaton site (fig. 4), the zone situated
between the points 34 and 36 is a well defined anomaly
for all the emanometry profiles but it is rather low for
the integrating techniques. So, we can release this
anomaly and its evolution by a repetition of emanometry
measures over a long duration (one year e.g.) but, in
this case, we can not release it by integration over a
month. This effect is not fully understood but shows
that direct measures repeated several times or integrating measurement give different results. Our experience
shows that, in fact, an implantation over a too long
period is not to be advised. On Blaton site, the detectors were buried in the overburden sand for a period of
43 days. It is too much for the influence of temperature and moisture which are, in that case, exaggerated
upon the absorption-desorption phenomenon of the detectors.

4. DEVELOPMENT OF A FULL PROGRAM OF RADON MEASURES IN
   THE MONS BASIN.

        In view to study the parameters influencing the
distribution and the evolution of the radon concentration in the overburden gas, a full program of measures
has been carried out since June 1983 (✕).
        At first, we performed an investigation of the sui-
table conditions for the detectors implantation. We

(✕) with the support of the "Ministère de l'Emploi et
    du Travail" T.C.T. N° 2067.

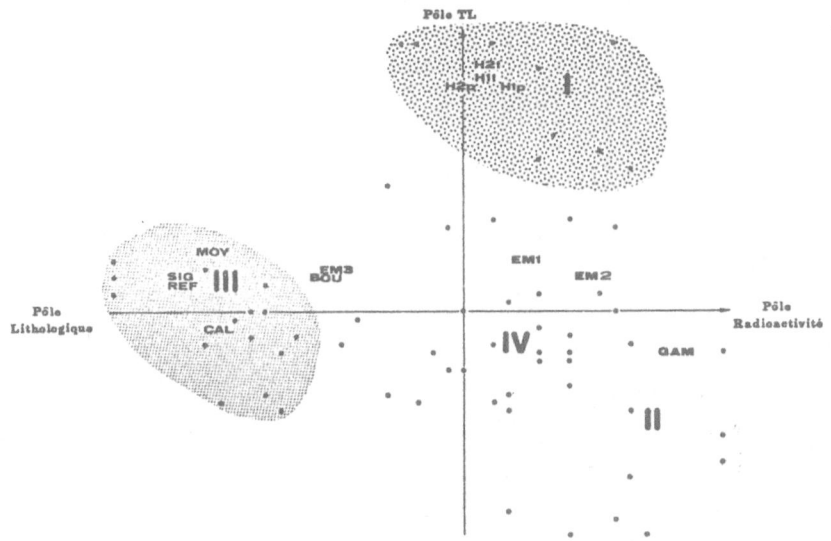

Fig. 3. - Results of the fraction analysis
        (principal axis) on the Ciply site.
        variables : BOU : Boukoals, CAL :
        calcimetry on loam fraction, EM1,
        2, 3 : emanometries at different
        dates, GAM : gamma activity of the
        loams, H1p, H1t, H2p, H2t : thermo-
        luminescence intensities, MOY : mean
        value of loam fraction, REF : the
        whithold at 80 μ, SIG : standard
        deviation.

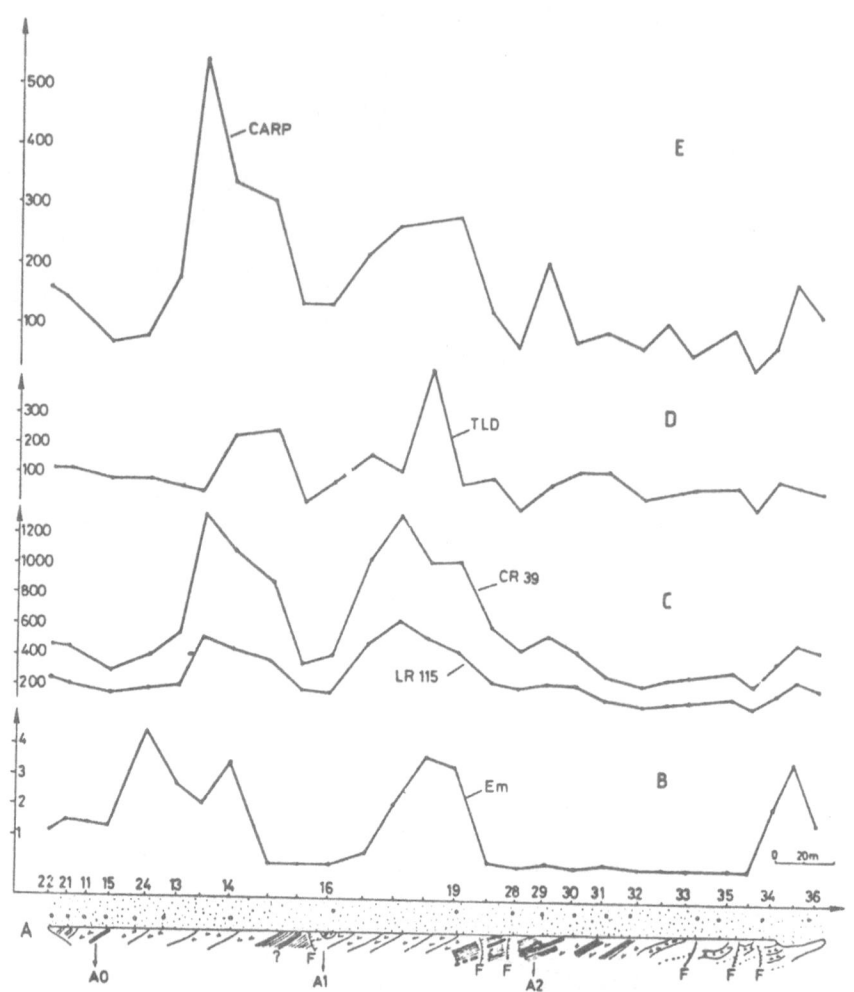

Fig. 4. - Results of different radon detection
          techniques on the Blaton profile.
          E : activated charcoal (total gamma
              count rates, counts/min).
          D : thermoluminescent dosimeters
              (arbitrary unit).
          C : alpha track plastic detectors
              (tracks/mm2).
          B : mean emanometry (count/sec).

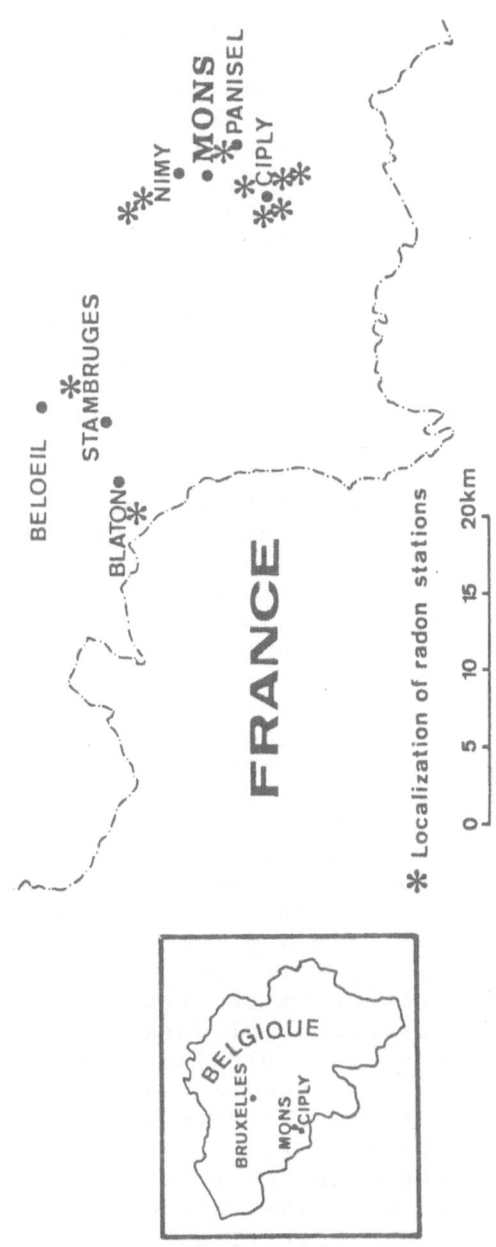

Fig.5 - Localization of radon stations

are using the system described on fig. 1. A radon detector is deposited at the bottom of the holes protected
with a plastic tube and closed at the top with a plug.

Afterwards, we selected forty two sites in the seismic active zone of the Mons basin (54). These sites are
distributed depend of the localization of radon anomalies
in relation with the superficial or deep sources, the
lithological features of the geological formations and
the microclimatological conditions. Two zones can be
distinguished (fig. 5.). The first one is situated in
the Southern part of the Mons basin (from Ciply to Mons)
where we find Cretaceous, Tertiary and Quaternary formations which are not very consolidated and not folded.
These post Paleozoic formations are thick of a few hundred meters.They are varied : Quaternary loams, Tertiary
sands, clays and calcarenite, Cretaceous chalks and
uraniferous phosphatic chalks. These Cretaceous formations are fractured and faulted. Besides, they are affected by dissolution pipes.
The second is situated in the Northern part of the basin
where the Visean limestones occur, only overlied by a
few meters of Tertiary sands. The Visean formations
are folded and faulted. Besides, many water springs
with gas emissions (nitrogen, CO_2) are located
in the Northern part of the Mons basin where some geologists have described great tectonic accidents. Those
springs are probably in relation with the geothermal
system discovered by the St. Ghislain boring at a depth
of about 2,500 meters. Around those springs, we have
detected a relative high concentration of radon in the
overburden gases. So, we have a good sampling of various
conditions allowing the study of the origin and the variations of the radon concentration in soils and waters.

A first complete field survey has been performed
since January 1984. We are using charcoal detectors
which are recovered all the fortnight. Up-to-now, a
"data base" is constituted with the results of the radon
survey and with the meteorological data (maximum and
minimum temperature, pression, pluviometry).

As soon as possible, we hope to add the data concerning the seismic activity in the Mons basin. Several
mobile stations should be established in collaboration
with Dr. MELCHIOR in the zones where radon migration is
studying. Besides, it would be interesting to dispose
of some field meteorological stations. Our purpose is
the measure on each site of all the microclimatological
parameters influencing the radon concentration in the
overburden.

CONCLUSIONS.

A complete bibliography about the radon methods and their applications in earthquake prediction shows that radon concentration in overburden can be affected by various parameters. In view to make conspicuous the relation between radon and sismic activity, it is, at first, necessary to release the external factors influencing the radon emission. We have demonstrated by some examples that it can be possible to distinguish a superficial source from a deep radon source by using several radon techniques with different integrating periods. We discussed of the advantages and disadvantages of four techniques developed at the "Faculté Polytechnique de Mons" : emanometry, charcoal detectors (Boukoal), alpha track plastic detectors, thermoluminescence dosimetry and thermoluminescence of the superficial formations. These techniques are being applied on several sites of a seismic active zone in Western Europe : the Mons basin.

From a statistical study of various parameters : radon and radiometry, geology, climatology and seismic activity, we hope to explain the variations of the radon concentration in space and time. It is a first step for a best exploration of the relation between radon and seismic activity.

ACKNOWLEDGMENTS.

The authors are grateful to the technical and administrative staff of the project T.C.T. N° 2065 (Ministère de l'Emploi et du Travail, Belgium).

BIBLIOGRAPHIE.

(1)   Andrews, J.N., Wood, D.F. 1972, Institution of Mining and Metallurgy Transactions Bulletin 81, pp. 198-209.
(2)   Fleischer, R.L., Mogro-Campero, A. 1978, J. of Geophys. Research, 83, n° B7, pp. 3539-3549.
(3)   Fleischer, R.L., Mogro-Campero, A. 1979, Geophys. Research Letters, 6, n° 5, pp. 361-364.
(4)   Hatuda, Z. 1953, Memoirs of the College of Science, University of Kyoto, Series B, Vol. XX, n° 4, pp. 285-306.
(5)   Holub, R.F., Brady, B.T. 1981, J. of Geophy. Research, 86, n° B3, pp. 1776-1784.

(6)   Melvin, J.D., Shapiro, M.H., Copping, N.A. 1978
      Nuclear Instrum. Methods, 153, pp. 239-251.
(7)   Tanner, A.B. 1978, Open File Rep. 78-1050, U.S.
      Geol. Surv., Reston, Va., pp. 5-56.
Location of active faults and application in volcanology.
(8)   Crenshaw, W.B., Williams, S.N., Stoiber, R.E. 1982,
      Nature, 300, pp. 345-346.
(9)   Gasparini, P., Mantouani, M.S.M. 1978, J. of Vol-
      canol. and Geotherm. Research, 3, pp. 325-341.
(10) King, C.Y. 1975, E.O.S., Trans. A.G.U., 56, n°12,
      p. 1019.
(11) King, C.Y. 1976, E.O.S., Trans. A.G.U., 57, p. 957
(12) King, C.Y. 1977, E.O.S., Trans. A.G.U., p. 434.
(13) King, C.Y. 1978, Nature, 271, pp. 516-519.
Earthquake prediction.
(14) Birchard, G.F., Libby, W.F. 1976, E.O.S., Trans.
      A.G.U., 57, p. 957.
(15) Birchard, G.F., Libby, W.F. 1980, J. of Geophys.
      Research., 85, n° B6, pp. 3100-3106.
(16) Chalov, P.I., Tuzova, T.V., Alekhina, U.M. 1976
      Doklady Akad. Nauk SSSR, 231, pp. 26-28.
(17) Fleischer, R.L. 1980, General Electric Compagny,
      Report n° 80 CRD 253, 6 p.
(18) Friedmann, H., Herneger, F. 1978, Geophys. Research.
      Letters, 5, n° 7, pp. 565-568.
(19) Friedmann, H., Aric, K., Duma, G., Gutdeutsch, R.,
      1982 Inst Meteorol. Geophy. Wien, pp. 427-433.
(20) Hauksson, E., Godard, J.G., Palsson, S.E. 1978,
      E.O.S., Trans. A.G.U., 59, p. 1196.
(21) Hauksson, E., Godard, J.G., J. of Geophys. Research
      86, n° B8, pp. 7037-7054.
(22) King, C.Y. 1980, J. of Geophys. Research, 85,
      n° B6, p. 3051.
(23) King, C.Y. 1980, J. of Geophys. Research, 85
      n° B6, p. 3065-3078.
(24) King, C.Y. 1981, Nature, 293, p. 262.
(25) King, C.Y., Wakita, H. 1981, Eartquake Notes
      ISSN0012-8287, USA, 52, n° 1 - 71 (seismologica
      Society of America, 76th Annual meeting, 1981-
      03-23, Berkeley.
(26) Klusman, R.W., Webster, J.D. 1981, Bull. of Seis-
      mol. Soc. of America, 71, n° 1, pp. 211-222.
(27) Lomnitz, O., Lomnitz, L. 1978, Nature, 271, pp.
      109-111.
(28) Melvin, V.I., Shapiro, M.H., Copping, N.A. 1978,
      Nuclear Instrum. Methods, 153, pp. 239-251.
(29) Mjachkin, V.I., Brace, W.F., Sobolev, G.A.,
      Dieterich, J.H. 1975, Pageogh., 113, pp. 169-
      181.

(30) Mogro-Campero, A., Fleischer, R.L., Likes, R.S. 1980, J. of geophys. Research, 85, n° B6, pp. 3053-3057.

(31) Moore, W.S., Chiang, S.H., Talwani, P., Stevenson, D.A. 1977, E.O.S. Trans. A.G., p. 434.

(32) Noguchi, M., Wakita, H. 1977, J. of Geophys. Research, 82, n° 8, pp. 1353-1357.

(33) Noguchi, M., Wakita, H. 1977, J. Soc. Instrum. Control E.N.G.R.S., 16, n° 9, pp. 700-706.

(34) Oddou, A. 1981, Thèse doctorat, Univ. Nice, lab. Géologie et Géochimie.

(35) Okabe, S. 1956, Mem. of the College of Science, Univ. of Kyoto, Series A, 28, n° 2, pp. 99-115.

(36) Press, F., Bullock, M., Hamilton, R.M., Brace, W.F., Kisslinger, C., Bonilla, M.G., Allen, C.R., Sykes, L.R., Raleigh, C.B., Knopoff, L., Clough, R.W. 1975, E.O.S., Trans. A.G.U., 56, pp. 838-881.

(37) Press, F. 1975, Scientific American, 232, n° 5, pp. 14-23.

(38) Raleigh, B., Bennett, G., Craig, H., Hanks, T., Molnar, P., Nur, A., Savage, J., Scholtz, C., Turner, R., Wu, F. 1977, E.O.S., Trans. A.G.U., 58, n° 5, pp. 236-272.

(39) Scholz, C.H., Sykes, L.R., Aggarwal, Y.P. 1973, Science, 181, n° 4102, pp. 803-810.

(40) Seidel, J.L. 1982, Thèse de 3ème cycle, Univ. Clermond-Ferrand II, U.E.R. R.S.T.

(41) Shapiro, M.H., Melvin, J.D., Tombrello, T.A., Mendenhall, M.M. 1981, J. of Geophys. Research, 86, n° B3, pp. 1725-1730.

(42) Shishkevich, C. 1971, Geoscience Bulletin, Scr. A., 2, n° 8-10, pp. 69-91.

(43) Smith, A.R., Wollenberg, H.A., Mosier, D.F. 1978, Open File Rep. 78-1050, U.S. Geol. Surv., Reston, Va., pp. 154-174.

(44) Talwani, P. 1979, Physics of earth and planetary interiors, 18, p. 1196.

(45) Talwani, P., Moore, W.S., Chiang, J. 1980, J. of Geophys. Research, 85, n° B6, pp. 3079-3088.

(46) Ulomov, V.I., Mavashev, B.Z. 1967, Doklady Akad. Nauk. S.S.S.R., vol. 176, n° 2, pp. 319-321.

(47) Wakita, H., Nakamura, Y., Notsu, K., Noguchi, M., Asada, T. 1980, Science, vol. 207, pp. 882-883.
Radon detection techniques and applications for the Mons basin.

(48) Telford, W.M. 1983, In Developments in geophysical exploration methods, 4, pp. 155-194. Ed. by A.A. Fitch.

(49) Mc. Laughlin, J.P., Charlet, J.M., Dupuis, Ch.,
     Quinif, Y., Bouko, Ph., Ramu, J.P. 1982, Ann.
     Soc. Geol. Nord, 105, pp. 211-221.
(50) Hambleton-Jones, B.B., Smit, M.C.B. 1980, Atomic
     Energy Board, South Africa, PER-48, pp. 1-24.
(51) Pacer, J.C. 1982, Uranium Exploration methods,
     Review of the NEA/IAEA R-D programme, OCDE, Paris,
     1st-4th June 1982, pp.495-504.
(52) Charlet, J.M., Dupuis, Ch., Quinif, Y., Bouko,
     Ph., Lair, Ph. 1982, Uranium Exploration methods,
     Review of the NEA/IAEA R-D programme, OCDE, Paris,
     1st-4th June 1982, pp. 545-555.
(53) Charlet, J.M., Quinif, Y., Bouko, Ph. 1984, Ann.
     Soc. Geol. Nord, France, in preparation (paper
     presented under date of 7th March 1984).
(54) Colbeaux, J.P., Dupuis, Ch. 1985, Nato Workshop
     243/94, Brussels, March 20-22, 1984.

CLUSTERING OF SEISMICITY IN THE REGIONS OF BRABANT AND ARDENNES
MASSIVES BEFORE THE LIEGE EARTHQUAKE OF NOV. 8, 1983.

M. De Becker, T. Camelbeeck, C. Poitevin

Centre de Géophysique Interne
Observatoire Royal de Belgique
3, Avenue Circulaire
B - 1180 Bruxelles

1. INTRODUCTION

Premonitory seismicity patterns before this earthquake are of
interest by two reasons :

a. Theoretical reason - their relevance to the study of similarity
   in the occurrence of earthquakes.

This earthquake occurred as a consequence of the relative movement of the massives which compose the European platform. On the
other hand most of premonitory seismicity patterns were established in areas with much higher tectonic activity and accordingly
higher magnitudes of the earthquakes. It would be of obvious theoretical interest to check whether a kind of similarity exists between the occurrence of earthquakes in the regions of high and
small seismicity.
There are evidences in favour of such similarity : fundamental
properties of earthquake sequences such as linear magnitude-
frequency-of-occurrence relation and clustering seem to be rather
similar in very diverse seismotectonic environments and there are
no a priori reasons to believe that premonitory seismicity patterns would have nothing in common in different environments. Indeed, several premonitory seismicity patterns after normalization
to the seismic activity seem to be applicable in rather wide magnitude range {1, 2, 3, 7}; so far, however, it is a reasonable
hope but not a firmly established fact.

b. Practical reason - the possible damage from the earthquakes in
   the area under consideration is rapidly increasing due to appearance of high-risk objects (dams, nuclear power plants, ...)

175

P. Melchior (ed.), Seismic Activity in Western Europe, 175–187.
© 1985 by D. Reidel Publishing Company.

and increase of vulnerability of some common constructions and
lifelines.

This is a common situation for all industrialized areas exposed
to the earthquakes, but for areas of low seismicity the growth of
seismic risk just begins to be recognized. Therefore the practical
aspect of our problem is also of rather general importance.

2. DATA

We used the earthquake's data of the Royal Observatory of
Belgium actually revised by the Centre de Géophysique Interne {4}.
The basic statistics of this catalog are shown in the table 1.
We see from this table that the data seems reasonably complete

$$\text{for } M_L \geq 4 \quad \text{since 1930}$$

$$\text{for } M_L \geq 3 \quad \text{since 1946}$$

$$\text{for } M_L \geq 2 \quad \text{since 1964}$$

The earthquakes with unidentified magnitudes in 1931-1960 seem to
have $M_L \geq 2$. This is suggested by the fact, that average annual
number of earthquakes is about the same for all earthquakes in
1931-1960 (2.8 per year) and for $M_L \geq 2$ in 1961-1981 (2.5 per year).

The territorial distribution of seismicity is discussed in (5,
figure 1). We can see from this figure three active areas :
Hainaut, Liège and the ridge of Brabant massif.
The distribution of seismicity between these areas is shown in
table 2.-4.
The remarquable trait is the quiescence between 1973 and 1981.
In Brabant, this quiescence has started in 1958.

3. CLUSTERS

Definitions

It was found in many studies {1-3, 6-8} that the strongest
earthquakes of a region are often preceded by an abnormaly large
cluster of earthquakes in medium magnitude range. Cluster means
a group of earthquakes in a narrow time-space domain. Several
overlapping formal definitions of a cluster were suggested.

The difference in the definitions is not essential for the region considered here. We will assume for the beginning the sim-
pliest definition used in consideration of premonitory pattern b_g
{6}. The size of the cluster b_g is the number of the earthquakes

within a space-time window d(km) ✗ d(km) ✗ e(days), in the magnitude range from M_1 to M_2. According to the procedure suggested in {1, 6} we look for the clusters premonitory to strongest earthquakes of the area, with magnitude $\geq M_0$. The cluster generated by the aftershocks of these earthquakes are eliminated from consideration. The time of increased probability of a strong earthquake ("TIP") is diagnosed by condition that the size of the cluster exceeds some threshold B. The duration of a TIP was previously assumed 3 years, and for largest M_0 (around 8) - 4 years {3}. We assumed d = 20 km and e = 2 days (to check the stability of the results we will consider later on also d = 50 km, e = 0.25 days or 7 days).

The results for the 1st version are shown in Fig. 1 and table 5 {all explanations in the caption to table}.

The distribution of the size of the clusters (table 6) suggests the threshold B = 3 or 4 for identification of large clusters.

Figure 1 suggests that the strongest earthquakes are preceded by the largest clusters : 3 out of 4 clusters of size $b_g \geq 4$ and 4 out of 5 clusters of $b_g \geq 3$ were followed within 1 year by an earthquake of $M_L \geq 4.5$. We would probably accept as premonitory a cluster with a larger lead time, within a priori limit 3 years.

The total duration of such 3 years period before the earthquakes with $M_L \geq 4.5$ is 16 years or about 30 % of the total time considered. The probability to have 3 out of 4 clusters in this time by chance is about 8 %; for 4 out of 5 clusters it is about 3 %. These numbers cannot be accepted as a measure of confidence level because the threshold 4.5 was assumed a posteriori. Let us discuss now the stability of our conclusions to the arbitrary choice of the threshold.

The decrease of e to 0.25 days improves the result : the false alarm that is the cluster in 1940 which was not followed by a strong earthquake disappear. However we were resistant to assume this threshold since it breaks several obvious clusters. The increase of e to 7 days doesn't introduce significant changes: the number of earthquakes in the premonitory clusters increases but a new false alarm appears in 1944.

The change of radius up to 50 km leeds to no significant change. Our results are unstable only to one parameter, M_0 : assuming $M_0 = 4.4$, we will have only three clusters of size ≥ 3 and only two of them were followed by strong earthquakes within 3 years. In other words, only the two last strongest earthquakes would be preceeded by unusually large clusters. Similar but less definite results about low seismicity areas of Europe were obtained in {3}.

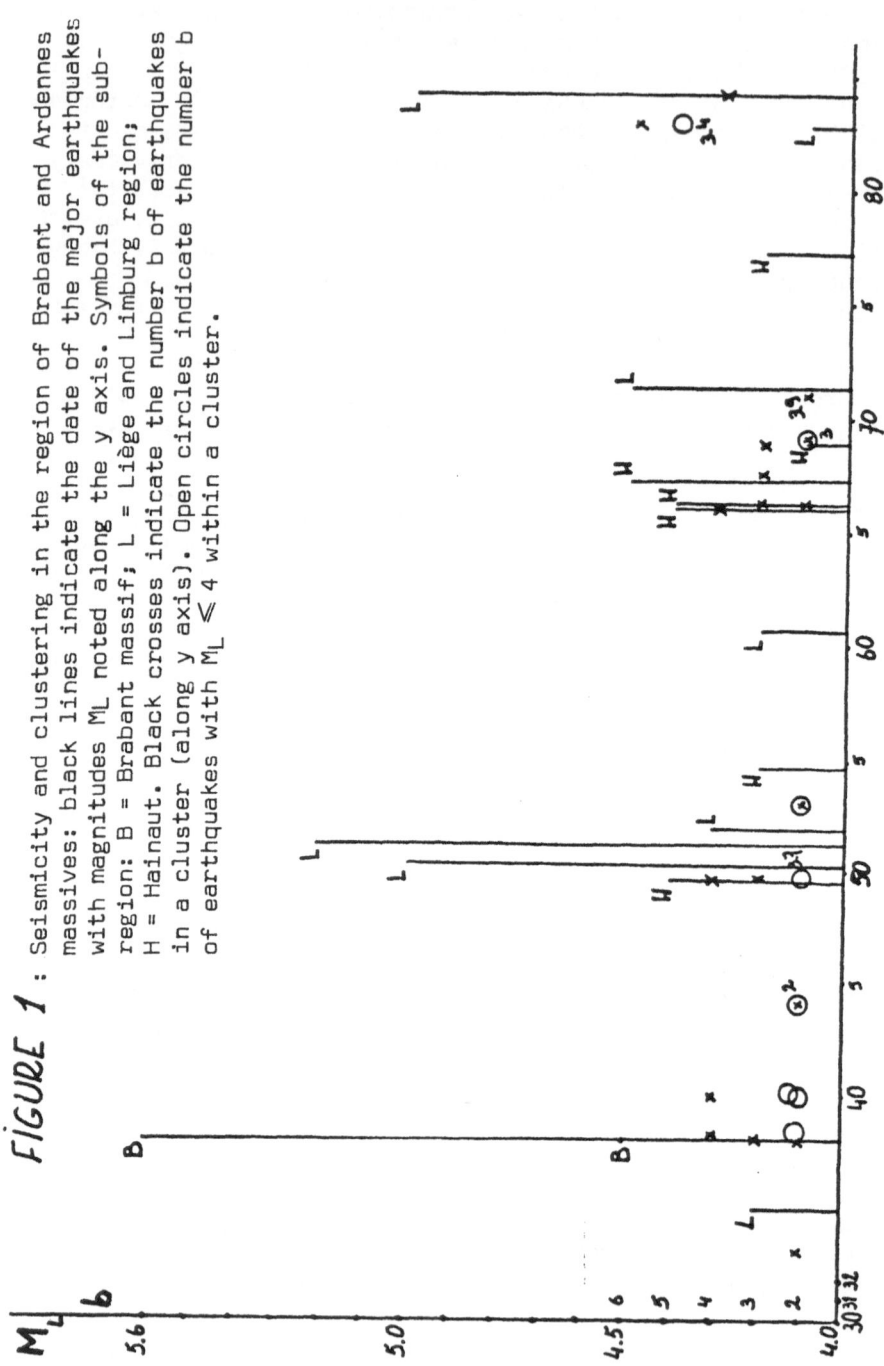

FIGURE 1 : Seismicity and clustering in the region of Brabant and Ardennes massives: black lines indicate the date of the major earthquakes with magnitudes M_L noted along the y axis. Symbols of the sub-region: B = Brabant massif; L = Liège and Limburg region; H = Hainaut. Black crosses indicate the number b of earthquakes in a cluster (along y axis). Open circles indicate the number b of earthquakes with $M_L \leqslant 4$ within a cluster.

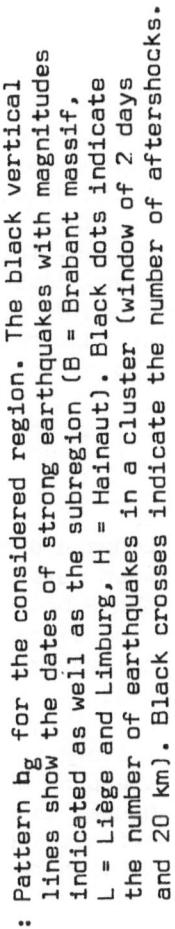

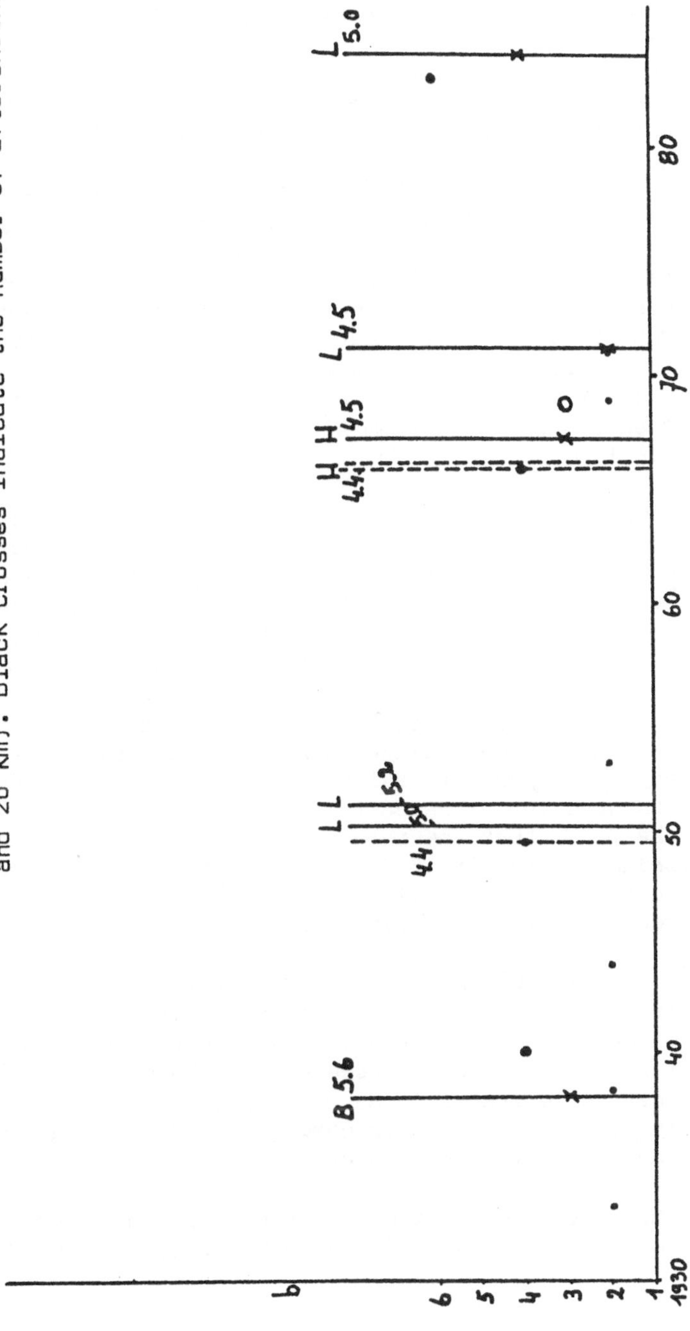

FIGURE 2 : Pattern b_g for the considered region. The black vertical lines show the dates of strong earthquakes with magnitudes indicated as well as the subregion (B = Brabant massif, L = Liège and Limburg, H = Hainaut). Black dots indicate the number of earthquakes in a cluster (window of 2 days and 20 km). Black crosses indicate the number of aftershocks.

Discussion

1. We considered together three active areas (Hainaut, Liège, Brabant) represented by isolated groups of epicenters. It is tempting to look for premonitory patterns in each area separately.
   However, the studies for many other regions indicate that at least long-term premonitory patterns are observed within comparatively large distances from the forthcoming strong earthquake : this is plausible from the geodynamical point of view. The earthquakes in Liège and Hainaut areas are the manifestation of the relative movement of the separate massives, into which this part of West-european platform is divided. Specifically Hainaut and Liège seismic areas are situated on different parts of the boundary between Brabant and Ardennes and the movements on different parts of this boundary have no reasons to be totally independent.
   Brabant area possibly may be considered separately; anyhow, its position in the system of the massives in this part of European platform is not clear.

2. Complete seismic quiescence between Hainaut and Liège areas (the segment of the above mentioned boundary between latitudes 4.5 and 5.5) deserves special attention. There should be some movement along this segment (while seismicity may be discontinuous along the boundary of two massives, the total movement along this boundary is certainly not, since the massives are comparatively rigid.
   There are two possibilities : either the movement along this segment may be realized through relatively slow creep or this segment represents seismic gap and a strong earthquake is not excluded in future on this segment. In favor of seismic gap is the fact, that the catalog {4} shows an earthquake with macroseismic intensity 7-8 in Tirlemont area, in 1828. However, due to the obvious unreliability of these data, the direct creep measurements could better resolve this alternative.

3. Physical explanation of premonitory clustering discussed here is connected with migration of fluids in the fault zone : both Liège and Hainaut area do have the sources of hydrothermal water. It is not clear however why the clustering is almost entirely concentrated in Hainaut area.
   It seems worthwhile to compare the fluctuation in water supply and seismicity in both areas.

4. Among other premonitory patterns, seismic quiescence was reported in many areas, with or without subsequent strong activation.

In this context, it is interesting to mention a remarkable quiescence since 1971, before the Liège earthquake in 1983. Catalog shows similar quiescence before the series of 2 strong earthquakes in 1950 and 1951. However the data for this period may be incomplete. Another area of complete quiescence since 1956 is Brabant area.

Conclusion

The hypothesis that strongest earthquakes of the area considered are preceded by unusually large clusters of earthquakes in medium magnitude range is in agreement with the available data on the seismicity of this area. The data are unsufficient for independent statistical estimation of this hypothesis. However since it was formulated a priori on the basis of the data of other areas this hypothesis deserves attention.

The final test would be the monitoring of the appearance of new clusters in order to check wether they will be followed by strong earthquakes.

To make such test significant, it is important to specify all parameters a priori. The study suggests that the time of increased probability of an earthquake with $M_L \geq 4.5$ can be diagnosed by the appearance of the cluster of ≥ 4 earthquakes with $M_L \geq 1.9$, within the square of 20×20 km^2 and the time interval 48 hours (fig. 2). This figure suggests the possibility to decrease the uncertainty in the time of occurrence of such earthquakes by factor 3 to 4. If successful such long term prediction could be useful for the search of medium and short-term precursors and for implementation of some important low key safety measures (12).

Acknowledgements :

The authors are very grateful to Professor V.I. Keilis-Borok (Institute of Physics of the Earth, USSR Academy of Sciences, Moscow) who has guided this work with his great competence and kindness.

82

Table 1.

Magnitudes M_L

Year	∿	1.9-2.4	2.5-2.9	3.0-3.4	3.5-3.9	4.0-4.4	4.5-4.9	≥ 5	Σ
1984 - 82		7	4	4	2	1	—	1 (5.0)	19
1981 - 79		—	—	—	—	. —	—		—
1978 - 76		1	—	—	—	—	—		1
1975 - 73		—	—	—	—	—	—		—
1972 - 70		—	—	2	2	0	1 (4.5)		5
1969 - 67		2	5	2	2	0	1 (4.5)		12
1966 - 64		3	6	2	2	3	—		16
1963 - 61		—	1	—	1	—	—		2
1960 - 58	1	—	—	—	—	1	—		2
1957 - 55	12	—	—	—	1	—	—		13
1954 - 52	12	—	—	—	1	2	—		15
1951 - 49	4	—	—	2	3	2	—	2 (5.1-5.0)	13
1948 - 46	1	—	—	1	1	—	—	—	3
1945 - 43	7	—	—	1	—	—	—	—	7
1942 - 40	5	—	—	—	—	—	—	—	5
1939 - 37	5	—	—	—	—	2	1 (4.5)	1 (5.6)	9
1936 - 34	2	—	1	1	—	1	—	—	5
1933 - 31	8	—	—	—	—	2	1 (4.6)	1 (5.0)	13
TOTAL		13	17	15	15	14	4	5	

∿ = number of earthquakes for which the magnitudes are not yet defined.

Seismicity of the region (longitude : 3.60 E to 6.75 E / latitude : 50.37 N to 51.87 N) during the period 1931-1984.
The numbers in brackets are the local magnitudes of the earthquakes considered.
The last column "Σ" gives the total of earthquakes which took place during the period of 3 years under consideration, including those for which there is no magnitude defined (∿).

Table 2

year	~	1.9-2.4	2.5-2.9	3. -3.4	3.5-3.9	4. -4.4	4.5-4.9	≥ 5	All	+0
1984 - 82		3	3	1	2	1		1	11	11
1981 - 79									0	0
1978 - 76									0	0
1975 - 73			1						1	1
1972 - 70				2			1		3	3
1969 - 67									0	0
1966 - 64			1			1			2	2
1963. - 61			1		1				2	2
1960 - 58						1			1	1
1957 - 55					1				1	1
1954 - 52	1					1			1	2
1951 - 49						1			1	1
1948 - 46									0	0
1945 - 43									0	0
1942 - 40									0	0
1939 - 37	2								0	2
1936 - 34				1		1			2	2
1933 - 31							1		1	1
TOTAL	3	3	6	4	4	6	2	1	26	29

Seismicity of the region of Liège and Limburg from 1931 until 1984.
longitude : 5.25 E - 6.75 E
latitude : 50.?5 N - 51.87 N
The column called "All" contains the sum of all the earthquakes
of known M_L, during the period of 3 years considered.
The column "+0" contains the sum of all the earthquakes occurred
during the 3 years considered, even those for which a magnitude
is not yet available.

Table 3

year	~	1.9-2.4	2.5-2.9	3. -3.4	3.5-3.9	4. -4.4	4.5-4.9	≥ 5	All	+ 0
1984 - 82										
1981 - 79										
1978 - 76										
1975 - 73										
1972 - 70										
1969 - 67										
1966 - 64										
1963 - 61										
1960 - 58										0
1957 - 55	3									3
1954 - 52	7									7
1951 - 49										0
1948 - 46										0
1945 - 43										0
1942 - 40	1									1
1939 - 37	3					2	2	1 (5.6)	5	8
1936 - 34										0
1933 - 31	1									1
TOTAL	15					2	2	1	5	20

Seismicity of the region of Brabant massive from 1931 until 1984.
longitude : 3.5 E - 5.0 E
latitude : 50.55 N - 51.50 N

Same remarks as in table 2.

Table 4

year	~	1.9-2.4	2.5-2.9	3. -3.4	3.5-3.9	4. -4.4	4.5-4.9	≥ 5	All	+ 0
1984 - 82		4	1	3					8	8
1981 - 79									—	—
1978 - 76						1			1	1
1975 - 73									—	—
1972 - 70			1	1	2				4	4
1969 - 67		2	4	2	2	1	1		12	12
1966 - 64		3	5	2	2	2			14	14
1963 - 61									—	—
1960 - 58	1								—	1
1957 - 55	10								—	10
1954 - 52	6				1	1			2	8
1951 - 49	8					1			1	9
1948 - 46	2			1					1	3
1945 - 43	7			1					1	8
1942 - 40	4								—	4
1939 - 37	1								—	1
1936 - 34	2		1						1	3
1933 - 31	6								—	6
TOTAL	47	9	12	10	7	6	1	0	45	92

Seismicity of the region of Hainaut from 1931 until 1984
longitude : 3.5 E - 5.0 E
latitude : 50.25 N - 51.55 N

Same remarks as in table 2.

Table 5

M_L	4.0	4.1	4.2	4.3	4.4	4.5	4.6	5.0	5.1	5.6
n	3	3	3	3	4	3	1	3	1	1

Number of strong earthquakes occurred in the considered region versus local magnitude M_L, during the period 1931 - 1984.

Table 6

M_L	~			≤ 3			3. - 3.4			3.5 - 3.9			4. - 4.4			4.5 - 4.9			≥ 5		
	H	L	B	H	L	B	H	L	B	H	L	B	H	L	B	H	L	B	H	L	B
1 (singlet)	44	3	14	21	8	—	9	4	—	5	4	—	2	6	2	1	4	—	1	3	—
2	3		1							2			1		1		2				1
3													1								
4													2							1	
5							1														
6																					

This table shows the amount of clusters with one to six earthquakes versus the highest M_L of the earthquakes forming the cluster.
The space window is 20 km and the time window is 2 days.
The statistics has been made separately for the three regions :

H (Hainaut)
L (Liège and Limburg)
B (Brabant massif)

BIBLIOGRAPHY

1. Keilis-Borok, V.I., Knopoff, L. and Rotwain, I.M. : Bursts
   of aftershocks, long-term precursors of strong earthquakes.
   Nature, Vol. 283, N° 5744, Jan. 17, 1980, pp. 259-263.

2. Keilis-Borok, V.I. : The worldwide test of three long-term
   premonitory seismicity patterns.
   Tectonophysics, 85, 1982, pp. 47-60.

3. Shaposhnikov, V. : Long-term premonitory seismicity patterns
   in Ardennes, Schwarzwald and Vienna plato.
   In Russian, Dakladi Akademii Nauk SSSR, V269 N1, 1983, pp.
   73-76.

4. Preliminary catalog of seismicity in Belgium.
   Royal Observatory of Belgium.

5. Melchior, P. : Problèmes actuels de la séismologie en Belgique
   in Seismic Activity in Western Europe, with particular consideration to the Liège earthquake of november 8, 1983.
   Proceedings of the Nato Workshop, Brussels, march 20-22, 1984.

6. Caputo, M., Console, R., Gabrielov, A.M., Keilis-Borok, V.,
   Sidorenko, T.V. : Pattern b_g and strong earthquakes of Italy,
   GJRAS, 1982.

7. Evison, F.F. : Fluctuation of seismicity before major earthquakes.
   Nature (London), V. 266 : 710-712, 1977.

8. Wyss, M. and Habermann, R.E. : Seismic quiescence precursory
   to a past and a future Kurill Island Earthquake.
   Pageoph. Vol. 117, Birkhauser Verlag, Basel, 1979.

CATALOGUE DES SÉISMES SURVENUS AU MOYEN AGE EN BELGIQUE ET DANS LES REGIONS VOISINES.

Pierre Alexandre
Assistant en Histoire
Université de Liège
(3, Place Cockerill, 4000 Liège)

L'histoire des tremblements de terre survenus en Belgique au Moyen Age et aux Temps Modernes reste encore à élaborer. En effet, les travaux qui ont servi jusqu'ici de base à l'étude de la séismicité historique de notre pays sont des compilations vieillies, établies sans le secours de la critique des sources. Dans un travail antérieur {1}, nous croyons avoir démontré que les plus connus de ces recueils, - ceux de PERREY pour la France, de TORFS et de LANCASTER pour la Belgique, de SIEBERG pour l'Allemagne {2}, - présentent une proportion d'erreurs qui, pour la période médiévale, s'élève à plus de 75 %. En effet, ces auteurs, ignorant tout des règles de la critique historique, ont mis sur le même pied des textes sans valeur (déformés, tronqués ou mal datés) écrits par divers compilateurs plusieurs siècles après les faits, et des observations sismiques originales rédigées par des chroniqueurs contemporains des faits. C'est dans ces ouvrages que l'on trouvera, par exemple, la mention de séismes à Tournai en 330, 502 et 630, et à Cambrai en 450 : ces événements, longtemps considérés comme les plus anciens tremblements de terre survenus dans nos régions, n'ont en réalité jamais eu lieu; aucune des rares sources de cette époque ne les mentionne.

Il faut donc renoncer à l'emploi de ces compilations anciennes et retourner directement aux sources écrites susceptibles de nous fournir des observations sismologiques. Etant donné que la séismicité instrumentale ne s'est développée qu'à partir du début du XXe siècle, les observations dont nous disposons pour la période antérieure sont de nature qualitative; elles figurent dans des textes anciens sujets à la critique des sources, et qui relèvent donc du métier d'historien.

P. Melchior (ed.), Seismic Activity in Western Europe, 189–203.
© *1985 by D. Reidel Publishing Company.*

Pour le Moyen Age et pour nos régions, la plus grande partie
des observations sismologiques conservées proviennent de sources
narratives composées de façon annalistique (annales, chroniques,
etc .); ce n'est qu'à partir des XVe-XVIe siècles que l'on peut
espérer en découvrir dans des documents d'une autre nature, par
exemple des registres de comptes, des registres de paroisse, etc.

Le catalogue que nous présentons ci-dessous concerne la
Belgique et les régions voisines (Rhénanie, Palatinat, Picardie,
Nord de la Champagne et de la Lorraine), pour la période qui va
des origines à 1400; il est le résultat d'un travail de critique
des sources narratives et de comparaison des textes : il ne con-
tient donc que des observations originales, émanant de contemporains des faits. Les autres textes : copies déformées et datées
de manière inexacte, ou même relations de tremblements de terre
imaginaires (comme le sont par exemple les séismes inventés par
le chroniqueur liégeois Jean d'Outremeuse), ont été définitivement
éliminés.

Le plus ancien tremblement de terre dont nous trouvons mention pour la zone étudiée ici est celui de 801; l'extrême pauvreté
des sources antérieures au IXe siècle, pour nos régions, a pour
conséquence que nous ignorons tout des séismes qui ont pu y sur-
venir dans l'Antiquité et à l'époque mérovingienne. Ce n'est
qu'avec les sources carolingiennes (essentiellement les *Annales
regni Francorum* et leurs continuations) que commence l'histoire
des séismes en Europe du Nord-Ouest. A partir de cette époque, les
sources écrites narratives sont de plus en plus abondantes, et
l'on peut espérer à bon droit qu'aucun séisme majeur n'a échappé
à l'attention des annalistes.

Nous n'avons pu reproduire ici les textes sismiques originaux
que nous avons relevés : on les trouvera dans les éditions de
sources mentionnées dans la bibliographie. L'analyse que nous don-
nons des événements est la plus complète possible et n'a laissé
de côté aucun élément susceptible de nous informer sur la date
exacte, sur le nombre de secousses, sur la localisation et l'importance des dégâts de chaque séisme. Certains textes ne con-
tiennent aucune mention des localités touchées par le tremblement
de terre; il pouvait être utile dès lors de signaler le lieu où
notre source a été rédigée, car il est probable dans ce cas que
le séisme a secoué la région dans laquelle a vécu l'annaliste ou
le chroniqueur qui nous rapporte l'événement.

REFERENCES

{1} P. ALEXANDRE, Problèmes de méthode relatifs à l'étude des
     séismes médiévaux, dans *Tremblements de terre, histoire et
     archéologie, Actes des IVèmes Rencontres internationales
     d'histoire et d'archéologie (Antibes, 2-4/11/1983)*, 1984.

{2} A. PERREY, *Mémoire sur les tremblements de terre ressentis en
     France, en Belgique et en Hollande depuis le IVe siècle
     jusqu'à nos jours (1843)*, Bruxelles, 1845.
     L. TORFS, *Fastes des calamités publiques survenues dans les
     Pays-Bas et particulièrement en Belgique, depuis les temps
     les plus reculés jusqu'à nos jours*, t. 2 : *Hivers-Tremblements
     de terre*, Tournai, 1862.
     A. LANCASTER, *Les tremblements de terre en Belgique*, dans
     Annuaire météorologique de l'Observatoire Royal de Belgique,
     1901, pp. 194-228.
     A. SIEBERG, *Beiträge zum Erdbebenkatalog Deutschlands und
     angrenzender Gebiete für die Jahre 58 bis 1799*, Berlin, 1940.

ABREVIATIONS

C.R.H., Série in -4°: Publication in -4° de la Commission
     royale d'histoire de Belgique.

M.G.H., SS: Monumenta Germaniæ historica, Scriptores.

Script. rer. Germ.: Scriptores rerum Germanicarum.

Soc. hist. Fr.: Publications de la Société de l'histoire de France.

CATALOGUE DES SEISMES SURVENUS EN BELGIQUE, EN RHENANIE ET DANS LE NORD DE LA FRANCE DE 800 A 1400

1. **801.** Séisme(s) "à certains endroits" dans la région du Rhin.

    Source originale (1) : *Annales regni Francorum.*

    Remarque : La même source signale également, en 801, des secousses sismiques en "Gaule" et en "Germanie", sans plus de précision, ainsi qu'un tremblement de terre le 30/4 en Italie, lors du séjour de Charlemagne à Spolète.

2. **803.** Séisme au début de l'année, en hiver, à Aix-la-Chapelle et dans les régions voisines.

    Source originale (2) : *Annales regni Francorum.*

3. **823.** Séisme à Aix-la-Chapelle.

    Source originale (3) : *Annales regni Francorum.*

4. **829.** Séisme peu de jours avant le 28/3, de nuit, à Aix-la-Chapelle.

    Source originale (4) : *Annales regni Francorum.*

5. **838.** Séisme le 18/1, à l'heure de vêpres, à Lorsch et dans les régions de Worms, de Spire et de Ladenburg.

    Sources originales : A. *Annales Fuldenses.* B. *Annales Xantenses.*

6. **845.** Séisme dans la région de Worms à deux reprises, dans la nuit du 22 au 23/3, et dans la nuit du 28 au 29/3.

    Source originale : *Annales Xantenses.*

7. **855.** Séisme à Mayence, à une vingtaine de reprises.

    Source originale : *Annales Fuldenses.*

8. **858.** Important séisme le 1/1 dans plusieurs régions, notamment à Worms, où une secousse a lieu à l'heure de matines, et surtout à Mayence, où 13 secousses ont lieu avant le point du jour. Dégâts importants à Mayence : des constructions anciennes sont fendues, et l'église St-Alban est secouée à tel point qu'un mur tombant du faîte détruit la chapelle St-Michel avec son toit et ses plafonds.

    Sources originales : A. *Annales Fuldenses.* B. *Annales Xantenses.* C. *Annales Bertiniani.*.

    Remarque : Les *Annales Fuldenses* et les *Annales Xantenses* placent le séisme de Mayence le 1/1/858, alors que les *Annales Bertiniani* situent le même événement le 25/12/857 (5) ; mais la première de ces dates est la meilleure, car les *Annales Fuldenses,* qui ont sans doute été rédigées à Mayence pour la partie qui va de 830 à 887, présentent une information plus sûre que celle des *Annales Bertiniani,* rédigées à Troyes (de 835 à 861), donc loin de la région touchée par le séisme.

9. **859.** Séismes pendant toute l'année dans la région de Mayence.

   Source originale : *Annales Fuldenses.*

10. **867.** Séisme le 9/10, `` à plusieurs endroits ``.

   Sources originales : A. *Annales Fuldenses.* B. *Annales Xantenses.*

   Lieu de rédaction : A. Mayence *(Annales Fuldenses)* B. Cologne ? (*Annales Xantenses,* de 861 à 873).

11. **870.** Séisme à Mayence, à deux reprises.

   Source originale : *Annales Fuldenses.*

   Lieu de rédaction : Mayence.

12. **872.** Séisme à Mayence le 3/12, à l'heure de prime.

   Source originale : *Annales Fuldenses.*

   Lieu de rédaction : Mayence.

13. **881.** Important séisme à Mayence le 30/12 (6), avant le point du jour ; les édifices sont secoués, les pots de terre se brisent en se heurtant les uns les autres.

   Source originale : *Annales Fuldenses.*

   Lieu de rédaction : Mayence.

14. **922.** Séisme dans la région de Cambrai ; des maisons sont renversées.

   Source originale (7) : Flodoard, *Annales.*

   Lieu de rédaction : Reims.

15. **939.** Séisme.

   Source originale : *Annales Colonienses breves.*

   Lieu de rédaction : Cologne.

16. **947.** Séisme à Reims.

   Source originale : Flodoard, *Annales.*

   Lieu de rédaction : Reims.

   Remarque : Le texte de Flodoard n'est pas clair : il y est question d'une `` tempête à Reims pendant toute une nuit, avec des éclairs continuels et un tremblement de terre ``.

17. **954.** Plusieurs séismes en `` Gaule `` et en `` Germanie ``, `` à certains endroits ``.

   Source originale (8) : *Annales Augienses.*

   Lieu de rédaction (pour les années 953-954) : Mayence.

18. **1000.** Important séisme le 29/3, dans plusieurs régions.

   Sources originales (9) : A. les *Annales Leodienses,* les *Annales S. Jacobi Leodiensis* et les *Annales Laubienses* dérivent d'une source originale, les annales perdues de Liège. B. *Annales Elnonenses.* C. *Annales S. Medardi Suessionensis.*

   Lieu de rédaction : A. Liège. B. St-Amand. C. Soissons.

19. **1013.** Séisme le 18/11, vers midi.

> **Sources** (10) : *Annales Leodienses* et *Annales S. Jacobi Leodiensis,* dérivant des annales perdues de Liège.
>
> **Lieu de rédaction** : Liège.

20. **1026.** Séisme le 16/2.

> **Source** : les *Annales Elmarenses,* qui dérivent d'une source originale, les annales perdues de St-Pierre de Gand.
>
> **Lieu de rédaction** : Gand.

21. **1080.** Séisme à Mayence le 1/12 ; les murs sont ébranlés ; au cimetière du monastère de St-Victor, près de Mayence, les cercueils sont projetés hors de terre.

> **Source originale** (11) : Marianus Scottus, *Chronicon*
>
> **Lieu de rédaction** : Mayence.

22. **1081.** Important séisme le 27/3, à la 1ère heure de la nuit.

> **Sources originales** (12) : A. *Annales Leodienses* et *Annales S. Jacobi Leodiensis,* dérivant des annales perdues de Liège. B. *Annales Laubienses.* C. Sigebert de Gembloux, *Chronographia.* D. *Annales Blandinienses.*
>
> **Lieu de rédaction** : A. Liège. B. Lobbes. C. Gembloux. D. Gand.

23. **1095.** Séisme le 10/9, vers minuit.

> **Source originale** : Sigebert de Gembloux, *Chronographia.*
>
> **Lieu de rédaction** : Gembloux.

24. **1112.** Séisme le 20/4.

> **Source originale** : *Annales Aquenses.*
>
> **Lieu de rédaction** : Aix-la-Chapelle.

25. **1117.** Important séisme le 3/1, à l'heure de vêpres, dans de nombreuses régions, "plus violent à certains endroits, plus clément à d'autres" ; à Liège, il n'y a pas de victime, "mais la crainte est grande" : dans la cathédrale, on voit bouger les vases, les crucifix et tout ce qui est suspendu (couronnes, lampes, etc. . .) ; à Reims, on voit bouger dans les églises les images et les objets suspendus. Près de l'abbaye de Susteren (?), la Meuse a été vue "quasi suspendue en l'air, laissant son lit à sec".

> **Sources originales** (13) : A. *Annales Leodienses.* B. *Chronicon rhythmicum Leodiense.* C. Anselme de Gembloux, *Continuatio.* D. *Annales Laubienses.* E. *Annales Rodenses.* F. *Annales Brunwilarenses.* G. *Auctarium Laudunense.* H. *Annales Mosomagenses.* I. *Annales S. Dionysii Remensis.* J. *Annales Catalaunenses.*
>
> **Lieu de rédaction** : A et B. Liège. C. Gembloux. D. Lobbes. E. Rolduc. F. Brauweiler. G. Laon. H. Mouzon. I. Reims. J. Châlons.

Remarques :

1) Les *Annales Catalaunenses* mentionnent ce séisme le 30/12 et non le 3/1 ; mais il s'agit sûrement d'une erreur du copiste des annales, qui a inscrit "3 des Calendes de Janvier" (30/12) au lieu de "3 des Nones de Janvier" (3/1). Une erreur de transcription est sans doute aussi à l'origine de la date fournie par l'*Auctarium Laudunense*, qui signale l'événement "le mercredi 9/1"; or, le 9/1/1117 tombait un mardi, alors que le 3/1 était bien un mercredi.

2) Anselme de Gembloux rapporte que le 2 mai de la même année, il y eut à Liège un "violent orage avec un tremblement de terre"; mais ce texte est très suspect car d'autre part l'auteur du *Chronicon rhythmicum Leodiense*, un témoin de première main, nous donne une longue description de cet orage du 2/5 mais ne mentionne pas de séisme à cette date.

3) Ce tremblement de terre a été ressenti dans une grande partie de l'Europe ; il est également mentionné par des sources de Dijon, d'Einsiedeln (Suisse alémanique) et de Bonnevaux (Dauphiné), et surtout par un grand nombre de sources de Lombardie, où plusieurs villes furent détruites le 3/1/1117.

4) Dans les sources de nos régions, seules Liège et Reims sont mentionnées dans des textes rédigés dans ces deux villes ; mais il y a tout lieu de croire que le séisme a également été ressenti à Gembloux, à Lobbes, à Rolduc, à Brauweiler, à Laon, à Mouzon et à Châlons, puisqu'il est cité (sans indication précise de lieu) par des sources écrites dans ces villes ou ces abbayes.

5) L'étrange assèchement subit de la Meuse qui eut lieu à la même époque s'est déroulé près d'un endroit nommé "Sustula" par nos sources (identifié sans certitude avec Susteren, dans le Limbourg hollandais).

26. **1121.** Séisme le 10/12, à l'heure de tierce.

Sources originales (14) : A. Anselme de Gembloux, *Continuatio.* B. *Annales Aquenses.* C. les *Annales S. Pantaleonis Coloniensis,* qui dérivent d'une source originale, des annales perdues de Cologne.

Lieu de rédaction : A. Gembloux. B. Aix-la-Chapelle. C. Cologne.

Remarque : Les *Annales S. Pantaleonis Coloniensis* mentionnent le fait en 1120, mais il doit s'agir d'une erreur de transcription, car il ne s'agit ici que d'un témoignage de seconde main.

27. **1134.** Assèchement subit de la Sambre à Namur, pendant un jour.

Source originale : *Annales Fossenses.*

Lieu de rédaction : Fosses.

possible de dire lequel de ces deux millésimes est exact.

28. **1141.** Séisme le 24/4.

    Source originale : *Annales Aquenses.*
    Lieu de rédaction : Aix-la-Chapelle.

29. **1146.** Séisme à une quinzaine de reprises.

    Source originale : *Annales Disibodenbergenses.*
    Lieu de rédaction : Disibodenberg.

30. **1167.** Séisme le 20/1, vers minuit.

    Source originale (15) : *Annales Colonienses maximi.*
    Lieu de rédaction : Cologne.

31. **1179** ou **1180.** Important séisme le 1/8, à la 4e veille de la nuit.

    Sources originales (16) : A. Lambert le Petit, *Annales S. Jacobi Leodiensis.* B. *Annales Floreffienses.* C. *Annales Elmarenses,* dérivant des annales perdues de St-Pierre de Gand. D. *Annales Aquenses.* E. *Annales Brunwilarenses.* F. *Annales Colonienses maximi.*
    Lieu de rédaction : A. Liège. B. Floreffe. C. Gand. D. Aix-la-Chapelle. E. Brauweiler. F. Cologne.
    Remarques :

        1) Sur six sources qui relatent ce séisme du 1/8, quatre d'entre elles placent le fait en 1180 et deux seulement (*Annales Brunwilarenses* et *Annales Colonienses maximi*) le situent en 1179 : c'est pourtant la date de 1179 qui paraît exacte, car d'une part Lambert le Petit et l'auteur des *Annales Aquenses* ont décalé d'un an par erreur plusieurs événements mentionnés dans leurs annales, tandis que les *Annales Elmarenses* et les *Annales Floreffienses* ne sont pas très sûres ; d'autre part les *Annales Colonienses maximi* signalent que " le même mois eut lieu une éclipse de lune " : or cette éclipse eut lieu le 19/8/1179.

        2) Les *Annales Elmarenses,* une source de seconde main, placent le tremblement de terre le 19/8 ; le copiste de ces annales paraît donc avoir non seulement décalé les faits d'un an, mais aussi confondu les dates du séisme et de l'éclipse lunaire.

32. **1189.** Séisme dans la nuit du 27/2 au 28/2, vers minuit.

    Sources originales : A. *Annales Disibodenbergenses.* B. *Annales S. Pauli Virdunensis.*
    Lieu de rédaction : A. Disibodenberg. B. Verdun.

33. **1214** ou **1215.** Séisme le 29/8 à l'heure de prime, surtout dans la région de Coblence.

    Sources originales : A. *Annales Colonienses maximi.* B. *Annales S. Pantaleonis Coloniensis.*
    Lieu de rédaction : A et B. Cologne.
    Remarque : Ce séisme est placé en 1214 par les *Annales Colonienses maximi* et en 1215 par les *Annales S. Pantaleonis Coloniensis ;* il n'est pas

34. **1223.** Important séisme le 11/1 à l'heure de prime, à Cologne et dans les environs ; les parois des édifices sont secouées et menacent ruine. Ensuite pendant environ deux semaines d'autres secousses sont ressenties à Aix-la-Chapelle.

   Sources originales : A. *Annales S. Pantaleonis Coloniensis.* B. Césaire de Heisterbach, *Dialogus miraculorum.*

   Lieu de rédaction : A. Cologne. B. Heisterbach.

   Remarques :

   1) Nos deux sources placent ces événements du 11/1 à l'année 1222 mais font débuter l'année à Pâques (le **style** de Pâques est habituel dans la région de Cologne au XIIIe siècle) ; il importe donc de corriger d'un an le millésime.

   2) Quelques jours auparavant, à la Noël 1222 et dans les deux semaines qui ont suivi, les villes de Lombardie furent ravagées par d'importantes secousses sismiques.

35. **1259.** Séisme le 3/5.

   Source originale : *Annales Blandinienses.*

   Lieu de rédaction : Gand.

36. **1291.** Séisme le 10/9, au crépuscule, dans la région rhénane, notamment dans la région de Worms.

   Sources originales : A. *Annales Moguntini.* B. le *Chronicon pontificum et imperatorum Rhenense,* qui dérive d'une source originale, des annales perdues de Worms.

   Lieu de rédaction : A. Mayence. B. Worms.

   Remarque : Ce séisme est également mentionné par une source de Strasbourg.

37. **1316.** Séisme en septembre.

   Source originale : *Gesta Treverorum.*

   Lieu de rédaction : Trèves.

   Remarque : Cet événement, signalé en 1318 dans les *Gesta Treverorum*, une source à la chronologie peu sûre pour le début du XIVe siècle, est à rapprocher du séisme du 11/9/1316, mentionné dans des sources rédigées à Paris, à Tours et à St-Evroult (Normandie).

38. **1356.** Important séisme le 18/10, notamment dans la région de Metz.

   Sources originales : A. La *Chronique dite du doyen de St-Thiébaut de Metz*, la *Chronique* de Jacomin Husson et la *Chronique* de Philippe de Vigneulles dérivent d'une source originale, une chronique perdue de Metz. B. *Chronicon Moguntinum.*

   Lieu de rédaction : A. Metz. B. Mayence.

Remarque : Il s'agit du célèbre séisme du 18/10/1356, qui détruisit entièrement Bâle et la région avoisinante. Il est mentionné par beaucoup de sources de nos régions, mais qui ne font pas état de secousses ressenties ailleurs qu'à Bâle. Dans la zone que nous étudions, seuls les chroniqueurs de Metz signalent que ce séisme a également touché leur région ; c'est probablement aussi le cas à Mayence, mais le texte du *Chronicon Moguntinum* n'est pas suffisamment clair à cet égard. Enfin, une chronique parisienne, la *Continuation de la Chronique de Guillaume de Nangis,* rapporte que la secousse fut également ressentie à Reims.

39. **1357.** Séismes très fréquents entre le 9/4 et le 28/5 ; à Mayence, les faîtes de certains édifices s'écroulent. Important séisme en mai, notamment dans les régions de Spire et de Trèves.

   Sources originales (17) : A. *Chronique dite du doyen de St-Thiébaut de Metz,* dérivant d'une chronique perdue de Metz. B. *Chronicon Moguntinum.* C. Werner de Bonn, *Chronica.*

   Lieu de rédaction : A. Metz. B. Mayence. C. Bonn.

   Remarque : Cette série de secousses sismiques a également touché Bâle, pour la seconde année consécutive, le 6/4, ainsi que Strasbourg, le 9/5.

40. **1372.** Séisme le 1/6 dans la région de Metz.

   Sources : *Chronique dite du doyen de St-Thiébaut de Metz, Chronique* de Jacomin Husson et *Chronique* de Philippe de Vigneulles, dérivant d'une chronique perdue de Metz.

   Lieu de rédaction : Metz.

   Remarque : Ce séisme du 1/6/1372 fut également ressenti à Bâle et à Strasbourg.

41. **1382.** Important séisme le 21/5, dans l'après-midi, une heure avant vêpres, en Picardie, en Flandre, en Brabant et dans le pays de Liège (notamment à Liège et en Hesbaye) ; des pierres tombent des cheminées ; les édifices sont secoués, certains d'entre eux s'écroulent. Un second séisme a lieu le 24/5, à la 3e heure de la nuit.

   Sources originales : A. *Chronique liégeoise de 1402.* B. Jean de Stavelot, *Chronique.* C. la *Brabantsche Chronijk* et le *Breve Chronicon Brabantinum* d'Edmond de Dynter, qui dérivent d'une source originale, une chronique perdue du Brabant. D. *Chronicon comitum Flandrensium.* E. la *Chronique des Pays-Bas et de Tournai,* qui dérive d'une source originale, une chronique perdue dè Tournai.

   Lieu de rédaction : A et B. Liège. C. Brabant. D. Bruges ? E. Tournai.

   Remarque : La *Chronique des Pays-Bas et de Tournai* situe l'événement en 1380, mais la chronologie de cette source de seconde main est très imprécise.

42. **1395.** Important séisme le 11/6 au point du jour, notamment dans les régions de Liège et de Cologne (où cela dure le temps d'une patenôtre) ; à Liège, les pots, les plats et les chaudrons pendus aux murs sont violemment secoués ; à Cologne, les maisons sont secouées et les pots de terre s'entrechoquent.

Sources originales : A. Jean de Stavelot, *Chronique.*
B. *Cölner Jahrbücher.*

Lieu de rédaction : A. Liège. B. Cologne.

Remarque : Ce séisme du 11/6/1395 fut également ressenti à Limburg an der Lahn (comme en témoigne la *Limburger Chronik*).

1) Sources non originales : Annales Tiliani, Annales Mettenses priores, *chronique de Réginon de Prüm,* Annales Sithienses, Annales Fuldenses.

2) Sources non originales : Annales Mettenses posteriores, *chronique de Réginon de Prüm,* Annales Xantenses, Annales Maximiniani.

3) Source non originale : Annales Fuldenses.

4) Source non originale : Annales Fuldenses.

5) Le fait est placé à l'année 858 dans les Annales Bertiniani, *mais les sources de l'époque carolingienne font débuter l'année le 25/12 et non le 1/1.*

6) Contrairement à l'usage, habituel au IXe siècle, de faire débuter l'année le 25/12, l'auteur de cette source a placé cet événement survenu le 30/12 à la fin de l'année 881 et non au début de l'année 882.

7) Source non originale : Richer, Historiae.

8) Source non originale : Marianus Scottus, Chronicon, *et* Sigebert de Gembloux, Chronographia, *qui placent le fait à tort respectivement en 952 et en 950.*

9) Sources non originales : Sigebert de Gembloux, Chronographia ; Annales Parchenses, Annales Blandinienses, Annales Formoselenses, Annales S. Martini Tornacensis.

10) Source non originale : Sigebert de Gembloux, Chronographia; Annales Parchenses.

11) Source non originale : Sigebert de Gembloux, Chronographia.

12) Sources non originales : Annales Parchenses, Annales Elmarenses, Annales Formoselenses.

13) Sources non originales : Annales Remenses et Colonienses, Continuatio Praemonstratensis *(qui mentionne le fait à tort en 1118).*

14) Source non originale : Chronicon S. Bavonis Gandensis.

15) Source non originale : Annales S. Pantaleonis Coloniensis.

16) Source non originale : Annales S. Pantaleonis Coloniensis.

ANNEXE : LISTE DES FAUX SEISMES

Les séismes suivants, cités dans les compilations traditionnelles, mais qui ne figurent dans aucun document original, sont à rejeter définitivement. Ils peuvent être de trois types différents :

1) Séismes purement imaginaires, dont l'existence n'est due qu'à l'imagination trop fertile de chroniqueurs médiévaux ou modernes.

2) Séismes qui ont eu lieu, mais pas à la date indiquée par le compilateur, qui donne une année ou un jour de l'année faux.

3) Séismes qui ont eu lieu, mais pas aux endroits mentionnés par le compilateur, qui signale des destructions dans des localités qui ne sont pas citées dans nos sources.

330. Tournai. D'après l'ouvrage sans valeur de Haverlant (Torfs, Lancaster, qui mettent cependant le fait en doute).

450. Cambrai. D'après l'ouvrage sans valeur de Le Carpentier (Torfs, Lancaster).

502. Tournai. D'après Haverlant (Torfs, Lancaster).

630. Tournai. D'après Haverlant (Torfs, Lancaster).

749. Aix-la-Chapelle (Sieberg, qui ne cite aucune source originale et dont le travail est particulièrement fabuleux).

827. Spire (Sieberg).

834. Aix-la-Chapelle (Sieberg).

842. Coblence (Sieberg).

854. Tournai et Cambrai. D'après Haverlant et Le Carpentier (Torfs, Lancaster).

857. 15/9 ou 25/9. Cologne et Trèves (Sieberg).

858. 25/12. Mayence (Perrey, Sieberg, qui commettent une erreur de chronologie ; cfr. plus haut, année 858, note 5).

880. 1/1. Mayence (Perrey).

882. 29/12. Mayence (Perrey, qui met cependant le fait en doute). C'est en réalité celui de 881.

885. Mayence (Perrey, qui tient cependant le fait pour douteux).

950, 951 ou 952. "Gaule" et "Germanie" (Perrey, Torfs). C'est en fait celui de 954 (cfr. plus haut, année 954, note 8).

968. Cologne (Sieberg).

980. Mayence (Sieberg).

1000. 29/3. Lobbes, Tournai, St-Omer, Gand (Torfs). Ce séisme n'est attesté qu'à Liège, St-Amand et Soissons.

1001. Cambrai. D'après Le Carpentier (Torfs, qui tient cependant le fait pour douteux).

1005. Liège (Sieberg).

1009. Belgique (Sieberg).

1012. 19/10. (Sieberg).

1013. 18/9 (Perrey, Torfs, Lancaster, qui citent deux séismes, le 18/9 et le 18/11, alors que seul le second est réel).

1024. (Torfs).

1070. 11/5. Cologne. D'après l'ouvrage sans valeur de Von Hoff (Perrey, Torfs, Sieberg).

1080. Flandre (Torfs, d'après une chronique du XVIe siècle).

1081. Reims (Torfs).

1085. Lorraine (Perrey, Torfs, qui mettent en doute la localisation mais pas le fait lui-même ; Sieberg).

1086. Flandre (Torfs, d'après Jacques de Meyer, chroniqueur du XVIe siècle ; Lancaster).

1087. 14/7. Soissons (Perrey).

1089. Lorraine (Sieberg).

1094. Lorraine (Sieberg).

1095. Gand (Torfs, d'après la chronique de St-Bavon, une source sans valeur du XVe siècle).

1101. Pays de Liège (Torfs, Lancaster, qui mettent cependant le fait en doute).

1103. Mayence (Sieberg).

1108. Namurois (Torfs, Lancaster).

1109. Tournai (Torfs, Lancaster).

1112, 3/1. Liège (Perrey, Lancaster). Il s'agit évidemment du séisme de 1117.

1114. St-Omer (Torfs).

1116. Liège (Torfs, Lancaster, qui mettent cependant le fait en doute).

1118. Liège (Torfs, Lancaster, qui mettent cependant le fait en doute).

1120. Flandre (Torfs).

1122, 10/12. Cologne (Perrey, Sieberg). C'est en réalité celui de 1121.

1142, 2/11. Lorraine (Sieberg).

1166, 26/1. Cologne (Sieberg). C'est en réalité celui de 1167.

1173 ou 1174. Maestricht (Sieberg).

1181. Malines (Torfs, Lancaster, d'après une chronique malinoise du XVIe siècle).

1222, 11/1. Cologne (Perrey, Torfs, qui commettent une erreur de chronologie ; cfr. plus haut, année 1223, remarque 1).

1259. Flandre (Lancaster, qui met cependant le fait en doute).

1300. Reims. D'après Von Hoff (Perrey, pour une fois perspicace, a vu que Von Hoff confondait Reims avec Rieti, ravagée à cette date par un séisme).

1317, 14/8. Flandre (Ninove, Grammont, Renaix), Artois, Hainaut, Namurois (Torfs, d'après des . .. "Notules d'un manuscrit du XVIIIe siècle"!). Il s'agit peut-être du séisme de 1316 (ou 1318), mais qui n'est nullement attesté dans ces localités.

1342. (Lancaster).

1346. (Lancaster).

1350. (Lancaster).

1381, 21/5 (Lancaster). C'est en réalité le séisme de 1382.

1383. Anvers (Torfs, qui met en doute l'année, mais pas cette localisation qu'il tire d'une chronique anversoise du XVIe siècle).

1395. Anvers, Malines (Perrey, Torfs, Lancaster, d'après des chroniques sans valeur du XVIe siècle).

BIBLIOGRAPHIE

Annales Aquenses : G. WAITZ, dans M.G.H., SS, t. 24, 1987 :
34-39.

Annales Augienses : Ph. JAFFE, Bibliotheca rerum Germanicarum, t. 3, 1866 : 702-706.

Annales Bertiniani : F. GRAT - J. VIELLIARD - S. CLE-
MENCET, Annales de St-Bertin, Paris, 1964 (Soc. hist.
Fr.).

Annales Blandinienses : Ph. GRIERSON, Les Annales de St-
Pierre de Gand et de St-Amand, Bruxelles, 1937 : 1-73
(C.R.H., Série in-8º).

Annales Brunwilarenses : G.H. PERTZ, dans M.G.H., SS, t. 16,
1859 : 724-728.

Annales Catalaunenses : G.H. PERTZ, dans M.G.H., SS, t. 16,
1859 : 488-490.

Annales Colonienses breves : G.H. PERTZ, dans M.G.H., SS,
t. 16, 1859 : 730-731.

Annales Colonienses maximi : G. WAITZ, Chronica regia Coloniensis, 1880 (Script. rer. Germ.).

Annales Disibodenbergenses : G. WAITZ, dans M.G.H., SS.,
t. 17, 1861 : 6-30.

Annales Elmarenses : Ph. GRIERSON, op. cit. : 74-115.

Annales Elnonenses : Ph. GRIERSON, op. cit. : 132-175.

Annales Floreffienses : L. BETHMANN, dans M.G.H., SS,
t. 16, 1859 : 618-631.

Annales Fossenses : G.H. PERTZ, dans M.G.H., SS, t. 4, 1841 :
30-35.

Annales Fuldenses : Fr. KURZE, Annales Fuldenses, 1891
(Script. rer. Germ.).

Annales Laubienses : G.H. PERTZ, dans M.G.H., SS, t. 4, 1841:
9-28.

Annales Leodienses : G.H. PERTZ, dans M.G.H., SS, t. 4,
1841 : 9-30.

Annales Moguntini : G.H. PERTZ, dans M.G.H., SS, t. 17,
1861 : 1-3.

Annales Mosomagenses : G.H. PERTZ, dans M.G.H., SS, t. 3,
1839 : 160-166.

Annales regni Francorum : Fr. Kurze, Annales regni Francorum, 1895 (Script. rer. Germ.).

Annales Rodenses : P.C. BOEREN - G. PANHUYSEN, Annales
Rodenses, Assen, 1968.

Annales S. Dionysii Remensis : G. WAITZ, dans M.G.H., SS,
t. 13, 1881 : 82-84.

Annales S. Jacobi Leodiensis : J. ALEXANDRE - L. BETH-
MANN, Annales S. Iacobi Leodiensis, Liège, 1874 : 1-30.

Annales S. Medardi Suessionensis : L. D'ACHERY, Veterum
scriptorum Spicilegium, t. 2, 1723 : 486-492.

Annales S. Pantaleonis Coloniensis : G. WAITZ, Chronica regia
Coloniensis, 1880 (Script. rer. Germ.).

Annales S. Pauli Virdunensis : G.H. PERTZ, dans M.G.H., SS,
t. 16, 1859 : 500-502.

Annales Xantenses : B. VON SIMSON, Annales Xantenses et Annales Vedastini, 1909 (Script. rer. Germ.).

Anselme de Gembloux, Continuatio : L. BETHMANN, dans M.G.H., SS, t. 6, 1844 : 375-385.

Auctarium Laudunense : L. BETHMANN, dans M.G.H., SS, t. 6, 1844 : 445-447.

Brabantsche Chronijk : Ch. PIOT, Chroniques de Brabant et de Flandre, 1879 : 49-62 (C.R.H.).

Césaire de Heisterbach, Dialogus miraculorum : J. STRANGE, Cologne-Bonn, 1851, 2 vol.

Chronicon comitum Flandrensium : J.J. DE SMET - L.A. WARNKOENIG, Recueil des chroniques de Flandre, t. 1, 1837 : 223-257 (C.R.H.).

Chronicon Moguntinum : C. HEGEL, Chronicon Moguntinum, 1885 (Script. rer. Germ.).

Chronicon pontificum et imperatorum Rhenense : L.WEILAND, dans Neues Archiv der Gesellschaft für ältere deutsche Geschichtskunde, t. 4, 1879 : 74-85.

Chronicon rhythmicum Leodiense, C. DE CLERCQ, Reimbaldi Leodiensis opera omnia, Turnhout, 1966 : 122-140.

Chronique des Pays-Bas et de Tournai : J.J. DE SMET, Recueil des chroniques de Flandre, t. 3, 1856 : 115-570 (C.R.H.).

Chronique dite du doyen de St-Thiébaut de Metz : A. CALMET, Histoire de Lorraine, t. 5, 1745 : VII-CXVII.

Chronique liégeoise de 1402 : E. BACHA, La Chronique liégeoise de 1402, 1900 (C.R.H., Série in-8o).

Cölner Jahrbücher : H. CARDAUNS, Die Chroniker der niederrheinischen Städte. Cöln, t. 2, 1876 : 18-192 (Chroniken der deutschen Städte, t. 13).

Edmond de Dynter, Breve Chronicon Brabantinum : P. DERAM, Chronique des ducs de Brabant d'Edmond de Dynter, t. 1, 1854 : 57-60 (C.R.H.).

Flodoard, Annales : Ph. LAUER, Les Annales de Flodoard, 1905 (Collection de textes pour servir à l'étude de l'histoire, t. 39).

Gesta Treverorum : J.H. WYTTENBACH - M.F.J. MÜLLER, Gesta Trevirorum, t. 2, Trèves, 1838 : 179-271.

Jacomin Husson, Chronique de Metz : H. MICHELANT, Metz, 1870.

Jean de Stavelot, Chronique latine : S. BALAU, Chroniques liégeoises, t. 1, 1913 : 69-143 (C.R.H.).

Lambert le Petit, Annales S. Jacobi Leodiensis : J. ALEXANDRE - L. BETHMANN, op. cit. : 30-48.

Marianus Scottus, Chronicon : G. WAITZ, dans M.G.H., SS, t. 5, 1844 : 481-564.

Philippe de Vigneulles, Chronique : Ch. BRUNEAU, Metz, 1927-1938.

Sigebert de Gembloux, Chronographia : L. BETHMANN, dans M.G.H., SS, t. 6, 1844 : 300-374.

Werner de Bonn, Chronica : E. BALUZE - G. MOLLAT, Vitae paparum Avenionensium, t. 1, Paris, 1914 : 223-550.

PROBLÈMES DE SISMICITÉ HISTORIQUE : EXEMPLES DE FAUX SEISMES,
DE SEISMES MECONNUS ET DE SEISMES REINTERPRETES DANS L'ENSEMBLE
ALLEMAGNE/BELGIQUE/NORD-OUEST DE LA FRANCE/SUD DE GRANDE-BRETA-
GNE.

J. Vogt

Antenne Sismicité, Service Géologique Régional
(B.R.G.M.), Strasbourg 1).
avec la collaboration de B. Cadiot et J. Lambert

ABSTRACT
     Classical cataloguesare often unreliable. Until recently,
Perrey's has often been used in France in a most uncritical way.
Since 1975/76 has however been going on a painstaking revision
of the historical seismicity of the country and neighbouring
regions, as a sounder basis for the assessment of risk, zoning
and seismic engineering. Samples are given of a "fake quake"
(Avesnes, 1838), of an enigma (1456), of a hitherto scarcely
known quake (Ardennes, 1733), of a relocation of epicenter
(Channel, 1889), etc.. The need for such basic interdisciplinary
work is easily forgotten by sophisticated specialized research
often faithfully relying on uncritically fed data-banks.

1. INTRODUCTION.

     Il ne sera guère question de la sismicité de la région de
Liège pour la simple raison que le séisme majeur de 1692 a fait
l'objet d'une mise au point lors des "Rencontres d'Antibes" de
1983, consacrées à la sismicité historique. Par une coïncidence
fâcheuse ou heureuse - c'est affaire de point de vue - cette
présentation d'un "précédent" par excellence a précédé de quatre
jours le séisme liègeois du 8-11-1983.

     En soulignant *l'actualité de la sismicité historique*, cette
coïncidence invite à un rapide survol méthodologique. D'une manière générale, la définition souvent laborieuse du risque sismique et l'essor du génie sismique posent avec une acuité crois-
sante le problème de la *maîtrise de la sismicité historique*.

205

Sans revenir ici sur le cloisonnement des disciplines qui compromet souvent cette maîtrise 2), il convient de souligner les insuffisances criantes de nombreux catalogues classiques qui se reprennent volontiers les uns les autres. A cet égard, l'exemple de la France est particulièrement significatif : pour sa sismicité ancienne ne s'est-on pas longtemps reposé pour l'essentiel sur les travaux de Perrey, au milieu du XIXe siècle ? Il y a quelques années encore, n'a-t-il pas été déclaré que Perrey avait tout dit et qu'il serait non seulement superflu, mais aussi inconvenant d'entreprendre de nouvelles recherches de sismicité historique ? C'est méconnaître les préoccupations particulières et les conditions de travail de Perrey dont l'oeuvre, certes remarquable en son temps, ne répond en aucune manière aux besoins actuels 3). Aussi bien de tels points de vue relèvent-ils souvent à la fois du fatalisme et de l'irresponsabilité.

Bien plus, le mal est parfois aggravé par une *informatisation intempestive* de tels catalogues, mis en oeuvre tels quels, sans la discussion serrée qu'ils appellent. Ainsi sont créés des outils qui ne présentent que les *apparences de la rigueur*. Faute d'être avertis de ce processus, les utilisateurs de tels outils participent parfois à leur corps défendant à l'engrenage de l'irresponsabilité. Ce n'est cependant pas le lieu de développer ces considérations négatives, d'ordre épistémologique et éthique.

Heureusement, cette situation paradoxale a suscité progressivement de saines réactions dont l'oeuvre multiforme du Professeur Ambraseys est le meilleur exemple.

En France, les exigences du programme nucléaire ont conduit de proche en proche à une prise de conscience de ces problèmes. En commençant par un retour, riche en surprises, aux propres sources de Perrey, elle a conduit, avec l'aide d'historiens - soulignons, pour le Moyen Age, l'apport de P. Alexandre - et d'archivistes à une large révision de sismicité historique de son territoire et des régions voisines. Imprévisible au départ pour tout un chacun, cette révision représente en fin de compte l'apport le plus original et le plus substantiel du "Projet Sismotectonique" (1976-78) et de ses prolongements dans le cadre de contrats conclus par le C.E.A. et l'E.D.F. avec le B.R.G.M. 4).

Pour l'essentiel, ce débroussaillage a été assuré en trois ans (1976-78), au pas de course, par une équipe cohérente, démantelée depuis lors. Actuellement, cette tâche se poursuit au ralenti, en mettant l'accent sur la mise en valeur de riches matériaux, en majeure partie inédits 5), des recherches approfondies, à propos de cas d'espèce, et la réexploitation des archives du Bureau Central Sismologique Français qui couvrent un demi-siècle.

C'est à la suite de la première étape de cette révision qu'a été créé en 1979 au B.R.G.M. un fichier-informatique très élaboré faisant grand cas du *degré de fiabilité* 5). Ce fichier s'enrichit régulièrement de données nouvelles, qu'il s'agisse de sismicité historique ou de macrosismicité actuelle.

Après cet indispensable survol examinons quelques cas, exemplaires, parmi bien d'autres.

2. UN FAUX SEISME : L'EXPLOSION DE HERNU DU 26 OCTOBRE 1838.

Un récent catalogue français, inédit, fait état, le 26 octobre 1838, d'une "secousse très forte à Avesnes", sans autre précision. Un tel événement présenterait certes un grand intérêt sismotectonique, dans un cadre très large. Le vague de cette information ponctuelle a cependant conduit à la contrôler.

Il apparaît qu'elle est empruntée au catalogue de Perrey, sans plus. Ce dernier ne fait état que d'une seule source, insolite, à savoir une compilation ... italienne. Par l'intermédiaire de Perrey, cette information a d'ailleurs aussi été reprise par les compilations de Douxami et Torfs. Ce dernier a le mérite de retracer le circuit de l'information : "... secousse très forte à Avesnes... M. Perrey la cite d'après le *Giornale* de Colla".

En fait, il s'agit de l'explosion d'une poudrière à Hornu. Cette explosion est présentée en grand détail par la presse régionale. La secousse est ressentie à quelque distance, par exemple à Maubeuge, à Ath et à Mons. A ce propos, il est parfois question de tremblement de terre. A Maubeuge, "... les uns l'attribuent à une explosion ; les autres, et c'est l'opinion la plus générale, l'expliquent par un tremblement de terre", écrit tel journal. Sans doute la grande presse s'est-elle bornée à donner la deuxième interprétation, reprise par la compilation italienne.

Il ne reste pas moins surprenant qu'en un siècle et demi les compilateurs successifs et surtout les compilateurs régionaux n'aient pas éprouvé le besoin d'étoffer la connaissance d'un tel événement, récent pour plusieurs d'entre eux, en faisant appel à des sources de première main. Ainsi est né un "faux séisme", un de plus. Sismicité historique, sismologie et sismotectonique ne cessent d'être à la merci de telles confusions.

3. L'ENIGME DU 26 AOUT 1456.

Un événement daté du 26 août 1456 pose des problèmes d'interprétation particulièrement ardus. Faisons le point d'après
les catalogues classiques. Perrey fait état d'un "tremblement
à Liège", d'après une seule source classique, le *Novus Thesau-
rus Anecdotorum* de Martène et Durand. Telle est aussi la source
de Torfs qui souligne que "en tout cas cet ébranlement... est
essentiellement distinct de ceux qui surviennent ... dans le
royaume de Naples". Sieberg n'évoque cette date qu'incidemment,
à propos d'un séisme signalé cette année à Gera, sans autre
précision, en s'interrogeant au sujet de ses relations possibles
avec des secousses à Liège, en esquissant une "enveloppe", as-
sortie d'interrogations, d'après ces seuls repères. Fort à propos, le classique bilan de la sismicité belge par Van Gils ne
s'attarde pas à cet événement.

Il est vrai que deux chroniques de Maastricht, dont le
contexte nous échappe encore, font état d'un tel séisme. En ne
donnant ni le mois, ni le jour, ni la source, l'une dit : "... In
ditzelfde jaar waren er aardbevingen" 7) ; indiquant le jour,
l'autre fait état de dégâts : "... was hier ene groote aartbe-
vinge, die vel schade heeft gedaen" 8). Cependant, une rapide
lecture de nombreuses chroniques belges, publiées ou inédites,
liégeoises en particulier, ne nous a pas permis jusqu'ici de
trouver trace de ce séisme, alors qu'elles ne manquent pas de
mentionner plusieurs événements antérieurs.

Tout porte à croire qu'une confusion au moins partielle
avec un événement de nature différente, lointain, n'est pas
exclue. C'est en août 1456 que plusieurs chroniques évoquent un
ouragan destructeur en Toscane. Ainsi lisons-nous dans un opus-
cule consacré précisément aux tremblements de terre : "Im
Augstmonat war ein so grosser Wind in der florentiner und vole-
taner Gegend, dass viel Dörfer danieder gerissen worden" 9). Une
telle "nouvelle", sommairement présentée par des échos contemporains, est susceptible d'être réinterprétée et localisée ailleurs
par des plagieurs pressés : Précisément un contrôle a porté sur
la source qui fait état de Gera et que Sieberg a utilisée. Nous
lisons : "1456 ? Gera : << Im Monat August sind schreckliche
Sturmwinde und Erdbeben, jedoch in andern Ländern heftiger als
hier gewesen>>" 10). Cet écho est certes mince. Si l'enquête
n'a pas été poursuivie, ces remarques suffisent à susciter un
doute au moins partiel.

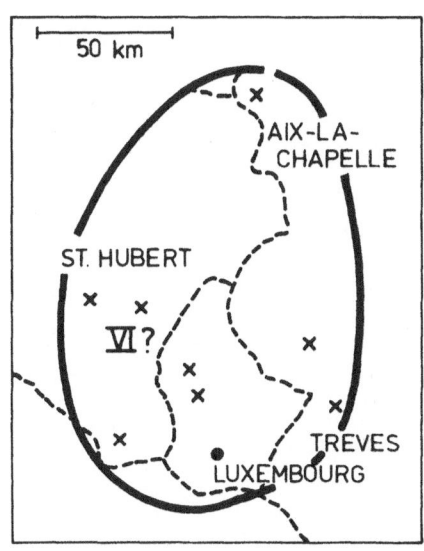

Enveloppe du séisme
ardennais méconnu du
13 avril 1733

4. UN SEISME MECONNU : LE TREM-BLEMENT DE TERRE ARDENNAIS DU 13 AVRIL 1733.

Ce séisme a échappé, semble-t-il, à Perrey. Sieberg l'évoque rapidement, vers 20 h 1/4, en un seul point, Aix-la-Chapelle, en suggérant une intensité III - IV. En fait, ce séisme affecte une aire notable. Au Sud, il est signalé en plusieurs points de l'ensemble Province du Luxembourg/Grand-Duché de Luxembourg/pays de Trèves. Dans le premier, il est mentionné à St. Hubert 11), St. Vincent 12) et, surtout, à Flamierge près Bastogne, vers 21 h : "tremblement de terre si violent que les cloches en ont sonné..." 13). Dans le second, il en est fait état en termes généraux par les archives paroissiales d'Insenborn : "1733 die aprilis quadrante ante horam 9 ^{nam} vespertinam factus est bis terrae motus in tota patria luxemburgensi" 14). Bien plus, une inscription en témoigne à Grosbous : "1733, 13 aprilis, terra tremuit" 15). En outre la secousse est notée à Bitburg - "... eine Erdtbibungh entstanden, aber gleich auffgehördt" 16) - et à Trèves 17).

S'il n'est pas destructeur à première vue, ce séisme n'est pas moins notable, à en juger par la nature des témoignages parmi lesquels figure, chose exceptionnelle, une source épigraphique. Il apparaît que son "enveloppe" englobe une partie des Ardennes et du Massif schisteux rhénan, avec un groupement de témoignages dans la partie centrale et méridionale. Par rapport à ce groupement, peut-être fortuit, Aix-la-Chapelle fait figure de point excentrique, avec une faible intensité. Il est donc plausible de songer à un épicentre dans la région de Bastogne.

On saisit l'intérêt de la discussion du contexte sismotectonique d'un tel événement, compte tenu de la connaissance très inégale d'autres séismes, mineurs, de ce domaine.

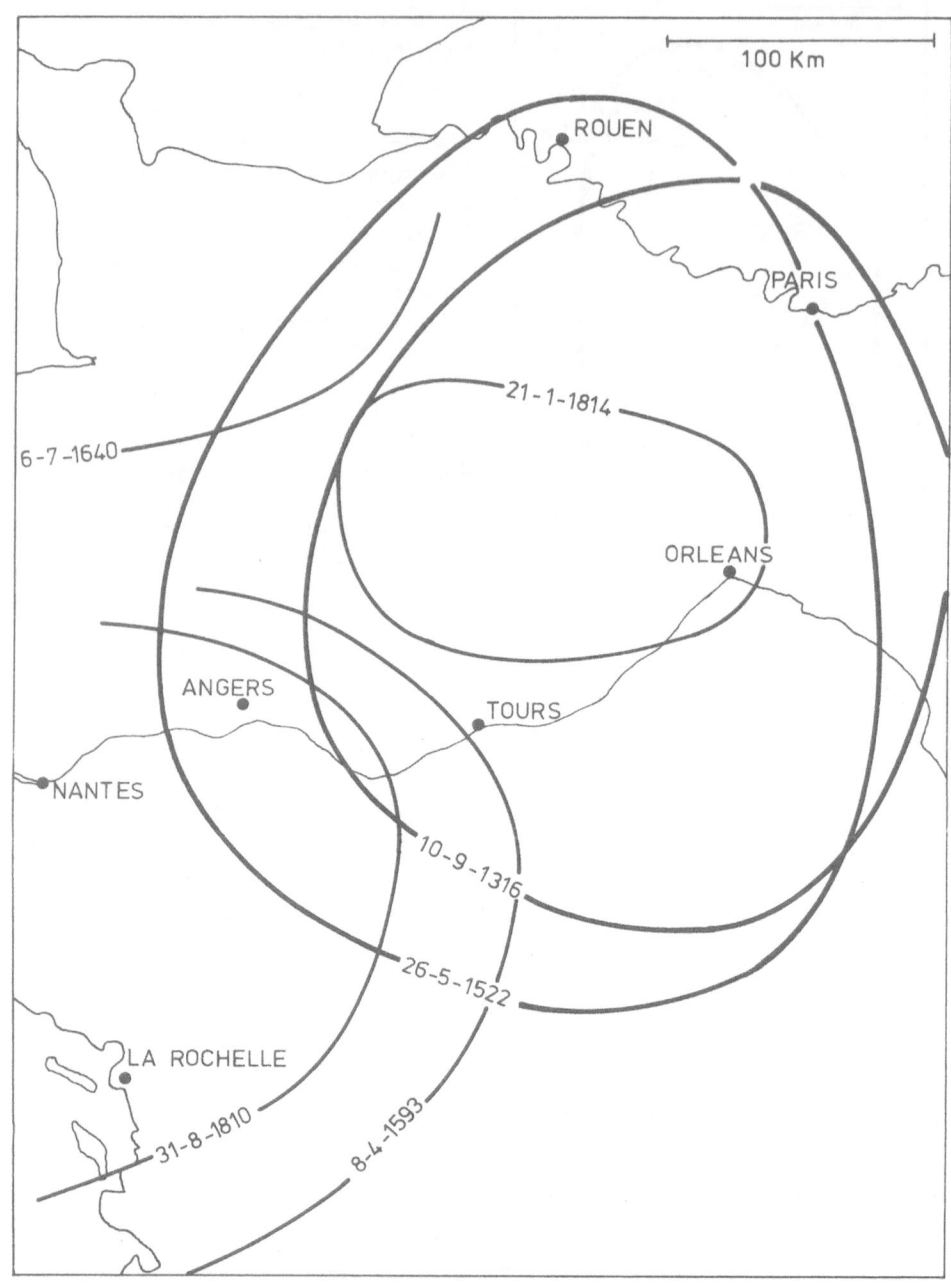

Enveloppes d'anciens séismes
majeurs de l'Ouest de la France, par
regroupement d'éléments émiettés

5. ESQUISSE D'ANCIENS SEISMES MAJEURS DE L'OUEST DE LA FRANCE.

En raison de problèmes chronologiques, les catalogues classiques ont fait "éclater" en éléments épars plusieurs séismes notables de l'Ouest de la France. L'analyse chronologique et de nouveaux apports se prêtent à des "reconstitutions" qui suggèrent des aires macrosismiques notables.

Pour commencer, tel est le cas du tremblement de terre du 10 septembre 1316. C'est d'une manière ponctuelle que Perrey et Y. Lacoste 18) mentionnent ce séisme, l'un à St. Denis et à Pontoise, l'autre en Berry, sans que ces indications aient été confrontées. D'autres sources font connaître cet événement à Orléans, en Touraine et dans la région de Laigle. Il en résulte une "enveloppe" d'une étendue insoupçonnée qui appelle de nouvelles recherches 19).

De la même manière apparaît l'importance du séisme du 26 juin 1522. A propos de l'Anjou, un catalogue récent le fait coïncider avec l'éclipse du 5 septembre. Ce propos résulte d'une mauvaise lecture d'une source, italienne, que Perrey donne *in extenso* et qui procède à une simple juxtaposition d'informations relatives à un séisme et à une éclipse. Or, c'est avec précision que de nouvelles recherches font connaître un séisme le 26 juin à Rouen et en Berry. Il est donc plausible de dater de ce même jour l'écho angevin. Anjou, Normandie, Berry, ce domaine rappelle celui du séisme de 1316.

De même, prend forme le séisme normand et breton du 6 juillet 1640. S'il échappe à Perrey, les compilateurs régionaux le découvrent chez les chroniqueurs du Mont-Saint-Michel. Par la suite les travaux de Mourant le font connaître avec quelque détail à Jersey. En outre, il est jalonné en Normandie et en Bretagne par des sources locales très précises dans les régions de Dinan, Vire et Argentan.

A trois séismes distincts des catalogues classiques la perspicacité de B. Cadiot a permis de substituer un seul, le 20 avril 1593, à Angers, Chinon, Chatellerault, L'Ile de Ré, La Rochelle et Saintes. Même processus, grâce à J. Lambert, pour les séismes du 31 août 1810 et du 21 janvier 1814, avec le regroupement de Saumur, Nantes et La Rochelle pour le premier, Orléans, Alençon et sans doute Le Mans, pour le second.

6. LA RELOCALISATION EN MER DU SEISME DU 30 MAI 1889.

Au cours du Projet Sismotectonique, il est apparu que la
sismicité de la Manche avait été méconnue par les catalogues
classiques, qu'il s'agisse des secousses des 5 novembre 1734,
28 janvier 1878 ou 30 mai 1889 20).

Le cas de ce dernier est particulièrement flagrant. En effet, il était admis, d'après les travaux de Mourant, spécialiste
de la sismicité des îles anglo-normandes, que l'épicentre était
situé entre Jersey et Cotentin. Une telle localisation était
cependant incompatible avec le tracé des isoséistes, caractérisés
en particulier, dans cette hypothèse, par le contournement d'une
grande partie du Cotentin, en raison de la densité des repères
au-delà. Sans doute cet épicentre excentrique avait-il été pro-
posé par analogie avec des séismes ultérieurs, par un processus
de "fixation" dont nous connaissons d'autres exemples malheureux.
Fort à propos, Lecornu, auteur contemporain, avait suggéré un
épicentre "en un point voisin de Guernsey et de Cherbourg". Une
enquête approfondie, en Normandie et en Grande-Bretagne, Wight
en particulier, a permis de conclure à la probabilité d'un épicentre en mer, sans doute au Nord ou au Nord-Est du Cotentin.

Si la sismicité des îles anglo-normandes est certes notable,
elle semble avoir exercé une singulière attraction, au point
d'éclipser une zone sismogène originale dont il reste à préciser
le cadre sismotectonique.

                                    X

Choisis parmi beaucoup d'autres, relativement simples, ces
"cas" ne sont certes pas replacés dans leur contexte, qu'il
s'agisse de l'histoire des séismes ou du cadre sismotectonique.
Néanmoins, cet échantillon permet de mesurer le travail accompli
et à accomplir pour l'élaboration, par un travail interdisciplinaire et transfrontalier, d'un *Corpus* de sismicité historique,
argumenté, solide 21). En fait, ce besoin fondamental est souvent
perdu de vue, éclipsé qu'il est par les thèmes à la mode, repris
avec un bel ensemble. A cet égard, fait souvent défaut, répétons-
le, la prise de conscience des insuffisances et des limites des
banques-informatique alimentées par des catalogues classiques.
Une banque fiable ne peut être que le reflet d'un *Corpus* solide,
permettant l'appréciation du processus d'interprétation, avec ses
forces et ses faiblesses, et surtout le retour, souvent indispensable, aux matériaux de base.

BIBLIOGRAPHY

1) Cette Antenne, créée en 1982, résulte d'une politique de
   décentralisation et de rapprochement avec le Bureau Central
   Sismologique Français et le Centre Sismologique Européo-
   Méditerranéen (Institut de Physique du Globe de Strasbourg).
   Elle assure en particulier la conservation et l'exploitation
   du patrimoine de sismicité historique et les enquêtes macrosismiques en France.

2) En dernier lieu, J. Vogt
   - "Problème de méthode de sismicité historique en France",
     Géologues (U.F.G.), n° 57, 1981.
   - "Quelques coupures entre démarches et entre spécialités
     des Sciences de la Terre", Géologues (U.F.G.), n° 63, 1983.
   - "Exposé introductif", à paraître en 1984 dans les Actes
     des Journées d'Antibes (1983).

3) Voir à ce sujet J. Vogt *et al.*, "Les Tremblements de Terre
   en France", Mémoire du B.R.G.M., n° 96, 1979.

4) Cf. J. Vogt, "Présentation de la carte sismotectonique de la
   France à 1/1 000 000" (Congrès Géol. Int. 1979), Bull. Int.
   Assoc. Eng. Geol., n° 23, 1981. D'une manière plus approfondie, l'ensemble de cette activité fera l'objet d'un chapitre
   par J. Goguel, J. Vogt et C. Weber dans un prochain ouvrage
   consacré au "Risque Sismique", sous la direction de V. Davi-
   dovici.

5) Outre quelques articles parus ou à paraître dans plusieurs
   revues internationales, nationales ou régionales, quelques
   rapides aperçus sont donnés depuis 1977 par les "Résumés des
   Principaux Résultats Scientifiques et Techniques du Service
   Géologique National" (B.R.G.M.).

6) Cf. P. Godefroy, S. Thirion, J. Lambert, B. Cadiot, "Infor-
   matisation du patrimoine de sismicité historique de la
   France", Bull. B.R.G.M., Section IV, n° 2, 1980/81 (26e
   Congrès Géol. Int.).

7) Kronick der Stad Maastricht, De Maasgouw, 1886/87.

8) Chronick der Stad Maastricht... , De Maasgouw, 1902.

9) M. Bernhertz, "Terraemotus...", Nuremberg, 1616.

10) "Chronik verschiedener Naturerscheinungen innerhalb Reussen-
    lands und insbesondere Gera's, bis 1862", Sechster Jahresbe-
    richt der Gesellschaft von Freunden der Naturwissenschaften
    in Gera..., 1863.

11) Registres paroissiaux, Archives de l'Etat, St. Hubert.

12) Justices subalternes, St. Vincent, Oeuvres de Loi, Archives de l'Etat à Arlon (aimable envoi de M. l'Archiviste).

13) Registres paroissiaux, Archives de l'Etat St. Hubert, source aimablement signalé par M. l'Archiviste.

14) Ardenner Zeitung du 24-3-1936.

15) Ardenner Zeitung du 14-4-1936.

16) J. Hainz *et al.*, Geschichte von Bitburg, Trier 1965.

17) Stadtarchiv Trier, Mst. 1615/411 (Clarisses).

18) Cf. Annales de l'Institut de Physique du Globe de Strasbourg, 1925, page 88.

19) P. Alexandre, J. Lambert, J. Vogt, "Deux séismes français majeurs au XIVe siècle", Rés. Sc., 1980 (cf. note 5).

20) J. Vogt, J. Delaunay, A. Levret, "Révision de sismicité historique du Cotentin et des Iles anglo-normandes", Rés. Sc., 1979 (cf. note 5).

21) Cf. N. Ambraseys *et al.*, "Notes on historical seismicity", Bull. Seism. Soc. America, t. 7, n° 6.

THE SEISMIC EVENT OF NOVEMBER 9, 1983
EARTHQUAKE OR SONIC BOOM

H.W. Haak

Royal Netherlands Meteorological Institute
De Bilt, The Netherlands

ABSTRACT

This study on the occurrence of unidentified infrasonic
signals was started after a mysterious acoustic event on
November 9, 1983, first misinterpreted as a Dutch earthquake.
It was found that the conditions for sound propagation in the
winter season can be very favourable, but no satisfactory
explanation could be constructed for the November 9 event. With
the aid of a simple microbarograph one can discriminate earthquakes against acoustic events.

INTRODUCTION

The day after the disturbing Liège earthquake a mysterious
event took place in the central part of the Netherlands. Judging
from the dozens of telephone reports we received at De Bilt the
effects of this event extended over 3500 km^2. People indoors felt
vibrations and heard rattling windows, outdoors heavy thunder -
like low noises were noticed. When the seismograms of the Dutch
network were analyzed it turned out that the event was a large
pressure wave in the atmosphere coming from NW direction.
Whatever the cause may be of the pressure wave, it was not an
earthquake as was published in all the major newspapers the
following day. Other events similar to the November 9 distur-
bance were noticed in the past, for example on March 30, 1976,
the same phenomena occurred in the northern part of the Netherlands. Up till now the event could not be attributed to a man-made
or natural cause.

P. Melchior (ed.), Seismic Activity in Western Europe, 215–222.
© *1985 by D. Reidel Publishing Company.*

Probably the low frequency of occurrence, one in two or three years, and subject lying outside the normal observatory practice, such pressure waves received only little attention from seismologists.

This short study reports on the findings of instrumental work after the November 9 event and an inventory of possible causes is given.

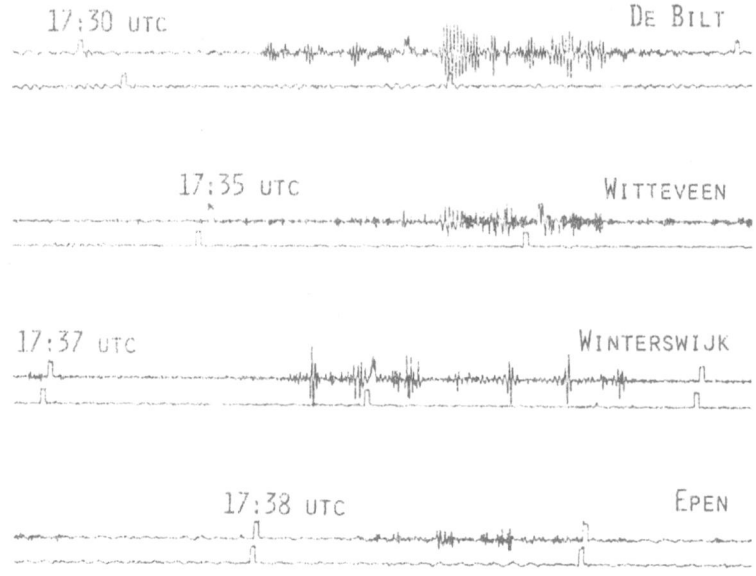

figure 1. Seismic records of the 1983 November 9 event. The top trace was recorded with the broadband Teledyne SL210 instrument, the other traces were recorded with short period Willmore MKII instruments.

THE NOVEMBER 9 EVENT

In figure 1 the seismograms are given recorded by a broadband Teledyne Geotech model SL210 and short period Willmore MKII seismographs. From the arrival time differences it is clear that the signals are due to a pressure wave propagating with sound velocity. The determination of the "epicenter" is severely hampered by the emergent onset of the signals especially on the

short period records. The propagating direction of the waves
was towards south-east . From figure 1 it can be seen that the
signals differ strongly amongst each other. Pulses are apparent
but the time distance between them is not the same for traces
of different stations. So a series of reflections is more likely
as the cause then a series of events. Although very small, the
signals are also to be found in the Galitzin seismograms (3
components) recorded at De Bilt. From the facts that horizontal
instruments deflected and the good sealing of the modern instruments from atmospheric pressure changes we conclude that an air
pressure wave will lead to a local ground motion. The ground
motion is too weak to generate seismic waves, that will reach
the other stations. It is interesting to see whether or not the
same mechanism is true for airwaves of longer periods, in the
order of minutes, from volcanic explosions or nuclear test
explosions in the atmosphere.

After the identification of the November 9 event as an airquake, a search for a possible man-made cause was made. Civil and
military airtraffic authorities in the region were checked and a
search was made for relevant publications in news papers. Until
today no man-made source was found. The high frequency content
of the signals and the quiet weather at the particular day indicates that the origin of the signal must be within 1000 km from
De Bilt, in the North Sea area. The same result was obtained from
a simple "epicenter" calculation with constant sound velocity
and zero wind velocity.

SOME HISTORY

It was not only the 1983 November 9 event which made us
decide to go into the affairs of this not strictly seismological
matter.

On March 30, 1976 the northern part of the Netherlands was
seriously disturbed by similar phenomena and again on September
7, 1978 and February 10, 1984. For none of these events
an explanation or reference to a man-made source could be found
with the exception of the last case the public in the first
instance identified the events as possible earthquakes, although
this part of the Netherlands has very low seismicity: no earthquake was ever detected instrumentaly in the region. In the
last 10 years at least 7 unidentified events of similar nature
took place in the Netherlands.

Surprisingly, almost in every continent there are places
where now and than or in certain periods "distant thunder" or
"canon fire" is heard; mist pouffers along the French-Belgian
coast, Sea Farts, Gouffres etc. are some name for such sounds;

often with long series of historic observations (ref. 1). Most
striking in this respect is a series of unusual noise events
reported along the east coast of Canada and the United States
in the period December 1977 to May 1978 (ref. 1). Two thirds
of the total of 594 of these events could be attributed to the
operation of supersonic aircraft. Most of the remaining 181
events are believed to have a natural, still unknown origin.

INSTRUMENTATION

    The detection of air pressure waves from various sources
such as volcanic, sonic boom, explosions, meteorites etc. with
a seismograph system is reported in a number of events. The
central problem is one of discrimination between airquakes and
true earthquakes. Therefore, a microbarograph was built with a
frequency response corresponding to that of a short period
seismograph i.e. an infrasound detector (ref. 3). The most
simple model was designed by Benioff and Gutenberg which is
shown in fig. 2 (ref. 2). It proved to be a useful concept as
long as the wind velocites were low. For higher wind velocities
the infrasound detector with noise reducing pipe array is probably more adequate (ref. 4, 5).

    The Benioff-Gutenberg instrument consists of a 200 liter
wooden box to which on one side a low frequency loudspeaker
(woofer) is mounted. Due to the low resonance frequency of 40 Hz
and the large cone area of 30 cm diameter, the sensitivity of
the system for low frequency (< 1 Hz) acoustic waves is high.

    The signal from the loudspeaker coil is amplified and
filtered to the required response by a first order low-pass
filter of 12 sec. With additional filtering it is possible to
reduce the long period > 10 sec. wind disturbances. In order to
facilitate the comparison with seismograms an identical recording
medium was chosen as for the short period seismometers: a recti-
linear ink recorder for a rotating paperband also known as
Karlsruhe recorder. The sensitivity on the barograms is 1 mm/μBar
over the band width from 10 Hz to 0.1 Hz. The time marks on the
records were incorporated in the conventional way.

    Since we were primarily interested in explaining some
spurious signals on seismograms it was not necessary to build
more than one infrasonic detector. With a set of three detectors
placed a few kilometers apart in a triangle the azimuthal angle
can be measured. In our case this information was already contained in the seismograms of the seismic network. The total
dimension of the wooden box and the loudspeaker can without
doubt be reduced when today's electronics is used.

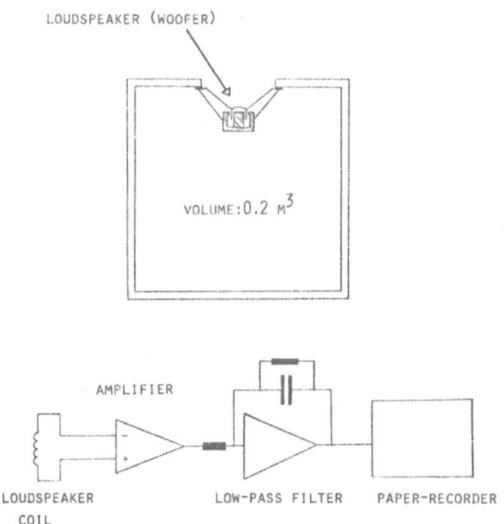

figure 2. Schematic diagram of the mechanical (top) and electronic
        (bottom) design of the Benioff-Gutenberg microbarograph.

RESULTS AND CONCLUSIONS

     During the three months January, February and March 1984
over 100 infrasonic events were clearly recorded, both on the
seismograms and the microbarograms (fig. 3). A large fraction
of these events were found to be related to the arrival of the
Concorde aircraft on the airports of London and Paris. The
signals were particularly large in periods with south-west wind.
In contrast to the event of November 9 we received no reaction
from the public on the Concorde events. The Concorde and other
infrasonic signals disappeared gradually in spring and early
summer, also under favourable wind conditions.

     On February 10, 1984 an unidentified event in the northern
part of Holland took place, which was heard by the public. Over
twenty telephone calls speaking of "booming noises" were
received. Around this date in mid-February, the Concorde signals
had large amplitudes. The same is true for the period in which
the November 9, 1983 event took place. This suggests that in the
winter season the·condition of the atmosphere with regard to
temperature and wind are favourable to sound propagation in the

infrasonic as well as in the low frequency audible regime. The
same conclusion can be drawn from the work of Claflin-Chalton
and MacDonald (ref. 1) concerning the North American events in
1978.

From the work of Donn and Rind (ref. 6) it is known that the
temperature profile up to 100 km altitude for a typical winter
situation differs considerably from that in the summer. The same
is true for the conditions for sound propagation. The other
parameter in the problem is the wind field which can be equally
important. These factors make it very hard to calculate the
exact "epicenter" of an airquake without knowledge of the local
sound velocity and wind profile. Due to the dynamic character
of all parameters the problem is essentially more complicated
than the seismic epicenter determination. Such an "epicenter"
determination could, however, give some insight in the possible
causes of the unidentified events.

The central problem still to be solved is the origin of the
unexplained events. The artificial man-made sources can be divided
in explosions of different origin and supersonic aircrafts.
Usually, the event is declared unidentified when all possible man-
made sources are ruled out by checking all the responsible authorities. Scientific rigor, however, is very hard to obtain in this
regard. Very little is known about the possible natural causes
such as meteorite impact, distant thunder and also earthquake
induced sounds. For the Netherlands region the last possibility
can certainly be ruled out. Since earthquakes can be very well
detected and localized in the Netherlands. Distant thunder can
easily be detected by the associated electro-magnetic pulses.
In the near future a network of lightning counters will be
installed in the Netherlands, by which dicrimination will be
possible. Meteorites as a possible cause can only be proved by
additional observations from an independent source such as visual
observation or ionospheric data. In the case of the November 9
event no such independent observation was available. Moreover,
only very little is known on the frequency of the meteorite
occurrence and on the effects of meteorite impact on the atmosphere
(ref. 7, 8).

In conclusion it can be said that 1) acoustic signals are
seen quite clearly on seismograms of short period vertical
seismometers; 2) that with the aid of a simple microbarograph
these signals can be discriminated against earthquake signals
and 3) that many cases of airquakes remain unexplained.

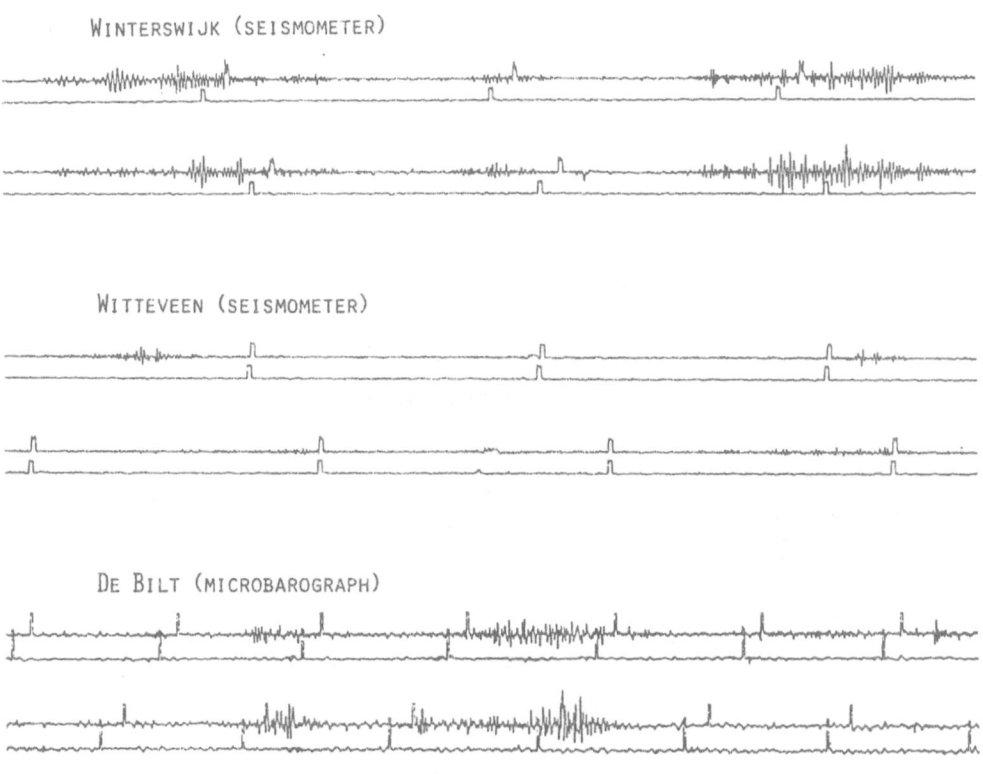

figure 3. Infrasonic signals due to the arrival of Concorde
flights. The three wave trains with approximately
one minute spacing are characteristic for these
signals. The time marks on the seismograms and the
microbarograms have a spacing of one minute.

REFERENCES

1.   Claflin-Chalton, S. and G.J. MacDonald (1978). Sound and
      light phenomena. The Mitre Corporation, McLean, Virginia
      22102, November 1978.

2.   Benioff, H. and B. Gutenberg. Bulletin Am. Meteor. Soc.,
      Vol. 20, No. 10. p. 421-426. (1939).

3.   Cook, R.K. and A.J. Bedard (1971). Geophys. J. R. Astr. Soc.
      26, 5-11.

4.   Grover, F.M. (1971). Geophys. J. R. Astr. Soc. 26, 41-52.

5.   Burridge, R. (1971). Geophys. J. R. Astr. Soc. 26, 53-69.

6.   W.L. Donn and D. Rind (1971). Geophys. J. R. Astr. Soc. 26,
      111-133.

7.   Jordon J.N. et al (1967). Earthquake information Bulletin,
      1, no. 3, p. 8-9.

8.   Quervain, A. de (1916). Ann. Schweiz. Meteor. Z. Aust.,
      p. 13-14.

L'ÉVÉNEMENT SISMIQUE DU 19 FÉVRIER 1984 EN PROVENCE
OCCIDENTALE : SEISME TECTONIQUE OU COUP DE TERRAIN MINIER ?

H. Haessler, P. Hoang-Trong et Y. Legros.

Institut de Physique du Globe de Strasbourg.

Abstract.

On February 19, 1984 the western Provence (France) has
been struck by a magnitude 4.5 seismic event. Largely felt from
Nice to Montpellier it was the biggest earthquake which occurred
in this area since the Lambesc destructive one of 1909. The tele-
metered seismological network installed by the IPG of Strasbourg
one year ago has allowed a detailed study of this event. The
focal depth is less than 2 km and the epicenter is located
within the coal mine of Gardanne (Bouches du Rhône). The focal
mechanism corresponds to a movement on a quasi-vertical E-W fault,
indicating a subsidence of the northern compartment. So , we are
at the limit of the inverse-normal case and this focal solution
does not inscribe very well in the global tectonic scheme of this
region. Furthermore, by a spectral analysis of aftershocks, it
was shown that the nature of corresponding signal is very close
to the one of a tectonic earthquake. Hence, although the contradiction between the focal mechanism and the regional stress
state, and the hypocentral location which are in favour of an
event associate to the mine exploitation, the February 19, 1984
shock was not a mine collapse.
It seems probably that this event takes its origin in a
local modification of the regional stress field, modification
induced by the mine exploitation.

Introduction.

Le 19 février 1984 à 21 h 14 (T.U.), un tremblement de
terre de magnitude 4.5 a secoué la Provence occidentale. Ce

P. Melchior (ed.), Seismic Activity in Western Europe, 223–232.

séisme, largement ressenti de Nice à Montpellier et de Marseille
à Digne, est le plus important qui se soit produit dans cette
région depuis l'événement destructeur de Lambesc de juin 1909.
Survenant dans une région à faible sismicité naturelle mais où
l'exploitation de la lignite du bassin de Gardanne provoque de
fréquents coups de terrain minier, la question de son origine,
naturelle ou artificielle, s'est alors posée immédiatement.

Situation tectonique.

      A l'Oligocène différents types de déformation se sont
succédés dans le domaine subalpin et provencal. Alors que les
Alpes méridionales externes étaient soumises à un régime compressif, une distension intracontinentale généralisée s'étendait
dans toute la Provence et le Languedoc. Au Miocène, le domaine
en compression limité jusque là au domaine subalpin interne, se
déplace vers l'ouest, le sud-ouest et le sud, en direction de la
Procence. Au fini-Miocène Pliocène, ce régime compressif s'in-
tensifie et couvre presque tout le domaine provencal. La Provence
est soumise à une déformation décrochante compressive compatible
avec une direction de raccourcissement horizontale subméridienne.
En particulier dans la région comprise entre le Lubéron et la
côte marseillaise, divers accidents tectoniques sont présents :
- les failles inverses et les chevauchements orientés approxima-
tivement E-O tels ceux du Lubéron, de la Chaîne des Costes, de
la Trévaresse, de la Sainte Victoire et du chaînon de l'Etoile,
et en arrière de celui-ci, la faille inverse de la Diote.
- les décrochements dextres orientés NO-SE et N-S comme celui
de Salon-Cavaillon, et senestres orientés NE-SO tel celui de la
Durance. En l'absence d'information précise, les caractères
tectoniques quaternaires dans cette région sont extrapolés de la
période précédente mio-pliocène (P. Combes, 1984).

Etat de la sismicité.

      Cette région est jalonnée dans le passé par les séismes
destructeurs d'Avignon (1769), Carpentras (1738), Beaumont de
Pertuis (1812), Manosque (1708), Chasteuil (1855, 1951) et de
Lambesc (1227, 1909). La Provence est ainsi considérée comme
l'une des zones les plus exposées aux risques sismiques de
France. Pendant longtemps la microsismicité de cette région a
été mal connue, c'est pourquoi l'Institut National d'Astronomie
et de Géophysique, en association avec l'Institut de Physique du
Globe de Strasbourg, a installé un réseau sismologique télémétré
opérationnel depuis novembre 1982 (fig.1). Parmi les événements
sismiques locaux antérieurs au 19.02.1984 et enregistrés par ce
réseau, quelques secousses de magnitude inférieure à 3 sont
localisées dans la région de Manosque, c'est-à-dire sur le
décrochement de la Durance, tous les autres ont leur épicentre

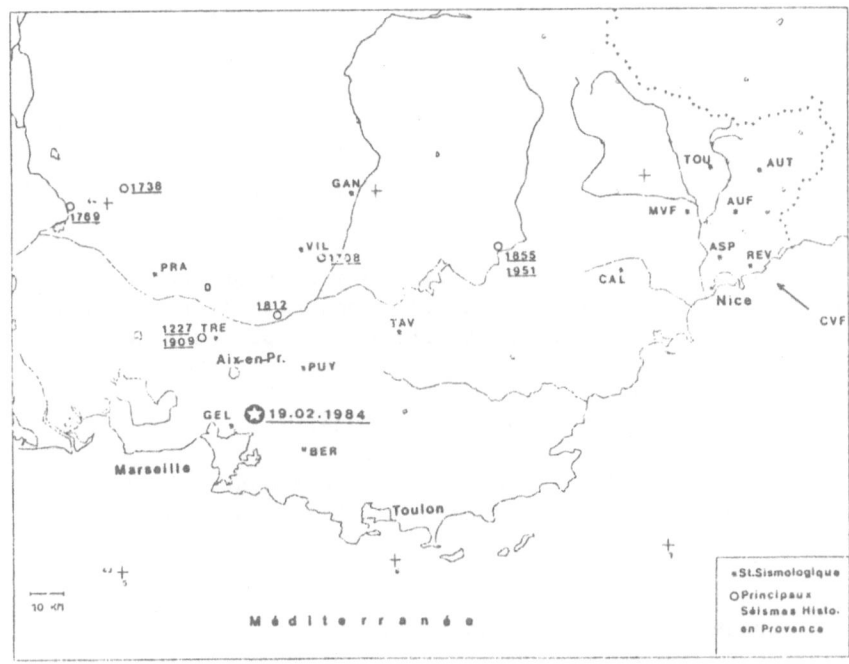

Fig. 1 : Les réseaux télémétrés de Provence et de la
            Côte d'Azur.

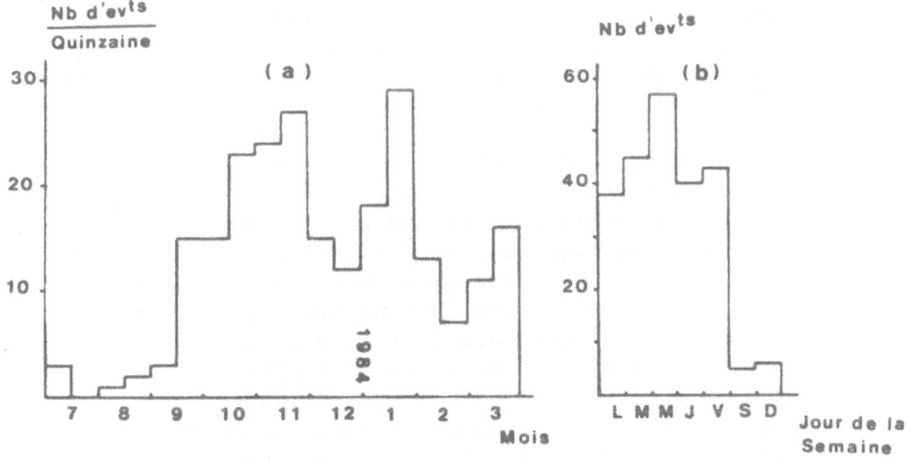

Fig. 2 : Histogrammes de la sismicité du bassin de
           Gardanne pour la période juillet 1983-mars 1984.

dans le bassin de Gardanne. Pour ceux-ci, les sismogrammes présentent un aspect typique par leur contenu spectral basse fréquence et par le développement remarquable des ondes de surface, ils sont ainsi reconnus depuis longtemps par la station sismologique de Cadarache située à 35 km comme étant dus à des coups de terrain minier liés à l'exploitation de la lignite. J.P. Rothé signale dès 1938 dans les Annales de l'Institut de Physique du Globe de Strasbourg l'existence de tels phénomènes. Par ailleurs, J. Vogt (1983) rappelle que certains de ces chocs sont régulièrement ressentis par la population. L'analyse statistique d'une série récente (juin 83 - février 84) montre que ces événements sont directement liés à l'exploitation puisqu'ils se produisent très rarement durant les week-ends et les périodes de congés (fig.2).

L'événement sismique du 19 février 1984.

        Les paramètres focaux de cet événement tels qu'ils ont été établis à partir des données du réseau local sont les suivants :
        latitude   : 43.41 N
        longitude   :   5.51 E
        heure origine : 21:14:36.57 T.U.

L'incertitude est de l'ordre de 500 m pour les coordonnées horizontales. Celle de la profondeur est un peu plus élevée car son calcul dépend fortement du modèle de vitesse utilisé. De toutes manières, quel que soit le modèle adopté dans nos calculs, la valeur obtenue pour la profondeur reste inférieure à 2 km. La magnitude de cet événement a été estimée à 4.5. Pendant les 12 heures qui ont suivi le choc principal, on a enregistré seulement deux répliques, le 20 février à 3 h 34 et à 8 h 49 T.U. (M = 3). Leur épicentre coincide aux erreurs de calcul près avec celui du choc principal. Ce faible nombre de répliques est inhabituel pour un séisme de cette magnitude.
        La carte macrosismique (fig.3) établie par J. Vogt indique une aire pléistoséiste correspondant à l'intensité V+. Cette aire englobe l'épicentre instrumental donné par le réseau local. Les isoséistes IV et V présentent une dissymétrie caractérisée par une extension importante vers le nord par rapport à la région épicentrale. L'origine de cette dissymétrie est à rechercher davantage dans le mécanisme à la source du séisme que dans la variété des terrains géologiques. Remarquons également l'étendue très grande de l'aire macrosismique pour un séisme de cette magnitude et de cette profondeur.

Mécanisme focal.

        La figure 4 représente les sens des premiers mouvements

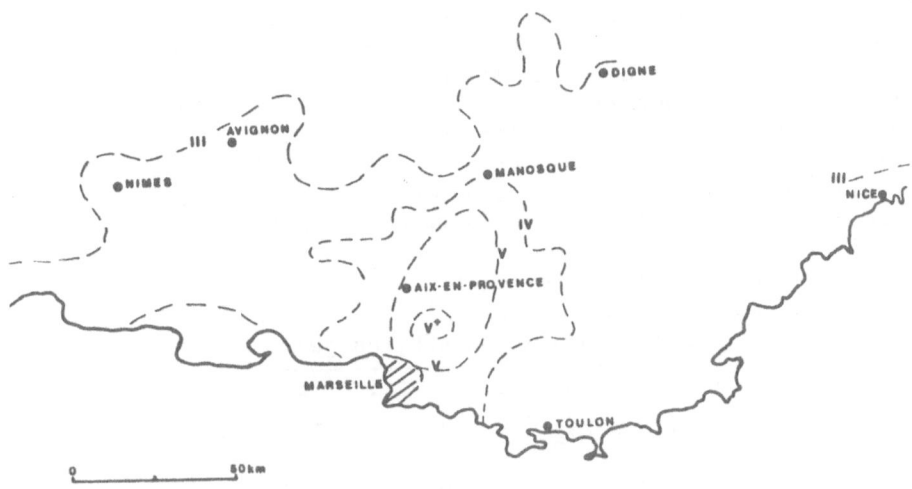

Fig. 3 : Carte macrosismique de l'événement du 19.02.1984.

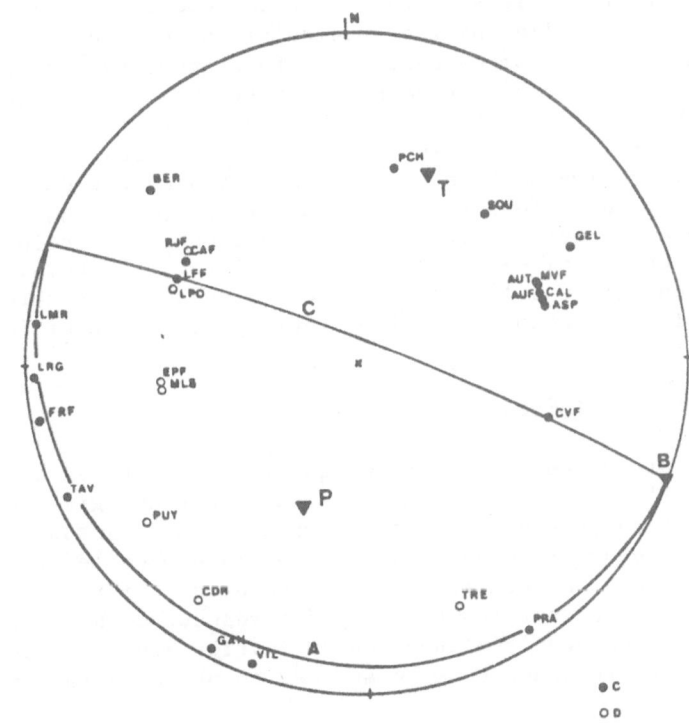

Fig. 4 : Solution focale de l'événement du 19.02.1984.

portés sur l'hémisphère focal inférieur, en projection stéréographique de Schmidt. Les lectures ont été effectuées sur les sismogrammes des stations proches et correspondent à des ondes Pg ou
Pn. Les ondes Pg traversant l'hémisphère supérieur sont représen-
tées par leurs symétriques par rapport au centre de la sphère. La
profondeur du foyer est suffisamment contrainte pour que l'incertitude sur l'incidence des ondes Pg soit faible. Grâce à la
qualité des données et à la bonne couverture azimutale, les plans
nodaux sont bien définis, le score étant de 96%.

Les paramètres du mécanisme sont :

	Plans Nodaux		Axes Principaux		
	A	C	P	T	B
Azimut	113	293	203	23	113
Pendage	5	85	50	40	0

Le vecteur glissement est quasiment vertical sur un plan à très
fort pendage nord. La solution indiquée correspond à une faille
normale. Cependant, elle se trouve dans la zone de transition
faille normale - faille inverse. Quoiqu'il en soit, distention ou
compression, ce mécanisme traduit un affaissement suivant un plan
quasi-vertical E-0 du bloc nord par rapport au bloc sud. Ce qui
pourrait expliquer l'extension dissymétrique vers le nord de
l'isoséïste IV-V. Ce mécanisme ne reflète pas le régime compressif
admis pour la région dans un cadre global et à partir de l'extrapolation à l'actuel d'observations mio-pliocène. Par contre il
semble en accord avec les mesures de nivellement qui indiquent
un affaissement au niveau du bassin de Gardanne, (P. Combes, 1984).
L'état de contrainte à l'origine de ce séisme pourrait être dû à
une combinaison du champ tectonique régional et de celui induit
par l'exploitation minière.

Examen du signal sismique.

   Les sismologues identifient qualitativement les coups de
terrain par l'examen de la forme du signal. Les enregistrements
correspondant au séisme du 19 février sont saturés jusqu'à 200 km
de distance, ce qui rend cet examen impossible pour le choc principal. Nous avons par contre effectué une étude comparative des
signaux enregistrés par les stations des réseaux de Nice et de
Provence pour les deux répliques du 20 février ainsi que pour
deux événements reconnus comme étant des coups de terrain miniers.

   Le tableau suivant donne les paramètres de ces quatre
événements. Nous avons choisi des événements de magnitude compara-
rable afin de s'affranchir des problèmes de facteur d'échelle.

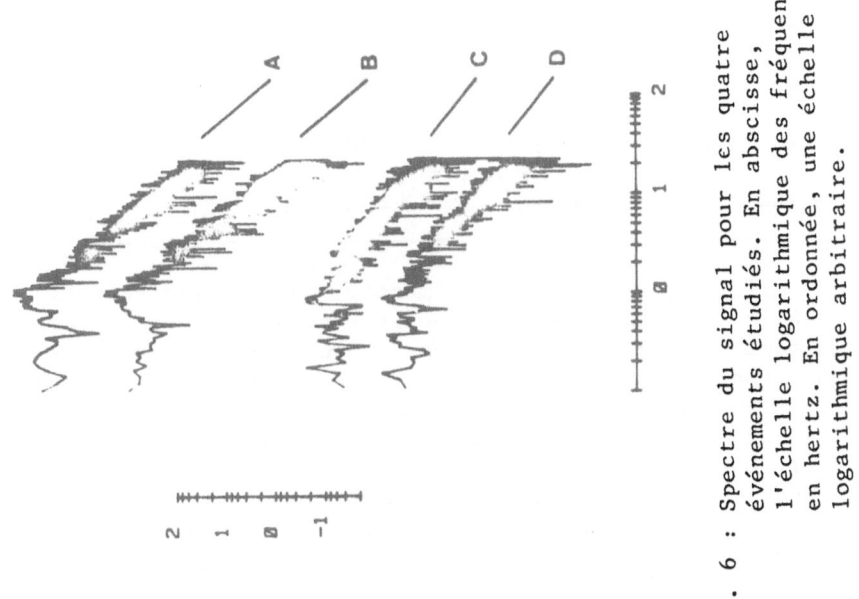

Fig. 6 : Spectre du signal pour les quatre événements étudiés. En abscisse, l'échelle logarithmique des fréquences en hertz. En ordonnée, une échelle logarithmique arbitraire.

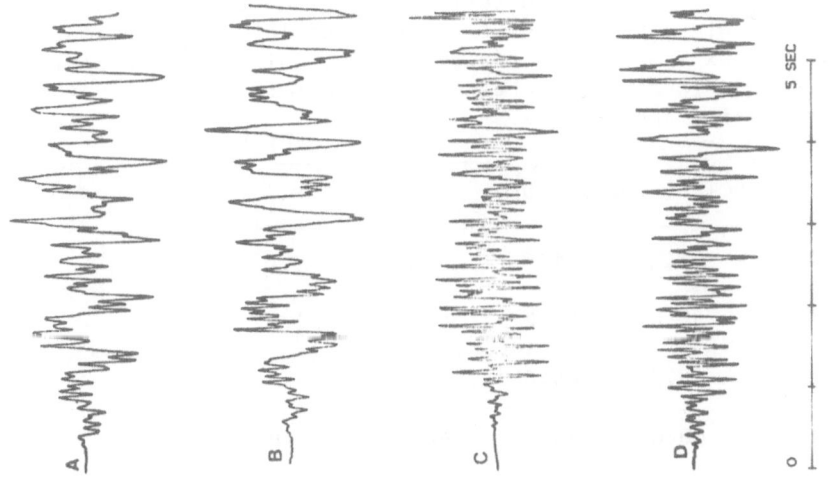

Fig. 5

	DATE	HEURE O.	LAT.	LONG	M
A	22/03/83	20:44:19.66	43.43N	5.50E	3.5
B	12/04/83	11:18:34.87	43.43N	5.47E	3.1
C	20/02/84	03:34:49.98	43.41N	5.51E	3.0
D	20/02/84	08:49:09.90	43.42N	5.52E	2.8

La figure 5 représente un montage des enregistrements obtenus à la station TRE pour les quatre événements concernés. On peut remarquer "de visu" que les signaux correspondant aux répliques sont plus riches en hautes fréquences que ceux qui se rapportent aux effondrements. Afin de préciser ces éléments qualitatifs nous avons procédé à une analyse spectrale de l'ensemble du signal. Les spectres ont été calculés à partir des échantillons de 45 secondes de signal numérisés à 45.45 points par seconde, soit au total 2048 points. Dans le domaine de fréquences considéré la réponse instrumentale est constante, ce qui nous permet de ne pas tenir compte de la correction d'appareillage. Ces spectres (fig.6) présentent une forme classique, c'est-à-dire, un niveau basse fréquence plat et une décroissance aux hautes fréquences. Les événements considérés sont localisés pratiquement au même endroit et enregistrés à une même station, si bien que l'effet du trajet est identique dans les deux cas. Les éventuelles différences doivent donc être imputées à la nature de la source. La comparaison porte sur la pente des spectres aux hautes fréquences et on trouve dans le tableau suivant le rapport (pente des répliques)/(pente des coups de terrain). L'événement A n'est pas enregistré à Nice. Remarquons toutefois que les rapports sont plus élevés aux stations du réseau de Nice qui sont plus éloignées de la région épicentrale. Cette différence est un effet typique de l'atténuation anélastique.

PENTE DES SPECTRES (DB/DECADE)

Evts	GAN	TRE	TAV	AUT	MVF	CAL
A	56.0	56.0	51.2	60.8	62.4	56.0
B	59.2	49.6	64.0	--	--	--
C	41.6	40.0	33.6	48.0	56.0	53.6
D	48.0	40.0	40.8	47.2	49.6	52.8

RAPPORT DES PENTES

Evts	GAN	TRE	TAV	AUT	MVF	CAL
C/A	0.74	0.71	0.66	0.79	0.90	0.96
C/B	0.70	0.81	0.53	--	--	--
D/A	0.86	0.71	0.80	0.78	0.79	0.94
D/B	0.81	0.81	0.64	--	--	--

Ce rapport est systématiquement inférieur à 1, indiquant que les pentes pour les répliques sont de 20 à 30 % inférieures à celles des coups de terrain minier. Les répliques ont donc un contenu spectral plus riche en hautes fréquences que les coups de terrain minier. Ce caractère dépend fortement de la fonction temporelle à la source, une rupture rapide comme un séisme engendrant plus de hautes fréquences qu'une rupture lente type effondrement.

Conclusions.

Etant donné que les signaux relatifs au choc principal ne peuvent se prêter à une analyse spectrale adequate et que le caractère "hautes fréquences" de ceux-ci peut néanmoins être apprécié de façon qualitative, on pourrait étendre les caractéristiques spectrales des répliques au choc principal. Dans ces conditions l'analyse de la forme du signal nous permettrait de dire que les événements du 19 et 20 février ne sont pas des coups de terrain minier. Leurs caractéristiques sont plus proches de celles de séismes tectoniques. Cependant, la solution focale présentée vient a contrario d'une interprétation évidente du rôle des accidents tectoniques dans la genèse de ces événements. De plus, la profondeur très faible du foyer et l'absence de répliques immédiates, sont plutôt en faveur d'un phénomène lié à l'exploitation minière. Il est par conséquent probable que le séisme du 19 février 1984 ne soit pas d'origine purement tectonique et qu'il puisse en partie résulter d'une modification locale du champ de contrainte, modification induite par l'exploitation minière.

Références.

Combes, P., 1984. La tectonique de la Provence Occidentale, microtectonique, caractéristiques dynamiques et cinématiques. Méthodologie de zonation tectonique et relations avec la sismicité.
Thèse de 3e cycle, Université Louis Pasteur, Strasbourg.

Combes, P., 1984. Compte rendu de mission du 23.02.84 au 29.02.84
        dans la région épicentrale du séisme du 19.02.84 : Analy-
        se du séisme et relation avec le cadre sismotectonique
        régional.
        Note technique SAER N 84/399 CEA.

Vogt, J., 1983. Rapport BRGM 83-RN-SIS-000.081.

THE EARTHQUAKES OF LIÈGE OF NOVEMBER 8, 1983 AND DECEMBER 21, 1965.

T. Camelbeeck and M. De Becker

Centre de Géophysique Interne
Observatoire Royal de Belgique
3, Av. Circulaire, 1180 Bruxelles.

ABSTRACT

The hypocenter of the earthquake of Liège of november 8, 1983 has been calculated using the data from 40 stations at an epicentral distance lower than 500 km :

Latitude	50.63°N
Longitude	5.50°E
Depth	4 km
Origin time	0h49m34.4s

The epicenter is close (2 km) to the more damaged area and to the epicenter of the M_L = 4.3 earthquake of december 21, 1965.
The mean value of the local magnitude determined by the coda duration measurement at the belgian seismological stations is M_L = 5.0 which is comparable to the values given by other networks.
Supposing the source being a circular crack, the radius has been estimated to 1200 m, the stress drop to 40 bars and the seismic moment 1.5 10^{23} dyne.cm.

1. INTRODUCTION

On november 8, 1983 at 0h49m an earthquake of moderate size (M_L = 5.0) was strongly felt in the region of Liège (Belgium). The cities of Saint-Nicolas, Glain and Montegnée have been particularly affected (intensity VII on the M.S.K. scale). The main shock was followed by some events of low magnitude. Three of them (having occurred on november 8) can be immediately associated with the main shock of 0h49m.

233

P. Melchior (ed.), Seismic Activity in Western Europe, 233–248.
© *1985 by D. Reidel Publishing Company.*

The table 1 contains the list of all the seismic events located in the Liège area since november 8, 1983 and studied using
the data from the belgian seismological stations.

Table 1. Seismic events localised in the Liège area since novem-
         ber 8, 1983.

Date	Origin time (TU)	Local magnitude (belgian network)
nov. 8 ,1983	0h49m	5.0
	0h55m	2.7
	1h25m	2.9
	2h13m	3.4
nov. 20,1983	8h03m	2.2
dec. 1 ,1983	21h53m	2.7
dec. 15,1983	15h17m	2.1
jan. 5 ,1984	12h38m	2.6
jan. 15,1984	13h22m	2.2
mar. 7 ,1984	16h	2.6
may 1 ,1984	11h31m	2.6

2. LOCATION

The location of the earthquakes of november 8, 1983 at
0h49m, 1h24m and 2h13m has been performed with the data provided
by 40 stations with an epicentral distance lower than 500 km.
Only P waves arrivals are considered. The crustal model used is
composed of a 30 km thick layer. The P-wave velocity is 6.1 km/s
in the crust and 8.1 km/s in the upper-mantle. In table 2 the
location of the different seismic events, the data used and the
temporal residues (O-C) for each data are indicated.
The error ellipses (of the epicenters) are calculated using
the χ_2^2 distribution at a confidence level of 95 % supposing the
standard deviation of each arrival time is 0.5 second.
These locations are featured in figure 1.

Table 2. Location of the earthquakes of nov. 8, 1983 in the Liège
area (from the preliminary report of the Royal Observatory of Belgium).

Principal event : The localisation was performed using 49 data from stations
with an epicentral distance lower than 500 km.

Latitude   50.63° N ± 0.60 km

Longitude   5.50° E ± 0.80 km

Depth   4 km   ± 2 km

Origin time   0h 49m 34.4s ± 0.2s

Residual   0.68s

Error ellipse data (95 % confidence) : semi major axis length   0.75 km
                                        semi-minor axis length   0.52 km
                                        azimuth of semi-major axis 145°

STATION	Δ°	Az°	1st Arrival		(O-C)	2nd Arrival		(O-C)
ENN	.30	62.8	$00^h49^m40^s2$	(Pg)	.26			
MEM	.32	93.5	40.5		.08			
JUE	.64	63.4	46.9		.63			
UCC	.74	283.5	48.1		-.03			
DOU	.79	227.7	49.0		-.10			
WLF	1.05	156.3	55.5		-.19			
BNS	1.11	71.9	55.0		-.01			
BGG	1.25	109.1	57.5	(Pn)	-.11			
ELG	1.25	109.1	57.8		.19			
BIR	1.37	132.2	59.5		-.35			
KOE	1.44	97.3	59.9		-.99			
DBN	1.49	352.3	50 01.5		.10			
ABH	1.51	118.9	01.4		-.54			
WTS	1.59	30.5	02.1		-.82			
KLT	2.13		09.5		-.80			
ROS	2.82	112.3	10.9		-.74			
WIT	2.30	17.9	16.1	(Pg)	1.18			
TRO	2.36	114.5	12.8	(Pn)	-.77			
HEI	2.42	119.4	13.8		-.57			
CDF	2.50	151.8	15.0	(Pn)	-.57	21.0	(Pg)	.09
BUH	2.65		17.6		.17			
HAU	2.68	167.8	17.0	(Pn)	-1.07	24.5	(Pg)	.22
BSF	2.92	162.7	20.7	(Pn)	-.67	29.1	(Pg)	.36
BFD	2.95	140.2	21.6		-.08			
STU	3.03	126.4	22.1		-.74			
FEL	3.21	148.2	24.9		-.40			
SLE	3.47	144.5	28.7		.12			
LOR	3.53	198.5	29.0		-.71	40.4		.40
ZUL	3.84	146.9	32.1		.49			
LBF	3.78	196.0	32.1		-1.05			
SSF	3.80	201.0	32.6		-.85	45.5		.46
AVF	4.09	201.1	36.2	(P.)	-1.24	50.8	(Pg)	.37
SMF	4.13	196.1	36.7	(Pn)	-1.28	51.6	(Pg)	.43
LDF	4.18	243.0	38.6	(Pn)	0.00	52.4	(Pg)	.40
SAX	4.22	141.7	40.4		1.20			
GRF	4.23	110.4	39.4		.20			
FLN	4.31	246.6	40.2	(Pn)	-.09			
LLS	4.43	147.2	43.5		1.40			
BGF	4.44	204.4	41.3		-.83	57.9		1.13
FUR	4.50	120.9	7		.97			

Second aftershock (11 data)

Nov. 8, 1983

Latitude   50.65° N ± 1.80 km

Longitude   5.55° E ± 2.00 km

Depth   0 → 4 km (not accurate determination)

Origin time   1h 24m 55.2s ± 0.4s

Residual   0.69s

Error ellipse data (95 % confidence) : semi-major axis length   2.03 km
                                        semi-minor axis length   0.89 km
                                        azimuth of semi-major axis 135°

Station	Δ°	Az°	1st Arrival		(O-C)	2nd Arrival		(O-C)
ENN	.26	63.6	$1^h25^m00.5^s$	(Pg)	.39			
MEM	.29	97.9	00.5	(Pg)	-.20			
DOU	.83	228.2	10.6	(Pg)	.08			
WTS	1.56	29.9	23.0	(Pn)	.14			
HAU	2.70	168.5	38.0	(Pn)	-.55	25 45.4	(Pg)	.14
LDR	3.56	198.9	50.7	(Pn)	.30	26 02.6	(Pg)	1.34
LBF	3.81	196.4	26 05.7	(Pg)	-.19			
SSF	3.83	201.4	25 53.1	(Pn)	-1.05	06.0	(Pg)	-.32

Third aftershock (21 data)

Nov. 8, 1983

Latitude   50.65° N ± 0.90 km

Longitude   5.48° E ± 1.20 km

Depth   4 km   ± 2 km

Origin time   2h 13m 22.2s ± 0.1s

Residual   0.49s

Error Ellipse data (95 % confidence) : Semi-major axis length   1.13 km
                                        Semi-minor axis length   0.75 km
                                        Azimuth of semi-major axis 147°

Station	Δ°	Az°	1st Arrival		(O-C)	2nd Arrival		(O-C)
ENN	.30	67.2	$2^h13^m26.3^s$	(Pg)	.40			
MEM	.34	96.8	28.2	(Pg)	-.32			
UCC	.73	282.2	35.4	(Pg)	-.21			
DOU	.79	226.0	36.8	(Pg)	-.06			
BNS	1.12	73.1	43.7	(Pg)	-.18			
BGG	1.27	109.8	45.5		-.24			
WTS	1.58	31.3	50.8	(Pn)	.57			
CDF	2.53	151.7	14 03.3	(Pn)	.14	14 08.8	(Pg)	-.24
HAU	2.70	167.6	04.6		-1.02	12.3		-.06
BSF	2.95	162.5	08.8		.14			
SLE	3.49	144.5	17.0		.52			
LDR	3.55	198.1	17.0		-.16			
LBF	3.80	195.7	20.2		-.41			
SSF	3.82	200.7	20.5		.04	33.0		-.38
AVF	4.11	200.8	24.2		1.16	39.5		-.67
SMF	4.15	195.8	24.9		.49	39.6		-.54

Figure 1.

The results are summarized and compared to the location of the
earthquake of dec. 21, 1965 in Table 3.

Table 3. Location of the Liège earthquakes of nov. 8, 1983 at
         0h49m, 1h24m and 2h13m and of dec. 21, 1965 at 10h.

	8.11.1983 main	8.11.1983 2nd aft.	8.11.1983 3th aft.	21.12.1965
Latitude	50.63°N±0,60km	50.65°N±1,80km	50.65°N±0,90km	50.62°N
Longitude	5.50°E±0,80km	5.55°E±2,00km	5.48°E±1,20km	5.53°E
Depth	4 km ± 2 km	0 → 4 km	4 km ± 2 km	0 → 4 km
Origin time(TU)	0h49m34.4s±0.1s	1h24m55.2s±0.4s	2h13m22.2s±0.1s	10h0m1.3s

The map of the epicentral area (figure 1) makes conspicuous
the link between the earthquakes of november 8, 1983 and december
21, 1965. The respective position of their macroseismic epicenters
(more damaged zone) in respect to the apparent trace in surface
of the geological faults is an argument to attribuate their origin
to a slippage in the faulting zone associated with the Saint-Gilles
fault.

3. FOCAL MECHANISM

Two kinds of fault-plane solutions are in agreement with the
P-wave first-motion data. The solutions presented here have been
calculated a few days after the main shock with the data available
at this time. More reliable solutions were obtained using all the
data from the nearest stations by Ahorner {1} and Bonjer {2}.

Solution A (figure 2) representents a strike-slip faulting
and is comparable to the solution of Ahorner showing a compressive
stress axis compatible with the regional tectonic stress in Central Europe {3}.

Solution B (figure 3) is a dip-slip solution which is close
to that formulated by Bonjer and confirmed by S. Faber {4} using
theoretical seismogram.

A more complete discussion of the focal mechanism is given
elsewhere {5}.

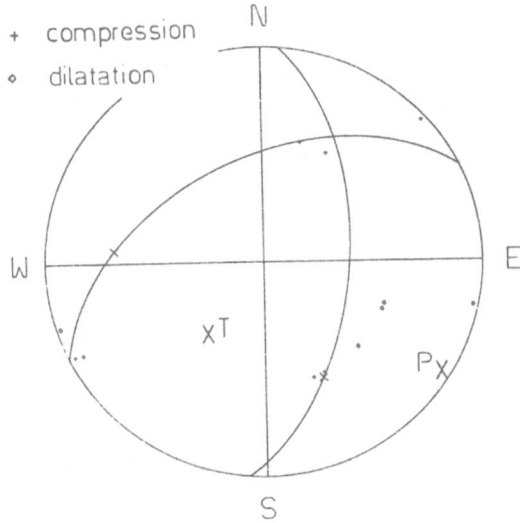

Figure 2. Focal mechanism for the Liège earthquake of november 8,
　　　　　1983　　0h49m　　Solution A.

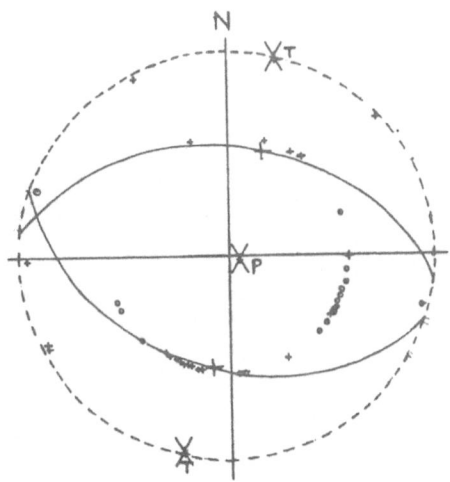

Figure 3. Focal mechanism of the earthquake of Liège of november
　　　　　8, 1983　　0h49m　　Solution B.

4. MAGNITUDE

     The main shock was too important to be recorded without satu-
ration on the paper recorders of the belgian stations. Thus, the
magnitude has been determined by the coda duration measurement on
the vertical short-period seismometers at the different stations.
     The relation between coda duration t (time since the first
arrival untill the amplitude of the decreasing coda becomes
twice the amplitude of the normal noise) and local magnitude is :

M_L = 2.50 logt - 1.67 for the vertical Benioff short-period seis-
     mometer recorded on the Helicorder at the station Dourbes
     (figure 4). The relation has been established with the data
     from near-earthquakes recorded between 1978 and 1982. No
     clear dependence of the distance appears up to 400 km.

M_L = 1.90 logt - 0.14 for the vertical Johnson-Matheson seismome-
     ter recorded on the Texas recorder at the station Membach.
     The formula is valuable to distances up to 200 km.

M_L = 1.78 logt + 0.37 for the vertical Johnson-Matheson seismome-
     ter recorded on the Brown recorder at the station Uccle
          (valid to distances up to 200 km).

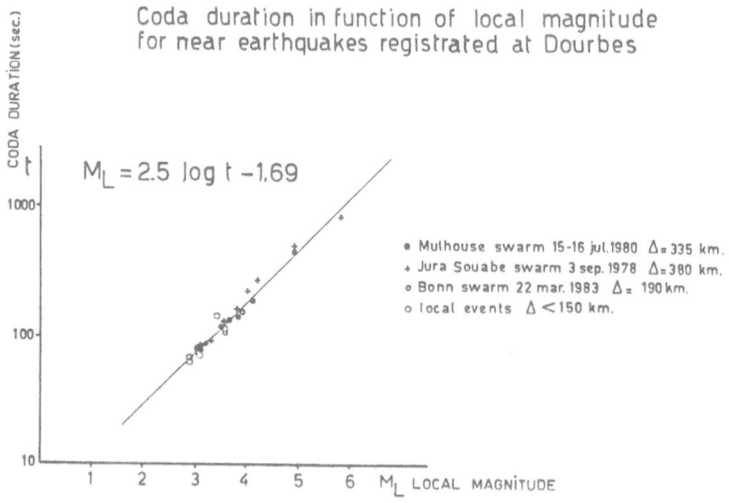

Figure 4. Coda duration in function of local magnitude on the
          vertical Benioff seismometer at the station Dourbes.

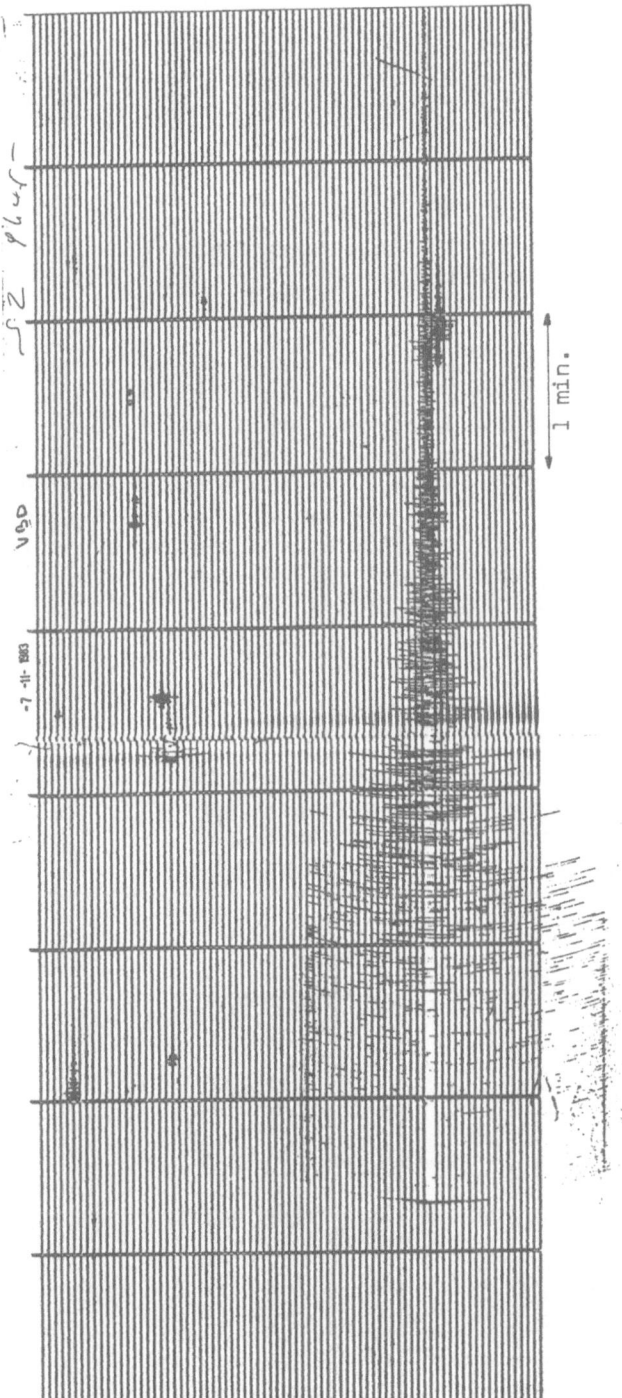

Figure 5. Seismogram from the station Dourbes (vertical Benioff short-period seismometer) of the earthquake of Liège of november 8, 1983 : 0h49m . The coda duration is t ≃ 460 s.

The measurements of t are :

t	=	460s	at	Dourbes	M_L	=	4.97
t	=	525s	at	Membach	M_L	=	5.02
t	=	420s	at	Uccle	M_L	=	5.03

The seismogram from the station Dourbes is given on the figure 5. The mean value of the magnitude is 5.0

The U.S.C.G.S. gives m_b = 5.0 (9 stations) whereas M_s is 4.3 (N.N. Ambraseys - personal communication).

5. SOURCE PARAMETERS

The seismograms from the broad-band stations GRF (Broad-band velocity .05 → 5 Hz) and KHC (broad-band velocity .01 → 3 Hz) indicate clearly (figure 6) that the main shock can be considered as a point source. The rise time of the P_n-wave is near 0.5 second in each case. Hence, an estimation of the source radius is possible using the pulse rise time measurement on the short-period seismograms at stations close to the epicenter.

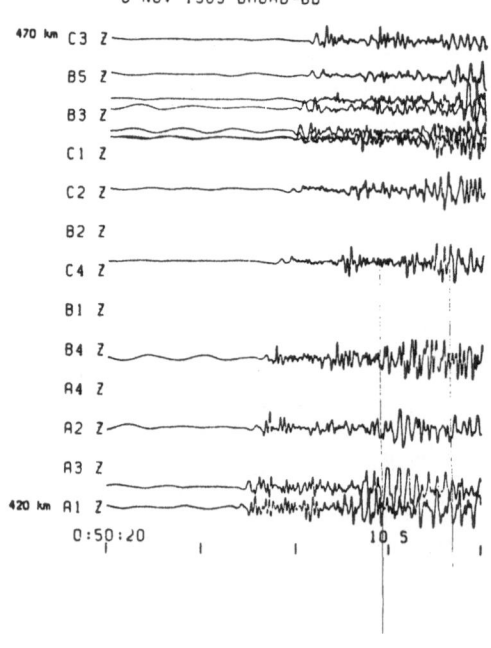

Figure 6a. Seismograms from the broad-band stations of the GRF array.

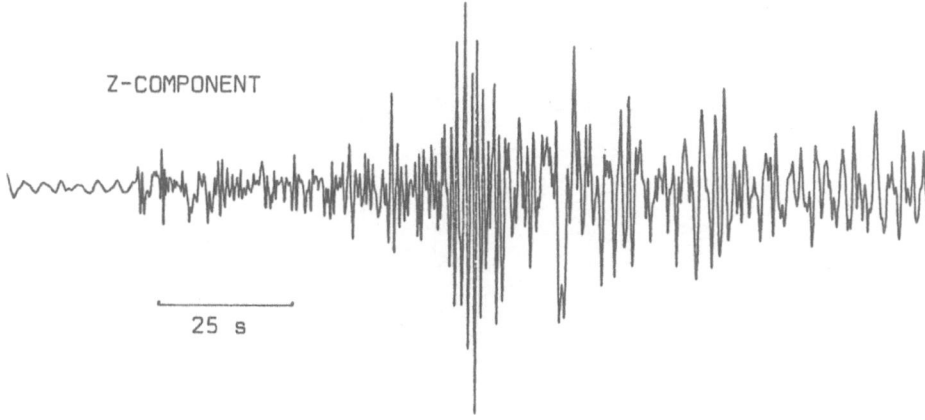

Figure 6b. Seismogram from the broad-band station KHC.

Boatwright (1) gives a relationship for a circular crack between
the pulse rise time τ_{12}, the source radius R rupture velocity v_r
and take off angle of the ray θ :

$$\tau_{12} = \frac{(13 - 12\,\xi)}{16}\,\frac{R}{v_r} \quad \text{with} \quad \xi = \frac{v_r}{v}\,\sin\theta \text{ (v is the S-wave velo-city).}$$

Averaging over all the focal sphere and supposing v_r = 0.8 V ; the
following relation has been obtained

$$R = 5.38 \quad \tau_{12} \qquad \begin{array}{l} R \text{ in km} \\ \tau_{12} \text{ in S} \end{array}$$

To give an estimation of R, the rise time has been determined with
the data from stations where it is possible to read with accuracy
the time scale (table 2 - figure 7). Considering seismometers as
ideal velocity-meters, only the wave attenuation must be corrected.
The relation between the time τ_{sis} from the first break to the
first zero crossing on the seismogram and τ_{12} is

$$\tau_{sis} = \tau_{12} + 0.5 \int_0^{\Delta} \frac{d\,X}{Q_\alpha\,v_\alpha} \quad \text{with} \begin{array}{l} Q_\alpha : \text{quality factor of } \alpha \text{ wave} \\ v_\alpha : \text{velocity of wave } \alpha \end{array}$$

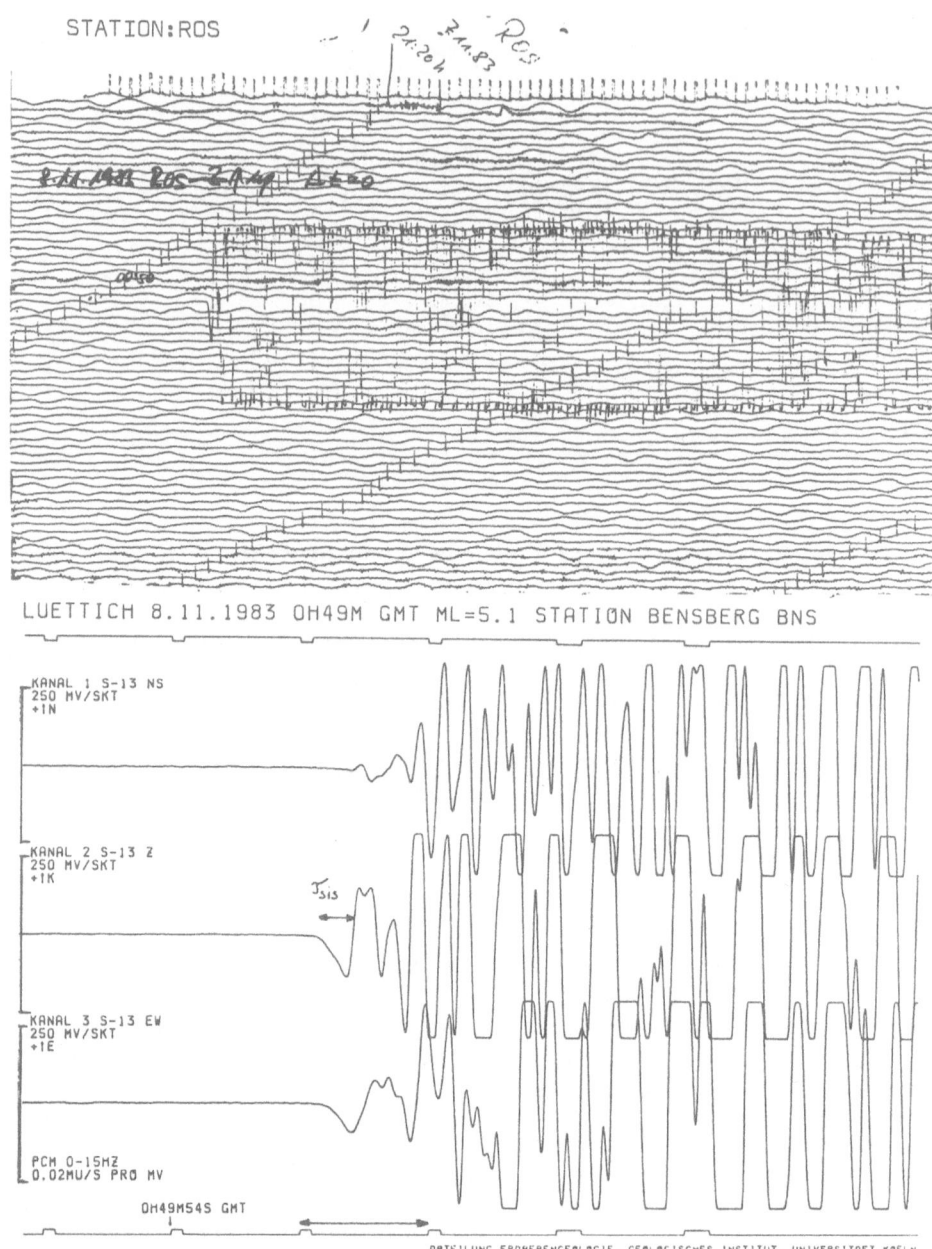

Figure 7. Seismograms of the main shock at the stations ROS
          and BNS.

Table **4.** Rise time determination at some stations close to the
epicenter.

STATION	Δ(km)	arrival	Q	τ_{sis}(s)	τ_{12}	R
ELG	139	Pg	500	0.26	0.24	1290 m
BNS	123	Pg	500	0.28	0.26	1390 m
BIR	152	Pn	500	0.45	0.43	2310 m
ROS	313	Pn	500	0.5	0.45	2420 m

Two data sets figure in table **4** : ELG and BNS are stations with
digital recorders whereas the measurement of τ_{sis} at stations BIR
and ROS are made on the data from paper recorders.
Certainly an upper bound of R is 2,4 km but a more reasonable estimation is comprised between 1,2 and 1,4 km.
The relation (figure 8) between local magnitude M_L , source radius R
and stress drop $\Delta\sigma$ for earthquakes in Hainaut {7} can be applied
to the earthquake of Liège. (The source characteristics and the
geological structure being approximatively the same in the two
zones).

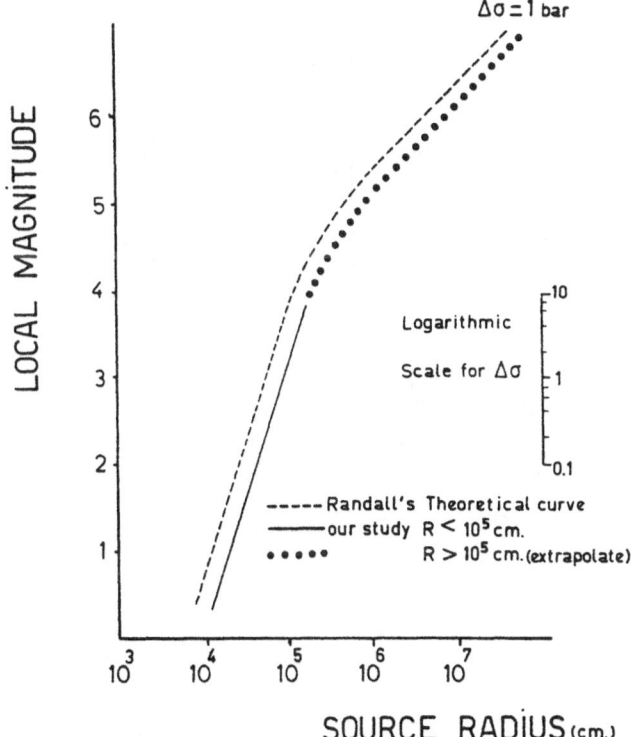

Figure 8. Relation between source parameters and local magnitude
for belgian earthquakes.

The stress drop calculated is 40 bars. The seismic moment
is 1.5 10^{23} dyne.cm and the average dislocation on the fault plane
9 cm.

6. SEISMIC SEQUENCE

The data from the station MEM indicates that four small
events of low magnitude ($M_L \sim 2.2$) have occurred between
october 20, 1983 and the date of the main shock. The lack of information about them makes impossible a detailed study.

The sequence of earthquakes located in the Liège area since
november 8, 1983 is presented in table 1. The three aftershocks
of november 8 can be directly associated with the principal event.
Their location is given in table 2. The magnitude has been determined using the Lg-wave amplitude measurement at a period of one
second on the vertical short-period seismometers from the belgian
stations (table 5).

Table 5. Magnitude determination for the aftershocks of november
         8, 1983 (A_{max}(Lg) is expressed in μm).

EVENT STATION	Nov. 8 0h55m A_{max}(Lg)	m_{bLg}	Nov. 8 1h24m A_{max}(Lg)	m_{bLg}	Nov. 8 2h13m A_{max}(Lg)	m_{bLg}
DOU	0.12	2.7	0.19	2.9	1.29	3.6
MEM	0.44	2.8	0.48	2.9	1.03	3.2
UCC					0.96	3.5
m_{bLg}		2.7		2.9		3.4

The presence of Rg-waves (figure 8) in regional short-period records for some of the events having occurred since november 20,
1983 indicates clearly their shallowness {8} (probably lower than
1 km).

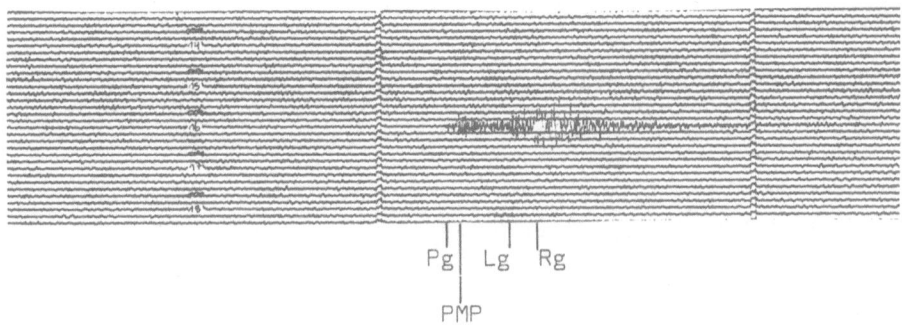

Figure 8. Seismogram from the station Dourbes of the earthquake
          of march 7, 1984 (16h).
          The different phases are indicated.

7. CONCLUSIONS

     The earthquake of november 8, 1983 in the Liège area has
shown the vulnerability of the highly industrialised and populated
regions of Northwestern Europe to an event of moderate size
(M_L = 5.0 - M_S = 4.3). This dramatic experience indicates the
importance of a coherent study of the seismic hazard in Belgium.
     From the scientific point of view some remarks can be formu-
lated :
- The earthquake of november 8, 1983 is not an unexpected event.
  The historical and recent (earthquake of dec. 21, 1965) seismicity data show precedents. The position of Liège in a joint position between the two active seismotectonic zones of the Lower
  Rhine Graben and the Brabant Massif is probably at the origin
  of this seismicity.
- The seismic source was of very short extent with a high stress-
  drop. These characteristics explain the high acceleration from
  which indices exist in some particular points in the epicentral
  zone.
- The attenuation of the intensity in the epicentral zone has been
  relatively important which is explained by the shallowness of
  the focus.

ACKNOWLEDGEMENTS

     We thank Dr Prof. Ahorner, Dr Bonjer, Dr Faber and Dr
Plesinger for providing the seismograms from the stations BNS and
ELG, BIR and ROS, GRF and KHC for the earthquake of Liège.

REFERENCES

{1} L. Ahorner, this volume.

{2} K. Bonjer, this volume.

{3} H.S. Liu, Physics of the Earth and Planetary Interiors, volume 32 (1983), n° 2, p. 146-159.

{4} S. Faber, this volume.

{5} Nato Advanced Study Institute Series - Reidel - **1985.** Seismic Activity in Western Europe ... (This volume)

{6} Boatwright - Bull. of Seism. Soc. of America, vol. 70 (1980), pp. 1-28.

{7} T. Camelbeeck, this volume.

{8} M. Both, Physics of the Earth and Planetary Interiors, volume 10 (1975), 369-376.

PHASE RECOGNITION AND INTERPRETATION AT REGIONAL DISTANCES FROM THE LIEGE EVENT OF NOVEMBER 8, 1983

Sonja Faber[1] and Klaus-Peter Bonjer[2]

1) Seismologisches Zentralobservatorium GRF,
Krankenhausstraße 1, 8520 Erlangen, Germany
2) Geophysikalisches Institut, Universität Karlsruhe,
Hertzstraße 16, 7500 Karlsruhe 21, Germany

ABSTRACT

Several dominant phases between P_n and P_g have been correlated in European records of the November 1983 Liège event. An extensive study using theoretical seismograms computed with the reflectivity method resulted in a successful interpretation of these phases and in determining a unique fault-plane solution which had been ambiguous from first motion data.

INTRODUCTION

The present investigation was primarily stimulated by the GRF-records of the November 1983 Liège earthquake. The GRF-array is located on the Franconian limestone-formation in Northern Bavaria and includes nineteen (13 vertical and 6 horizontal) broad-band Wielandt seismometers with a flat velocity response between 0.05 and 5 Hz (Wielandt and Streckeisen, 1982). Digital data are available since 1976. A detailed description of the GRF-array has been published by Harjes and Seidl (1978).

The GRF-data of the P-wave group from the Liège earthquake are displayed in Fig. 1. The stations are ordered from bottom to top according to the epicentral distance, A1 being the closest station at 423 km and C3 the most distant station with an epicentral distance of 475 km. The traces are shifted in time in order to line up the P_n-phase (phase 1) along a vertical line. The broad-band data clearly show a rather complicated P-wave signal. A second phase (2 in Fig. 1) with a higher frequency-

249

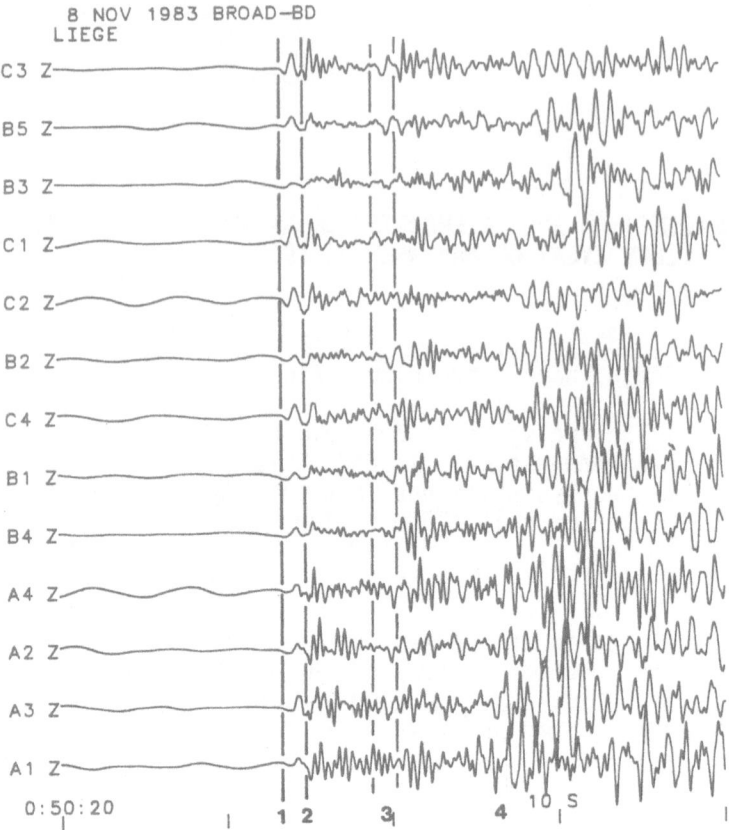

Fig. 1: Broad-band GRF data (vertical components) of the P-wave group from the November 1983 Liège earthquake. The distance ranges from 420 km (A1) to 470 km (C3); the average azimuth is 105°. Several phases (1 to 4) have been identified. Seismogram traces are shifted in time to line up for phase 1 (P_n).

content than the first onset of the P_n (1) appears on all records with variable relative amplitudes; the delay of phase 2 to the first onset is about 1.4 sec. Another dominant phase (3) arrives with a delay with respect to 1 of nearly 6 sec. The signal characteristic of the first two phases is reproduced (1 low frequency compared to 2); this may be especially recognized at station C3. Phase 4 is the crustal P_g-phase. One might think of several explanations for phase 2:

- source activity (aftershock or stopping phase)

- depth phase pP_n or sP_n

- near critical reflection or refraction from lower lithospheric structural contrasts.

In order to determine the nature of the phases in between P_n and P_g numerous records from European stations were collected. First-arrival times and first motions were read which allowed us to determine an accurate hypocenter-location as well as to investigate the focal mechanism. However, the final results were obtained through the computation of theoretical seismograms. The comparison of synthetic and observed P-wave groups for different possible fault-plane solutions enabled us to give a reliable explanation for the observed phases, and to solve the ambiguity of the possible fault-plane solutions which resulted from first-motion studies.

DATA

In order to make a proper interpretation of the phases correlated in the complex P-wave pattern of the GRF-data, it was first of all necessary to determine the hypocenter location and the fault-plane solution of the Liège event. Using a modified version of HYPO 71 (Gelbke, 1978) we determined the following focal parameters:

T_o = 00:49:34.61 UT, DEPTH = 4.0 km

LAT = 50° 38.45'N, LON = 5° 29.40'E.

The epicentral distances and azimuths of the stations used are assembled in Fig. 2. The trouble with the fault-plane determination was that it turned out to be impossible to deduce from first motion data a unique solution. The two possible fault-plane orientations, which are nearly equivalent as far as first motion data are concerned, are shown in Fig. 3, one being nearly a strike-slip (entitled KA2 here), the other one being nearly a dip-slip with normal faulting (KA1).

Most of the data available to us were short-period data (except for some few stations besides GRF). Their higher frequency content compared to the broad-band data of Fig. 1 and thereby the higher resolution of the source process and of small local heterogeneities in the crust, made it sometimes difficult to recognize the dominant phases which we correlated in the GRF-data. Also, it is not possible to determine the apparent velocity of a phase by interpreting single seismograms. To

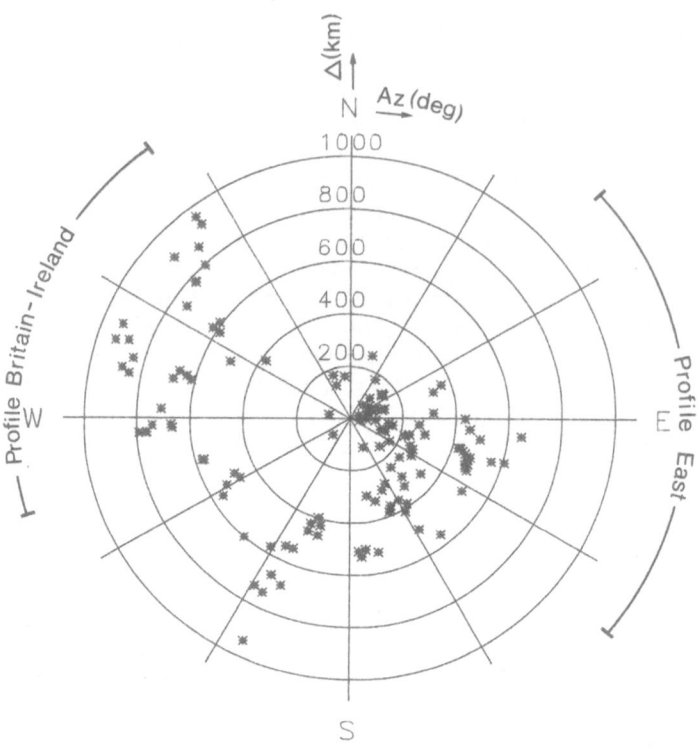

‾Fig. 2: Diagram showing distance (Δ) and azimuth (Az) of the
          stations used in the present study.

overcome these problems we decided to digitize seismograms of
stations in two different azimuthal directions and to construct
seismogram sections for one profile to the east (Fig. 4a) and
one to the west (Fig. 4b). The azimuthal ranges of these two
profiles are marked in Fig. 2. The seismograms plotted in Fig. 4
were shifted in time according to the travel-time of the first
onset through a model (Fig. 6), which for the crustal part
consists of a simplified version of the model used for
hypocenter location, and for the lower lithosphere depth range
represents an average velocity structure for the observed mantle
velocities. It should be mentioned at this point that mantle
velocities vary greatly with azimuth and that detailed veloci-
ty-depth models for different directions are still under
construction. The delays applied to the seismograms in Fig. 4
correct for lateral velocity variations and line up P_n. This
technique is permitted here, since we do not investigate

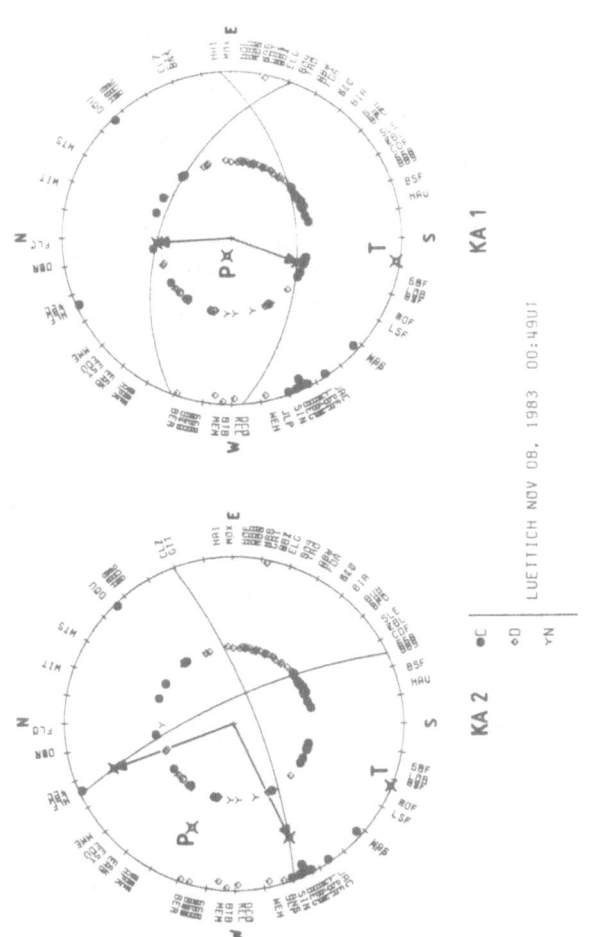

Fig. 3: Possible fault plane solutions for the Liêge earthquake deduced from first-motion investigations. The degree of inconsistency is comparable small for both solutions, called KA1 and KA2 here.

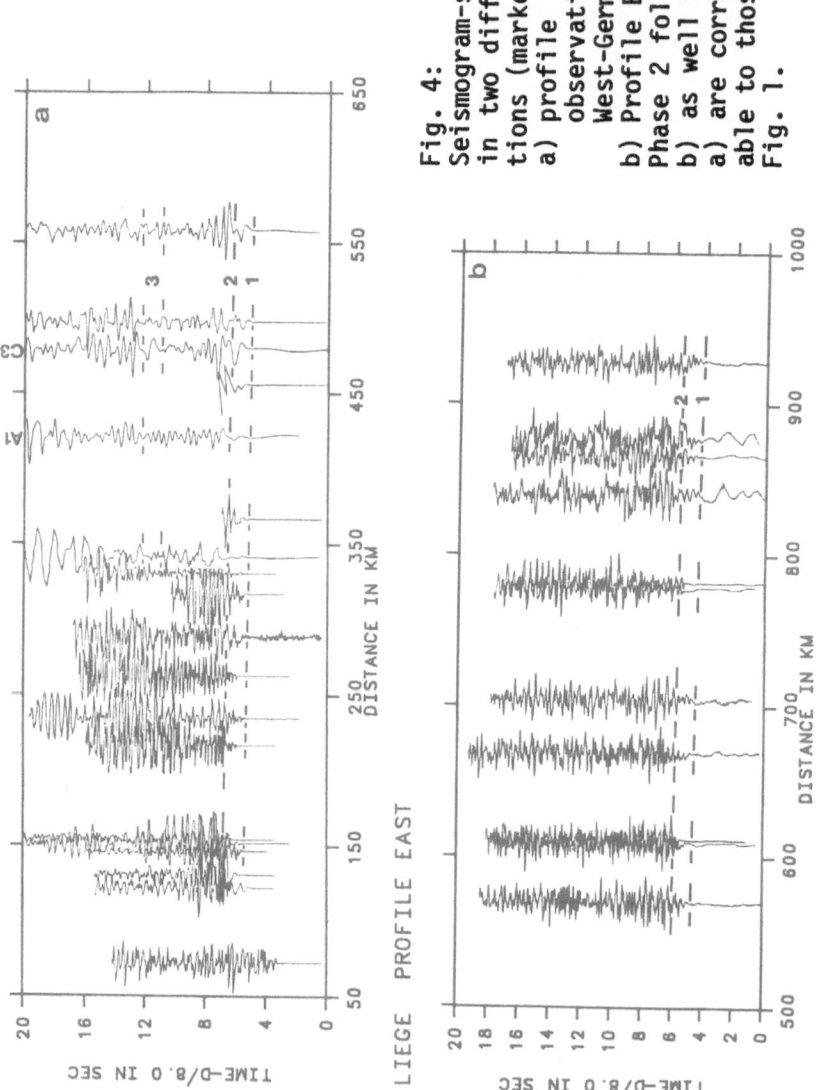

Fig. 4:
Seismogram-sections including data in two different azimuthal directions (marked in Fig. 2):
a) profile East includes only observations from stations in West-Germany.
b) Profile Britain-Ireland
Phase 2 following P (1) in a) and b) as well as the wave-group 3 in a) are correlated and are comparable to those phases correlated in Fig. 1.

absolute travel times in the present study. We are interested in relative times and amplitudes of the correlated phases with reference to P_n. The seismograms were plotted in normalized sections; true amplitude record sections could not be produced since the magnification factors of most analogue recordings were not at our disposal. On the other hand, true amplitude plots could not have resulted in more information, since we used these sections solely for recognition and relative amplitude studies of the dominant phases correlated in Fig. 1.

In both seismogram sections of Fig. 4 phase 2, with the same time delay to $P_n(1)$ as observed in the GRF-data (Fig. 1), and travelling with the same apparent velocity as P_n, can be clearly identified. Since this phase is parallel to P_n over several hundreds of kilometers, even in different azimuthal ranges, the possibility that this phase is a near critical reflection or refraction from lower lithospheric structure can be ruled out. The fact that phase 2 is clearly parallel to phase P_n indicates that this phase is bottoming at the same depth as P_n. In addition to this we have to report that in the whole compilation of the data available for all distances we could not find any clear evidence for interpreting phase 2 as source activity. The most plausible interpretation is therefore that this phase is a depth phase. This interpretation is in good agreement with a focal depth of about 4 km which resulted from our hypocenter determination and which predicts sP_n arriving with approximately the time delay to P_n picked for phase 2. Fig. 5 shows a few other examples of phase 1 (P_n) and 2 (sP_n) for several stations in azimuthal ranges which have not been covered by the seismogram section of Fig. 4. Again phase 2, being delayed to P_n by nearly 1.5 sec, is clearly identified on all of the records.

All of these observations leads to the statement that phase 2 which is most probably the depth phase sP_n is radiated with rather large energy from the focus in all azimuthal directions.

THEORETICAL INVESTIGATIONS

Synthetic seismograms were calculated with the reflectivity method. This method originally developed by Fuchs (1968) has been modified several times (Kind and Müller, 1975; Kind, 1978). It computes complete theoretical seismograms for a point source (dislocation or explosion source) buried in a layered medium. In the present investigation we used a dislocation point source, with the orientation inferred from the respective fault-plane solution studied. The source time function is identical to the one described by Brüstle and Müller (1983) and has the form:

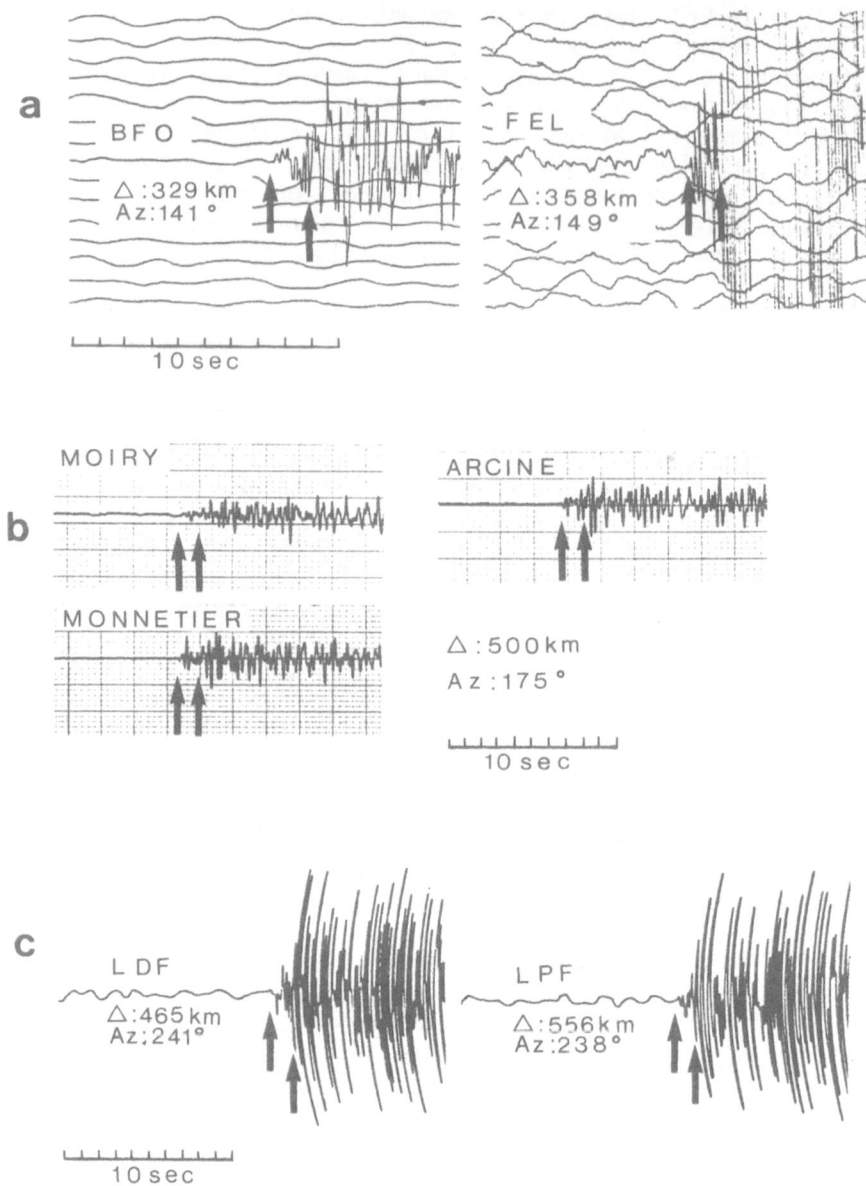

Fig. 5: P-wave groups from stations not included in the
         azimuthal range of the profiles shown in Fig. 4.
         Location of the stations:
         a) Rhinegraben (Black Forest); b) near Lake Geneva (SW
         Switzerland); c) Bretagne (NW France)
         A dominant second phase comparable to that one in Figs.
         1 and 4 can be identified on all records.

$$M(t) = \begin{cases} \dfrac{9M_0}{16} \left[1 - \cos \dfrac{\pi t}{T} + \dfrac{1}{9} \left(\cos \dfrac{3\pi t}{T} - 1 \right) \right] & , \ 0 \le t \le T \\ \\ M_0 & , \ t > T \end{cases}$$

with the final moment value M_0 and the risetime T. Attenuation in the average-velocity model (Fig. 6) was taken into account using a Q value of 1000 for P-waves and 200 for S-waves.

The normalized seismogram sections shown in Fig. 7 have been calculated for the fault-plane solutions KA1 (top) and KA2 (bottom); these were the two possible fault-plane orientations deduced as equivalent solutions from first motion investigations. Each seismogram in both sections is calculated at an epicentral distance of 500 km; the azimuth varies clockwise over 360°, starting at 0° (north) with one seismogram being computed every 15°. The phases which have been correlated in the data appear clearly in these synthetic plots, namely phase 1: P_n, phase 2: sP_n being preceeded by pP_n in some azimuthal

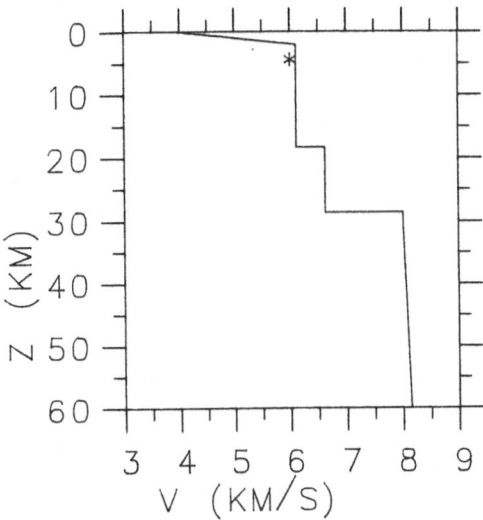

Fig. 6: Earth model used for the calculation of theoretical seismograms and of time shifts applied to the traces in the observed seismogram sections of Figs. 4, 8, and 9.

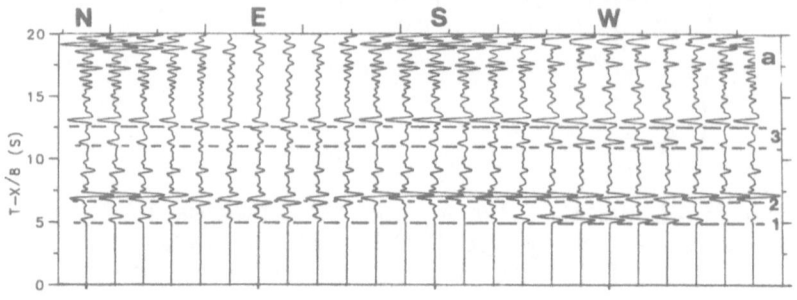

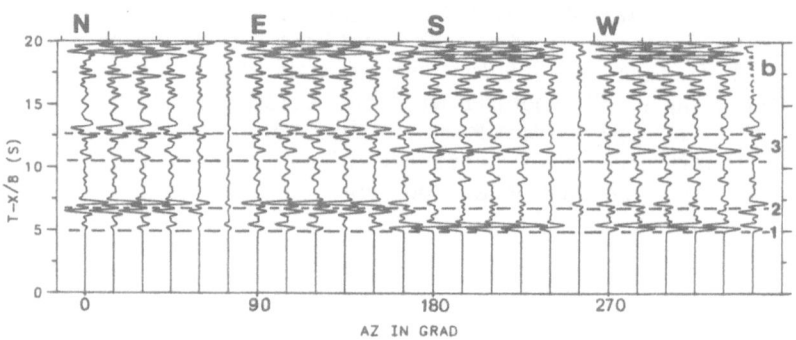

Fig. 7:   Theoretical seismograms (vertical component) calculated
          for fault plane solutions KA1 and KA2. Each seismogram
          is calculated at a distance of 500 km; the azimuth
          varies from 0°(N) to 345° with one seismogram every
          15°.

directions, phase 3: first crustal multiple of P_n and of the
depth phases. The following conclusion can be drawn from this
calculation of theoretical seismograms for an uniform azimuthal
coverage: the focal mechanism defined by fault plane solution
KA2 does not radiate energetic depth phases in all azimuthal
directions. The only focal mechanism which generates strong
depth phases with comparable amplitude ratios as have been
observed in all azimuthal directions is KA1. By this the
ambiguity of the focal mechanisms could be resolved and the
conclusion that the most likely explanation for phase 2 is that
it is the depth phase sP_n was confirmed.

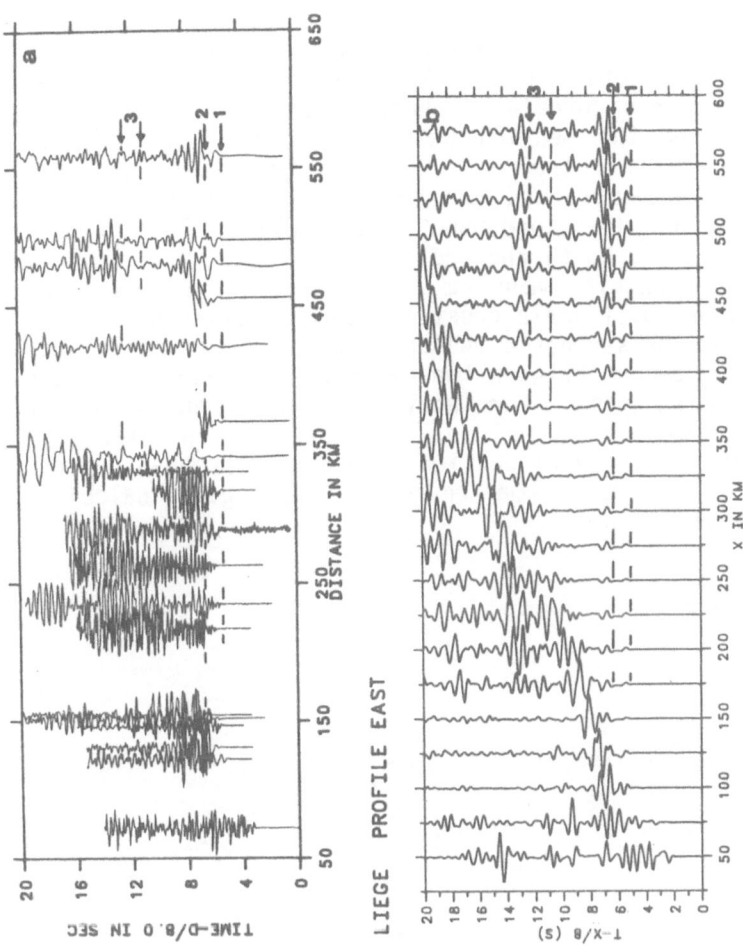

LIEGE PROFILE EAST

Fig. 8: Theoretical seismogram section (b) computed at an azimuth of 105° for the distance range from 50 km to 575 km to be compared with the observed profile East (a).

Fig. 8 shows, for better comparison with the data, a synthetic seismogram section calculated for the azimuth of 105° which is the average azimuth of the GRF-array and close to the average azimuth of the digitized profile to the east. The distances of the synthetic seismograms range from 50 to 575 km, comparable to the observed section (top of Fig. 8). The time scales of the synthetic and observed sections are identical. The frequency content of the theoretical seismograms is comparable to the broad-band data. The input-velocity model was the same as used before (Fig. 6). The synthetic section is again a normalized one, which means that the amplitude of each seismogram is normalized to the maximum amplitude of this trace in the plotted time window. This is the main reason why the amplitudes of the correlated phases 1 through 3 get much larger to the end of the section where the crustal phases with the strongest energy disappear out of the time window. The theoretical and observed seismograms are in good agreement, especially for the broad-band data included in the distance interval from 340 km to the end of the profile. The short-period seismograms contain many more reverberations due to the resolution of small-scale crustal heterogeneities which make the signal-shapes of synthetic and observed data look different. The excellent agreement between the observed broad-band data and the synthetic seismograms, of relative onset-times and signal-shapes of the correlated phases, is demonstrated in Fig. 9 which includes in a seismogram section of observed broad-band data one theoretical seismogram (marked by T) from Fig. 8, calculated at a distance of 525 km.

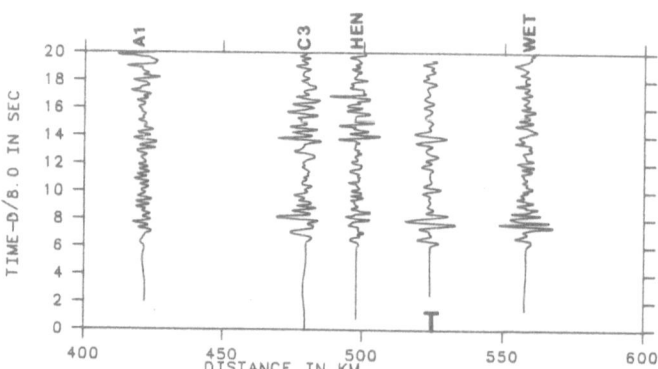

Fig. 9:   Comparison between observed broad-band records and synthetic seismogram (T) computed at a distance of 525 km and at the azimuth of the observing stations.

CONCLUSIONS

A successful interpretation of the dominant phases arriving in between P_n and P_g has been achieved using synthetic seismograms. Considering the good agreement between theoretical calculations and observations demonstrated above, it is clear that the rupture process of the Liège earthquake represents mainly dip-slip faulting with normal movement (solution KA1).

We were not able to model the differences in the frequency content between the P_n phase and the depth phases, repeated in the crustal multiples of these phases, as has been observed in the broad-band data of the GRF-array (Fig. 1). The most likely explanation for the relative high-frequency content of the depth phases is that it originates from a layered medium above the focus which is passed by the upgoing part of the phase before it gets reflected at the surface. Unfortunately, the version of the reflectivity method which we used for the calculation of the synthetic seismograms, can only be applied to lateral homogeneous structures, which means that we could not take into account in our calculations a different structure beneath the focal and the station site, thus testing quantitatively the suggestion made above.

ACKNOWLEDGMENTS

We are grateful to all colleagues who made data from their stations available. We thank especially T. Camelbeeck and M. de Becker who additionally provided us with data they had collected. The computations were carried out on the HP 1000 at GRF observatory and on the UNIVAC 1100 at the computer center of the University of Karlsruhe. J. Mechie and A. Gruber read the manuscript. We thank G. Bartman for typing the manuscript.

REFERENCES

Brüstle, W. and G. Müller (1983). Moment and duration of shallow earthquakes from Love-wave modelling for regional distances, Phys. Earth Planet. Inter., 32, 312-324.
Fuchs, K. (1968). The reflection of spherical waves from transition zones with arbitrary depth-dependent elastic moduli and density, J. Phys. Earth 16, 27-41.
Gelbke, C. (1978). Lokalisierung von Erdbeben in Medien mit beliebiger Geschwindigkeits-Tiefen-Verteilung unter Einschluß späterer Einsätze und die Hypozentren im Bereich des südlichen Oberrheingrabens von 1971 bis 1975, Dissertation, Geophys. Inst. Univ. Karlsruhe.

Harjes, H.P. and D. Seidl (1978). Digital recording and analysis
    of broad-band seismic data at the Graefenberg (GRF)-array,
    J. Geophys. 44, 511-523.
Kind, R. and G. Müller (1975). Computations of SV-waves in
    realistic earth models, J. Geophys. 41, 149-172.
Kind, R. (1978). The reflectivity method for a buried source, J.
    Geophys. 44, 603-612.
Wielandt, E. and G. Streckeisen (1982). The leaf-spring
    seismometer: design and performance, Bull. Seism. Soc. Am.
    72, 2349-2367.

THE SOURCE CHARACTERISTICS OF THE LIÈGE EARTHQUAKE ON NOVEMBER 8, 1983,   FROM DIGITAL RECORDINGS IN WEST GERMANY

L. Ahorner *,   R. Pelzing **

*)   Abteilung für Erdbebengeologie, Geologisches Institut, Universität zu Köln, Federal Republik of Germany

**)   Geologisches Landesamt Nordrhein-Westfalen, Krefeld, Federal Republik of Germany

Abstract

The instrumental hypocenters of the Liège mainshock on November, 8, 1983, at 0:49 GMT (with macroseismic intensity VII MSK-scale and local magnitude 5.1) and its largest aftershock at 2:13 GMT (magni= tude 3.0) have been calculated using weighted arrival times of direct P and S waves from 21 seismic stations in the distance range up to 144 km. A comparison of hypocenter results calculated with various velocity models shows the possibility of determining the epicentral coordinates with a precision of up to $\pm$ 1 to 2 km for the best fitting crustal model. The favoured solution for the mainshock is: Origin time 00:49:34.4 GMT, Longitude 5°30.8' E, Latitude 50°37.9' N, Depth 6 km. The instrumental epicenter is less than 1 km from the macroseismic epicenter at St.Nicolas.

From digital recordings in West Germany the seismotectonic source parameters of the Liège earthquakes were estimated using S wave displacement spectra of six stations and the Brune source model. In addition, the seismic energy released at the source was calculated from the time integrals of squared ground velocity seismograms. For the mainshock the following values have been found:

Seismic Moment M_O	1.6×10^{23} dyne cm
Source Radius r_O	870 m
Average Dislocation d_O	0.27 m
Stress Drop p_O	156 bar
Seismic Energy E_{SO}	1.3×10^{19} erg

The fault plane solution reveals a strike-slip dislocation, either right-lateral along a plane trending N 71°E, or left-lateral along a plane trending N 167°E. The P-axis is arranged NW-SE and nearly horizontal.

P. Melchior (ed.), Seismic Activity in Western Europe, 263–289.
© 1985 by D. Reidel Publishing Company.

1. Introduction

A damaging earthquake with intensity VII MSK and local magni=
tude 5.1 occurred on November 8, 1983, at 0:49 GMT near the Belgium
town of Liège in the border region between Belgium, Germany and the
Netherlands. This most severe earthquake in the northwestern part of
Central Europe since more than thirty years has well been recorded by
the dense seismic station network in West Germany which consists of
more than twenty stations situated in the Lower Rhine Embayment and
in the adjoining Rhenish Massif. Most of the stations are equipped with
modern digital recording systems. Thus high-quality seismograms are
available from the region east and southeast of the epicenter covering
a distance range between 70 km and 144 km. From these data the
source characteristics of the Liège mainshock and of its largest after=
shock have been determined in order to get more information on
the seismotectonic processes causing the Liège events.

2. Seismic Station Network

During the period of 1976 to 1982 the seismic station network
in the Lower Rhine Embayment and in the Rhenish Massif was consi=
derably enlarged and mostly equipped with digital recording systems
(Figure 1). At present there are more than twenty stations in operation.
Most of them are operated by the Department of Earthquake Geology
of the Geological Institute of the University of Cologne and the Geo=
logical Survey of Nordrhein-Westfalen at Krefeld. Two stations at the
western border of the Neuwied Basin (west of Koblenz) were operated
at the time of the Liège earthquake by the Department of Applied
Geophysics of the Technical University of Aachen.

The coordinates of all stations are listed in Table 1. Most station
codes are only provisional and are not included in the NEIS seismic
station list.

The stations are equipped with short period seismometers of the
types Geospace HS-10, Geotech S-13, Mark L-4-C, and Sensonics Will=
more MK IIIA, respectively, which have natural periods of 0.8 to 1.5
seconds and velocity transducers. At the stations marked by black tri=
angles in Figure 1 the data are recorded digitally on magnetic tape
using pulse code modulation (PCM), either by continuous recording or
by event triggering. The analog seismometer signals are resolved into
12 bit data words, which means a dynamic range of 72 dB, or with
gain ranging, of 120 dB. The upper frequency limits are 2O Hz, 75 Hz,
or 150 Hz, depending on the number of channels and the type of
equipment.

Due to the thick cover of Tertiary and Quaternary sediments
the seismic noise in the Lower Rhine Embayment is considerably higher
than in the mountainous areas to the east and to the south. Therefore
at the Stations Jackerath (JAC) and Pulheim (PUL), operated by the

Geological Survey, the seismometers are installed in boreholes which are about 400 m deep and extend down to the consolidated rocks of the Hercynian basement. The boreholes have a waterproof steel casing with a diameter of 32 cm. The seismic noise at a depth of about 350m to 400 m is only one tenth of that at 50 m depth. Thus these bore= hole stations are comparable in their noise conditions with favorable rock sites in the mountainous regions bordering the Lower Rhine Embay= ment. The maximum noise level of about 0.1 µm/sec at the borehole stations and rock sites allows to detect earthquakes down to mangitude 1.5 at a distance of 100 km, with both, P and S waves being recogniz= able. As the Liège area is about 100 km from the center of our station network, we are sure, that all aftershocks with magnitudes greater than 1.5 have been detected by our instruments.

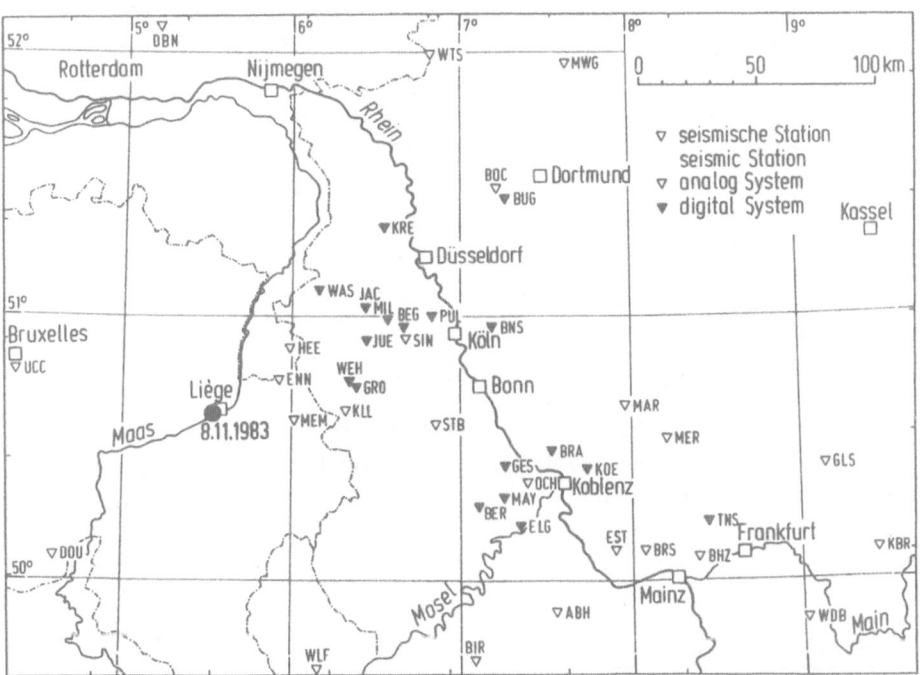

Figure 1.   Epicenter of the Liège earthquake (black point) and the seismic stations in the surrounding areas. Most of the stations are situated east and southeast of the epicenter in the Lower Rhine Embayment and in the Rhenish Massif.

Table 1.   Coordinates of seismic stations in West Germany

Station	Code	Geogr. Coordinates Long.E	Lat.N	UTM-Coordinates (Zone 32 U)		Elevation (m)	
Bensberg	BNS	7°10.53'	50°57.83'	371.94	5647.47	200	(1)
Kalltal	KLL	6°18.68'	50°38.79'	309.97	5614.06	390	(1)
Steinbach	STB	6°50.40'	50°35.72'	347.15	5607.13	270	(1)
Jülich	JUE	6°24.59'	50°54.83'	319.90	5643.30	91	(1)
Bergheim	BEG	6°39.34'	50°57.68'	335.29	5648.24	-95	B(1)
Sindorf	SIN	6°39.87'	50°55.12'	335.75	5643.48	-70	B(1)
Millendorf	MIL	6°33.55'	50°59.58'	328.63	5651.97	-25	B(1)
Jackerath	JAC	6°25.92'	51°02.18'	319.97	5657.04	-240	B(2)
Pulheim	PUL	6°49.21'	51°00.31'	347.06	5652.75	-300	B(2)
Großhau	GRO	6°22.64'	50°44.23'	314.96	5623.97	370	(2)
Wehebach	WEH	6°22.45'	50°45.44'	312.36	5626.08	250	(2)
Wassenberg	WAS	6°09.50'	51°06.13'	301.10	5665.05	60	(2)
Krefeld	KRE	6°32.25'	51°20.50'	328.50	5690.75	40	(2)
Braunsberg	BRA	7°31.39'	50°29.54'	395.20	5594.63	272	(1)
Köppel	KOE	7°43.98'	50°25.52'	409.92	5586.75	540	(1)
Burg Eltz	ELG	7°20.23'	50°12.36'	381.35	5562.95	140	(1)
Ochtendung	OCH	7°22.53'	50°22.25'	384.50	5581.25	120	(1)
Glees	GES	7°14.71'	50°26.10'	375.32	5588.68	240	(1)
Bermel	BER	7°05.79'	50°16.90'	364.32	5571.87	540	(3)
Mayen	MAY	7°14.49'	50°18.85'	374.74	5575.24	247	(3)

Operating Institutions:
(1)   Department of Earthquake Geology, Geological Institute, University of Cologne
(2)   Geological Survey Nordrhein-Westfalen, Krefeld
(3)   Department of Applied Geophysics, Technical University Aachen
B   means borehole station

3.   Example Seismograms

Typical seismograms of the Liège mainshock recorded digitally at our stations Wassenberg (WAS), Jülich (JUE) and Bensberg (BNS) are presented in Figures 2 to 4. The seismograms show the time history of ground velocity in three components. Following the mainshock three aftershocks were recorded at the German stations during the same night and another one on December 1. Seismograms of the strongest after= shock are displayed in Figure 5.

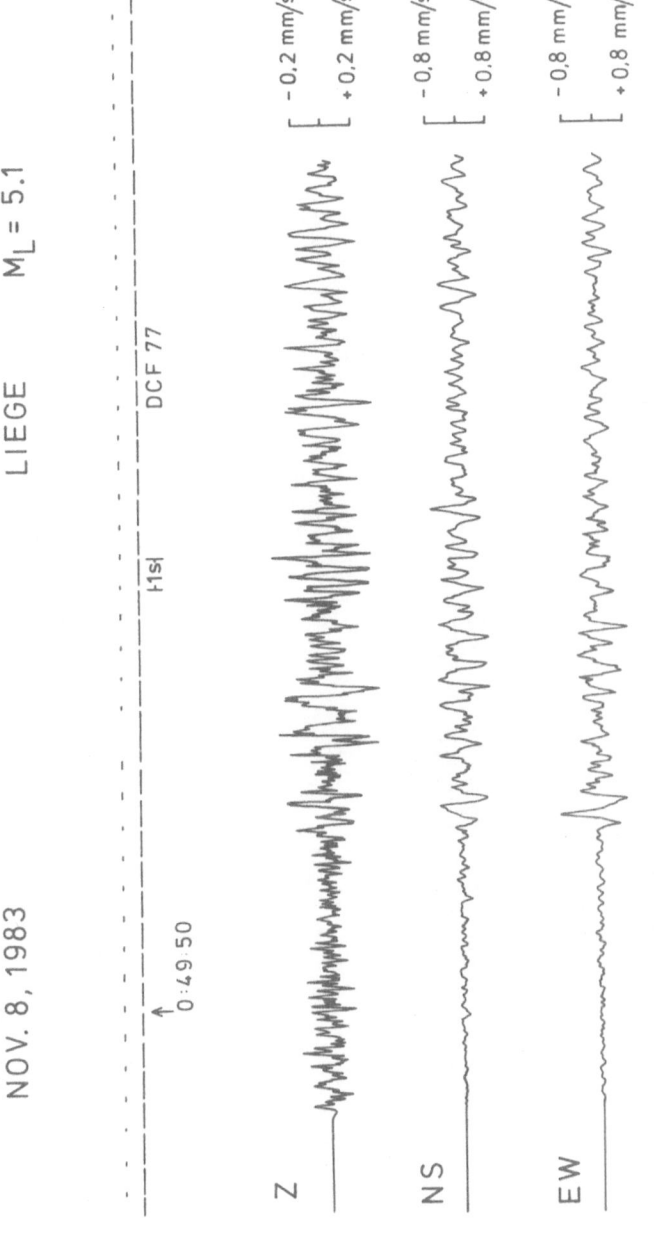

Figure 2. Seismogram of the Liège mainshock recorded at the Station Wassenberg (WAS). The hypocenter distance is 70 km. The recorded signal is proportional to ground velocity within the frequency range of 1 Hz to 150 Hz. The station underground consists of some hundred meters of soft soil (mainly fine sands of Tertiary age).

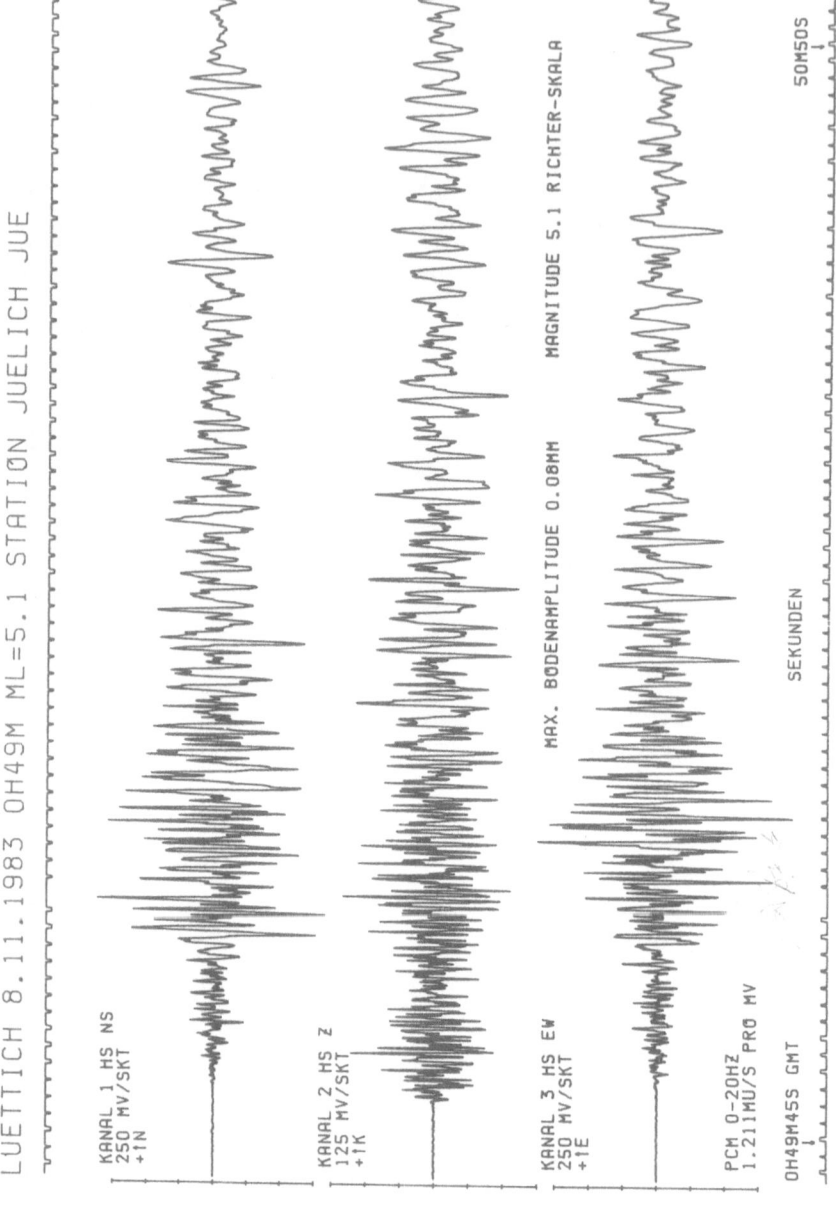

Figure 3.   Seismogram of the Liège mainshock recorded at station Jülich (JUE). Hypo-center distance 74 km. Station underground more than 1000 m soft soil. Note the long duration of the signal due to the underground conditions.

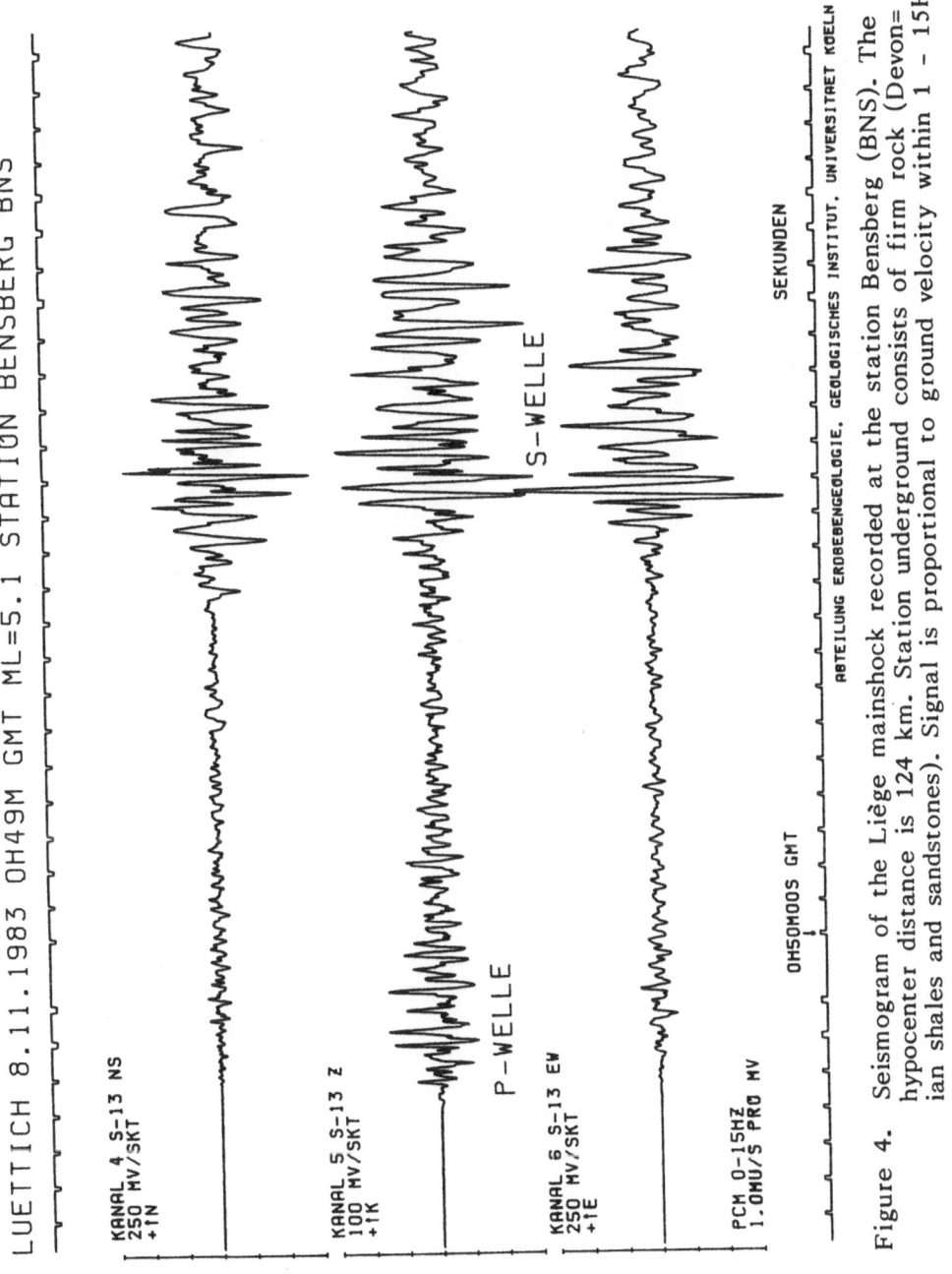

Figure 4. Seismogram of the Liège mainshock recorded at the station Bensberg (BNS). The hypocenter distance is 124 km. Station underground consists of firm rock (Devonian shales and sandstones). Signal is proportional to ground velocity within 1 – 15Hz.

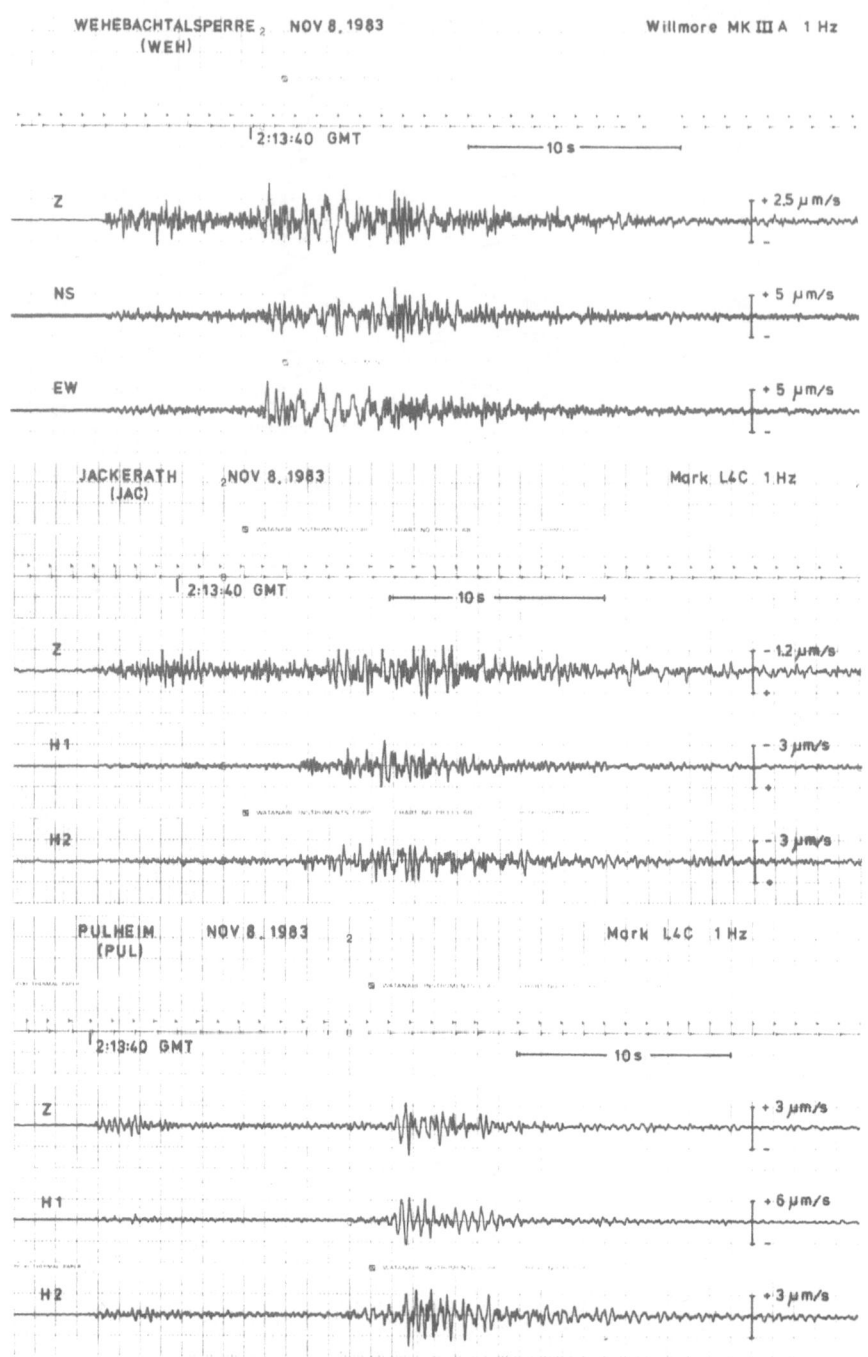

Figure 5. Seismograms of the strongest aftershock at 2:13 GMT, recorded at the Stations WEH, JAC, and PUL.

4. Magnitude Determination

The local magnitude M_L of the Liège earthquake and its after=
shocks has been calculated from S wave amplitudes A_h (horizontal
component of maximum ground displacement in μm) and hypocenter
distance R (in km) with the following formula deduced from Richters
original log A_0 values by converting Wood-Anderson amplitudes to
ground amplitudes (Ahorner 1983):

$$M_L = \log (A_h) + 1.90 \log (R) - 0.35$$

Following the procedure described by Richter in its original
paper the arithmetric mean of the two horizontal components was ta=
ken as local magnitude for a specific station. The formula given above
is valid for Central European earthquakes in the distance range of
10 km to 300 km.

Table 2 gives the basic data for the local magnitude calculation for
nine stations in the distance range of 70 km to 153 km. The corre=
sponding mean value for the local magnitude of the Liège mainshock
is:

November 8,   0:49 GMT M_L = 5.1 $\pm$ 0.2

This result is in good agreement with local magnitudes determined by
the Royal Observatory at Brussels with data from Belgian stations
(M_L=4.9), and with local magnitudes as reported by the European
Mediterranean Seismological Centre (EMSC) at Strasbourg (mean from
eight stations MAW = 5.1 $\pm$ 0.4).

Table 2. Local magnitude determination from the S wave amplitudes
of nine stations in West Germany for the Liège mainshock.

Station	Distance R (km)	Displacement Amplitude A_h(μm) NS-Component	EW-Component	Period T(sec)	Magnitude M_L
WAS	70	88.3	118.0	0.74	5.17
JUE	74	66.3	79.6	0.53	5.01
BEG	90	75.1	56.7	0.39	5.18
BER	120	46.8	49.1	0.25	5.29
BNS	124	38.3	66.9	0.39	5.32
MAY	129	17.2	18.9	0.49	4.93
ELG	139	7.4	9.4	0.14	4.65
BRA	144	15.4	20.8	0.32	5.03
BUG	153	27.0	27.0	0.51	5.23

Mean Value from 9 Stations: M_L = 5.09 $\pm$ 0.22

For the aftershocks the following magnitudes were calculated from the recordings of the German stations:

November 8,   0:55 GMT $M_L = 1.8 \pm 0.3$

                 1:25 GMT $M_L = 2.0 \pm 0.3$

                 2:13 GMT $M_L = 3.0 \pm 0.3$

December 1,   21:53 GMT $M_L = 1.8 \pm 0.3$

5. Location of Hypocenters

For locating the hypocenters of the Liège earthquakes only stations with direct P and S arrivals were used. The maximum epicentral distance to the stations used is about 140 km. Weights were given to the arrival times according to their accuracy of reading. Weights of stations with magnetic tape recording were twice as large as of stations recording on paper. In addition, weights of P arrivals were four times larger than those of S arrivals. All arrival times used are listed in Table 3, both for the mainshock at 0.49 GMT and the strongest aftershock at 2:13 GMT.

If P arrival times are plotted in a reduced time scale against di= stance to the macroseismic hypocenter, they can be approximated by four straight lines with velocities of 6.00, 6.25, 6.50, and 8.07 km/sec, respectively (Figure 6).

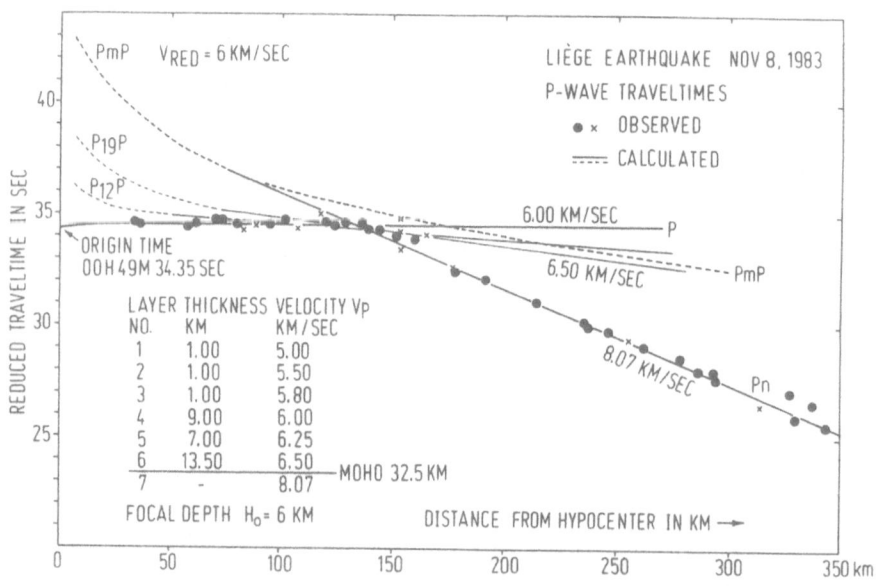

Figure 6.   Reduced travel time diagram for the Liège mainshock.

Table 3. P and S waves arrival times (in seconds) of the Liège main= shock and the largest aftershock

Station	First Motion	Mainshock 0:49 GMT t_P	t_S	Aftershock 2:13 GMT t_P	t_S
BNS	D	55.10	69.70	43.71	57.85
KLL	D	44.00	-	32.06	39.10
STB	D	50.40	-	38.65	51.00
JUE	C	47.00	54.14	-	-
SIN	C	50.00	-	-	-
MIL	C	49.01	-	-	-
ELG	D	57.48	-	45.50	62.85
JAC	C	47.75	57.50	36.00	45.74
PUL	C	51.76	64.20	39.84	52.52
WEH	D	44.75	52.15	32.91	40.67
WAS	C	46.36	55.92	-	-
KRE	-	52.30	-	-	-
MAY	D	56.07	71.10	44.50	60.00
BER	D	54.65	70.25	42.54	59.50
ENN	C	40.30	43.60	28.30	33.10
UCC	D	48.10	59.20	37.00	-
MEM	-	40.50	-	28.60	-
DOU	C	49.26	-	36.97	47.45
WLF	C	54.59	-	42.56	-
BRA	D	58.34	75.58	46.34	63.72
OCH	D	57.31	-	-	-

First Motion Direction of P waves for the Mainshock:

C = Compression, D = Dilatation

The velocity model given in Figure 6 fits the observed arrival times. It is a slight modification of a model derived earlier from earth= quakes in the eastern Lower Rhine Embayment, and therefore contains other velocity values as well. From previous works the velocity models shown in Figure 7 are also possible for the Liège area. Model A is an interpretation of refraction data in the Rhenish Massif by Mooney and Prodehl (1978). Model B is close to a Model given by Mechie, Prodehl and Fuchs (1983) for the area of the Hohes Venn, southeast of Liège, and is also based on refraction data. Model C was used by the authors as the best fitting model for two earthquakes in the German-Netherlands border region near Heinsberg (Ahorner and Pelzing 1983). Model D is the original Ahorner model, as deduced from earthquakes in the eastern part of the Lower Rhine Embayment, and model F its modification as shown in Figure 6. Model E is an approximation to the arrival times of waves from rock bursts in the Ruhr Coal District, which traverse the station network in the Lower Rhine Embayment with a mean velocity of about 6.2 km/s. Finally, model G is the simp= lest approximation possible to the Liège earthquake arrival times. By inverting hypocenter coordinates and the depth of the boundary between the 6.0 and 6.5 km/s layers simultaneously, a depth of 16 km was obtained (for the method see e.g. Crosson 1976 or Pelzing 1978).

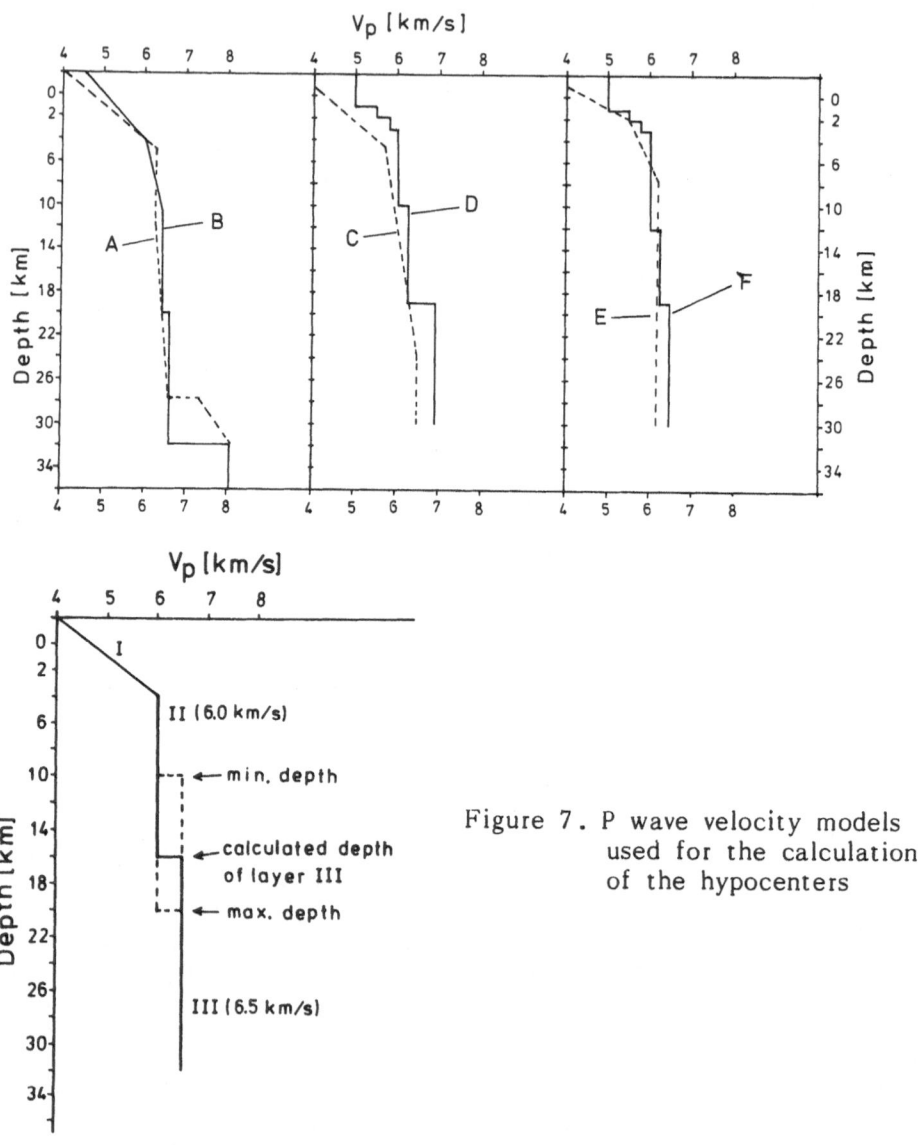

Figure 7. P wave velocity models
          used for the calculation
          of the hypocenters

For the calculation of the hypocenter coordinates an iterative location program, based on Geigers method (Pelzing 1978), was used. Only one velocity model was used at the same time for all stations.

The results of the hypocenter calculation are shown in Figure 8. At the left side of the diagram the hypocenter determination of the Liège mainshock with various velocity models is displayed. The macro= seismic epicenter is marked by a cross. The point chosen here is the intersection of Rue des Bons Buveurs and Rue Francisco Ferrer in St.Nicolas. The epicenters calculated with the seven velocity models are marked by black dots. The bars give the standard deviations of the epicentral coordinates. The all lie in the range of 1.5 to 2.5 km. The epicenters as reported bei CSEM and NEIS are also shown. They are situated about 12 to 15 km northwest of the macroseismic epicenter.

The best guess for the instrumental epicenter is the one calcu= lated with our velocity model G. The difference to the macroseismic epicenter is smaller than 1 km. However, the other solutions show, that the epicenter can be displaced by a few kilometers, if the velo= city model is varied between reasonable limits. According to the standard deviations the depth determination is less precise than the epicenter determination. The best guess for the mainshock is 6 ± 2 km. This value fits with velocity model G.

Of the aftershocks only the one at 2:13 GMT was recorded at a number of stations sufficient for a reliable determination of the hypo= center. The results are shown in the middle part of Figure 8. The overall pattern of epicenters (marked by black dots) is nearly the same as for the mainshock (marked by open circles), but is shifted somewhat to the south. Again there is a clustering of depth values around 6 km. The differences in the hypocentral coordinates of the two events are partly due to the fact that the number of observed arrival times is different in both cases (see Table 2). If only those arrival times are used which lead to an identical set of observations, the results are as shown in the right side of Figure 8. The southward deviation of the aftershock epicenter with respect to the mainshock is again visible. For the best fitting model G the aftershock epicenter is about 1.5 km south of that of the mainshock. Because the precision of depth deter= mination is relatively poor, it is assumed that the depth of both events is about 6 ± 2 km.

The final estimation of the hypocentral coordinates is shown in Figure 9 together with geographical features of the Liège area. The epicenters are given here in geographical coordinates. The instrumental epicenters for the mainshock and the strongest aftershock are relatively near to the macroseismic epicenter (distance smaller than 1 to 2 km).

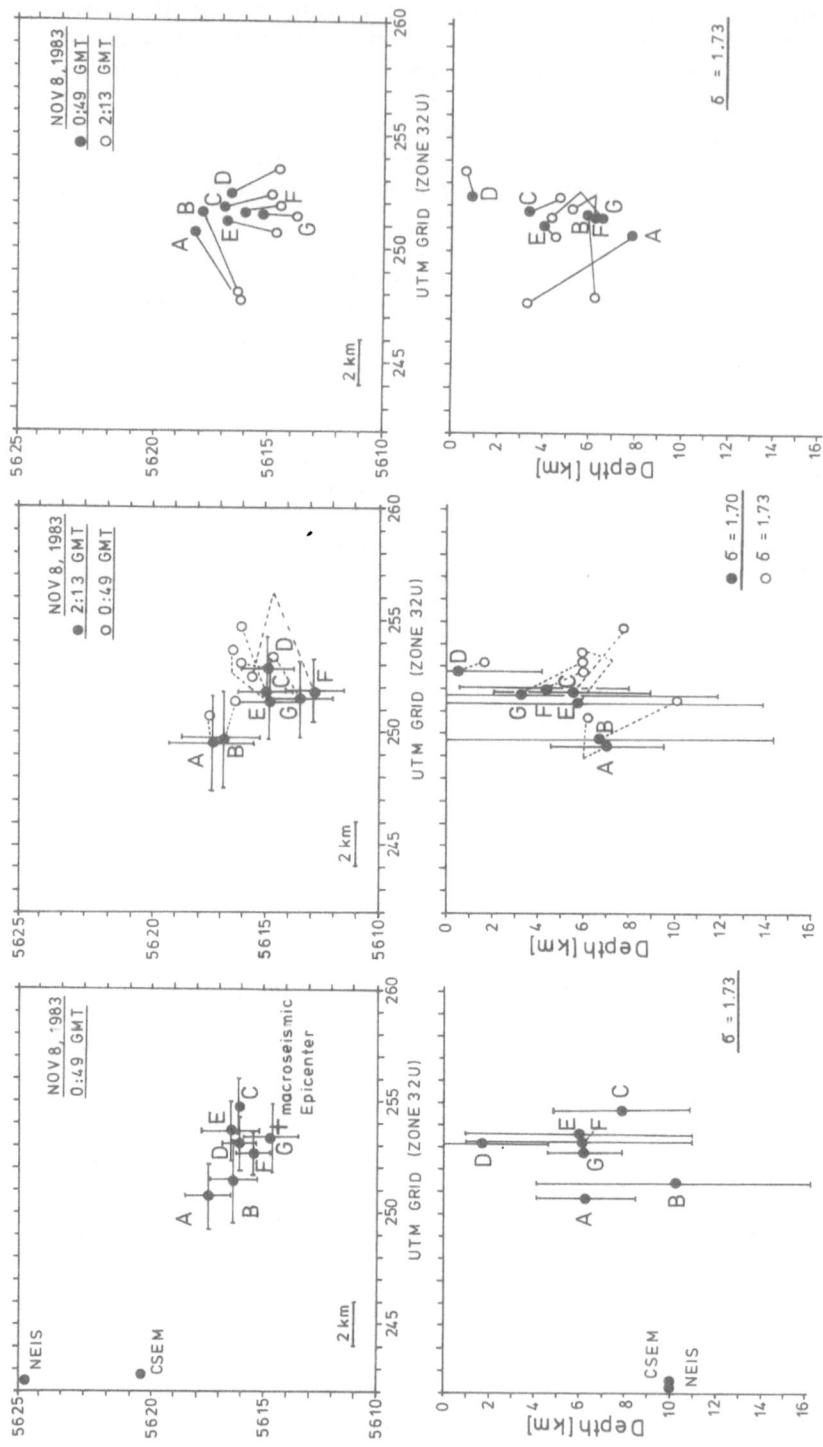

Figure 8. Epicenters (above) and focal depths (below) calculated with the velocity models Figure 7. The bars are the standard deviations of hypocentral coordinates (UTM grid, zone 32 U).

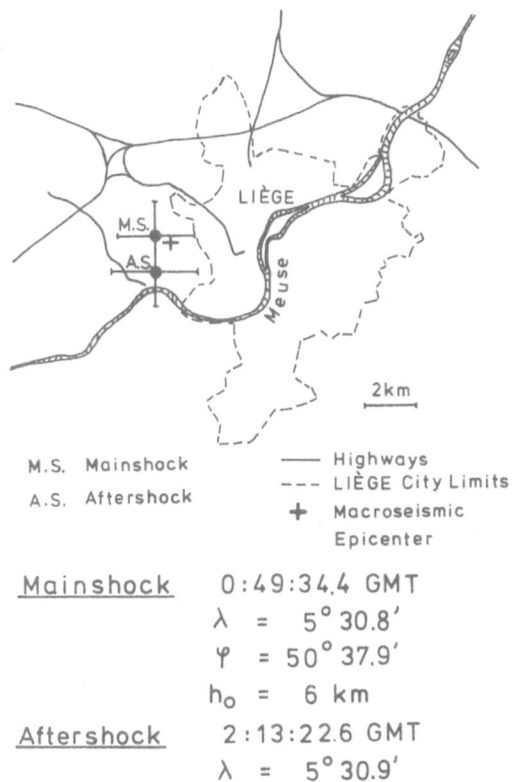

M.S. Mainshock
A.S. Aftershock

—— Highways
--- LIÈGE City Limits
✛ Macroseismic
    Epicenter

Figure 9.

Final positions of the instru=
mental epicenters of the Liège
mainshock and its largest
aftershock. Bars give standard
deviations of the epicenter
coordinates. The instrumental
epicenters are within 2 km
of the macroseismic epicenter
at St.Nicolas (cross).

Mainshock   0:49:34.4 GMT
                λ =   5° 30.8'
                φ = 50° 37.9'
                h_o =   6 km

Aftershock   2:13:22.6 GMT
                λ =   5° 30.9'
                φ = 50° 37.1'
                h_o =   6 km

6. Fault Plane Solution

A fault plane solution of the Liège mainshock based on first motions
of short period near stations given in Table 3 and, in addition, on many
other stations in Central Europe is displayed in Figure 10. The solution
is well defined and reveals a predominating strike-slip dislocation, com=
bined with a small thrust component. The seismotectonic bloc movement
is either right-lateral along a N 71°E trending and with 80° to the NW
diping plane, or left-lateral along a N 167°E trending and with 60° to
the NE diping plane. The P-axis is arranged nearly horizontal (dip 14°
to the NW) and strikes NW–SE (N 302°E). The T-axis is arranged
SW–NE (N 205°E) and dips with 29° to the SW. The direction of the
P-axis (pressure axis) is in agreement with the general NW–SE trend
of the compressive tectonic stress field in Central Europe (Ahorner 1975,
Ahorner, Baier and Bonjer 1983). Similar fault plane solutions to that
shown in Figure 10 have been worked out by some other investigators
(see e.g. Camelbeeck and De Becker, this volume). The seismotectonic
implications of the focal mechanism of the Liège earthquake are dis=
cussed in more detail in a special contribution (Ahorner, this volume).

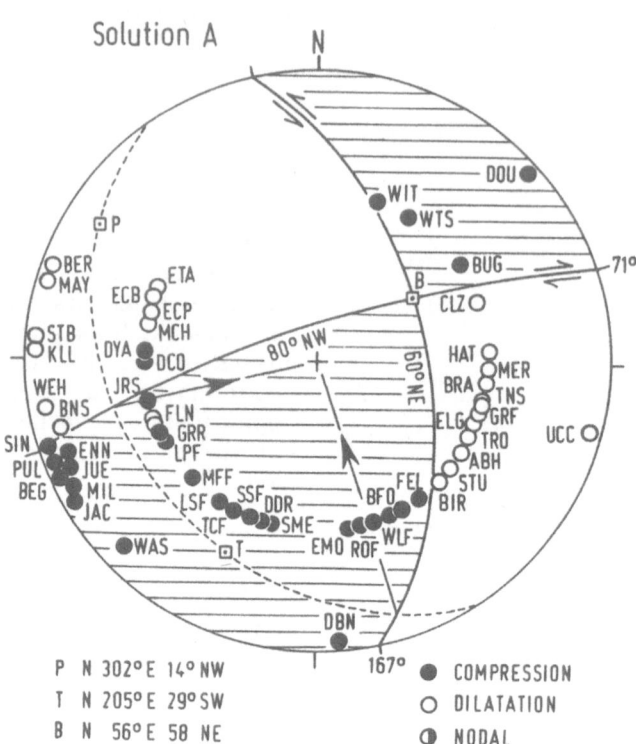

Figure 10.   Fault plane solution of the Liège mainshock on November 8,
1983, at 0:49 GMT, based on first motion of P waves.

7.   Seismotectonic Source Parameters

The seismotectonic source parameters of the Liège mainshock and
of its strongest aftershock were calculated from S wave displacement
spectra of digital stations using the Brune source model (Brune 1970).
The Brune model is based on a circular rupture plane and leads to
amplitude spectra which, in a biaxial logarithmic scale, have a plateau
at low frequencies and a fall-off to high frequencies with f^{-2}. The
plateau level Ω_0 and the corner frequency f_c seperating plateau and
high frequency fall-off are the input data for calculating the source
parameters with the following formulas (see e.g. Lee and Stewart 1981):

Seismic Moment $\qquad\qquad M_0 = \Omega_0\, 4\,\pi\, v^3\, R\quad C^{-1}$

Source Radius $\qquad\qquad r_0 = 2.34\quad v\quad (2\,\pi\, f_c)^{-1}$

Average Dislocation $\qquad\ d_0 = M_0\, (\pi\, \varrho\, v^2\quad r_0^2)^{-1}$

Stress Drop $\qquad\qquad\quad p_0 = 7\, M_0\, (16\, r_0^3)^{-1}$

where ϱ = density, v = S wave velocity, R = hypocenter distance in km,
C = term for correcting the average radiation pattern and the
free surface amplification.

The calculation was made in each case with ϱ = 2.7 g/cm^3, v = 3.5 km/s
and C = 0.85. The displacement spectra were corrected for instrumental
response down to about 0.2 Hz and for anelastic absorption with a
Q_S value of 500 to 750. Figure 11 shows typical displacement spectra
for the Liège mainshock and its strongest aftershock calculated from
digital recordings of the S waves at the Station Bensberg (BNS).

Due mainly to the radiation pattern the recordings of the different
stations lead to different values for the source parameters. The corres=
ponding values are given for six stations in the distance range of 70 km
to 144 km and in the azimut range of 44° to 107° in Table 4.

Table 4 . Seismotectonic source parameters for the Liège mainshock
from the recordings of six stations in West Germany

Station	R km	Az	Comp	$\log \Omega_0$ μ s	$\log M_0$ dyne cm	f_c Hz	r_0 m	d_0 cm	p_0 bar
WAS	70	44°	NS	2.40	23.48	1.48	880	37.7	195
WAS	70	44°	EW	2.50	23.58	1.32	987	37.8	174
JUE	74	67°	NS	2.55	23.65	0.98	1330	24.4	84
JUE	74	67°	EW	2.40	23.50	1.17	1114	24.7	101
BEG	90	68°	H1	2.00	23.19	2.24	581	43.8	342
BEG	90	68°	H2	2.05	23.24	1.68	775	27.6	162
BNS	124	74°	NS	1.80	23.13	2.09	623	33.1	241
BNS	124	74°	EW	1.90	23.23	2.04	638	40.0	285
MAY	129	107°	NS	1.30	22.65	1.64	794	6.8	39
MAY	129	107°	EW	1.40	22.75	1.70	766	9.2	55
BRA	144	97°	EW	1.55	22.95	1.23	1059	7.7	33

The mean values for the source parameters of the mainshock and
the strongest aftershock are presented in Table 5 together with the
standard deviations which give an impression of the accuracy of such
parameters. A comparison of the values for both shocks shows that
the difference in magnitude is mainly due to the higher stress drop
of the mainshock or, accordingly, to the larger dislocation. The stress
drop of about 156 bar for the mainshock is significantly higher than
for other earthquakes in the Lower Rhine Embayment and in the Rhenish
Massif which have been investigated previously (Ahorner 1983, Ahorner
and Pelzing 1983).

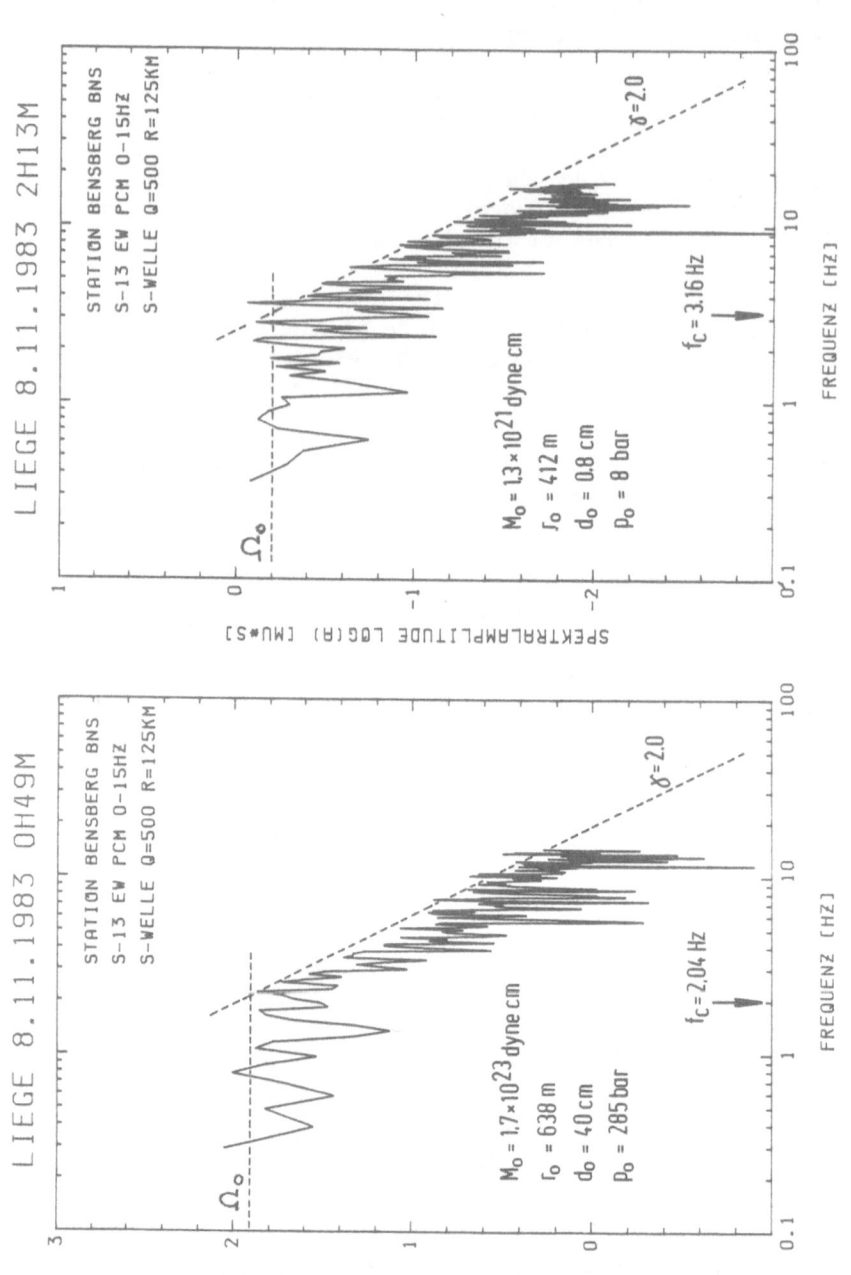

Figure 11. S wave displacement spectra for the Liège mainshock (left) and the strongest aftershock (right) calculated from the recordings at the station Bensberg (BNS). Hypocenter distance 124 km. The seismotectonic source parameters deduced from the spectra are given (Brune source model).

Table 5. Mean values with standard deviations of the seismotectonic source parameters for the Liège mainshock

		Mainshock 0:49 GMT	Aftershock 2:13 GMT
Magnitude	M_L	5.09 ± 0.22	3.00 ± 0.36
Seismic Moment	$\log M_o$ (dyne cm)	23.21 ± 0.33	20.76 ± 0.35
Corner Frequency	f_c (Hz)	1.60 ± 0.41	3.23 ± 1.08
Source Radius	r_o (m)	870 ± 230	460 ± 200
Average Dislocation	d_o (cm)	27 ± 14	0.37 ± 0.27
Stress Drop	p_o (bar)	156 ± 104	4.5 ± 3.8

The seismic moments of the Liège events are in good agreement with the general relationship between local magnitude and moment found by Ahorner (1983) from 26 small to moderate earthquakes in the Rhenish Massif and its vicinity (Figure 12). The relationship be= tween the average source dislocation and the local magnitude is more complex for the same data set, but it is quite clear that the source dislocation increases with magnitude from a few tenths of a mm in case of magnitude 1.0 to more than 100 mm in case of magnitude 5.0 (Figure 14). Again the values of the two Liège earthquakes, with a dislocation of 270 mm for the mainshock and 3.7 mm for the strongest aftershock, fit reasonably well to the Rhenish Massif data.

The P and S wave displacement spectra of the aftershock at 2:13 GMT are shown in Figure 13. Some spectra obviously have two corner frequencies (JAC, BNS, MAY, BER). This is an indication that the shape of the rupture plane is not circular as required for the Brune model, but elongated in one direction (unilateral rupture). Under the assumption that the focal mechanism of the aftershock is similar to that of the mainshock (see Ahorner, this volume) theoretical corner frequencies were calculated for rupture propagation on both fault planes (Table 6). The observed corner frequencies are better explained by an unilateral rup= ture on the NNW-SSE trending fault plane. The corresponding rupture length is about 500 m. The vector of the rupture propagation points to the south with an azimut of about 170° (with respect to north) and a dip of about 2°. However, regarding the geological situation in the focal region, the WSW-ENE trending fault plane is more convenient (see Ahorner, this volume).

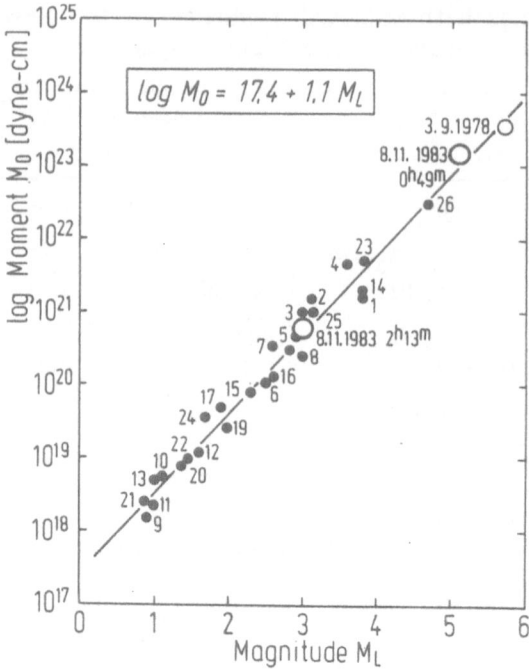

Figure 12. Relationship between seismic moment and local magnitude
found for 26 earthquakes in the Rhenish Massif (Ahorner
1983). The Liège data fit well to this relationship.

Table 6. Observed and theoretical corner frequencies and rupture para=
meters for an unilateral rupture on both fault planes of the
fault plane solution. Aftershock at 2:13 with magnitude 3.0

	Observed Corner Frequencies (Hz)		Theoretical Corner Frequencies (Hz)			
			Fault Plane I Strike 71° Dip 80°NW		Fault Plane II Strike 167° Dip 60°NE	
JAC	1.5	1.0	1.6	1.8	1.7	1.5
BNS	2.4	2.2	1.9	2.4	2.0	1.9
MAY	2.4	3.3	2.0	2.9	2.6	3.3
BER	2.6	-	2.1	2.9	2.7	3.5

Rupture Length	700 m	500 m
Azimut of Rupture Vector	120°	170°
Dip of Rupture Vector	-47°	-2°

The fitting equation: $\log M_0 = 17.4 + 1.1 M_L$

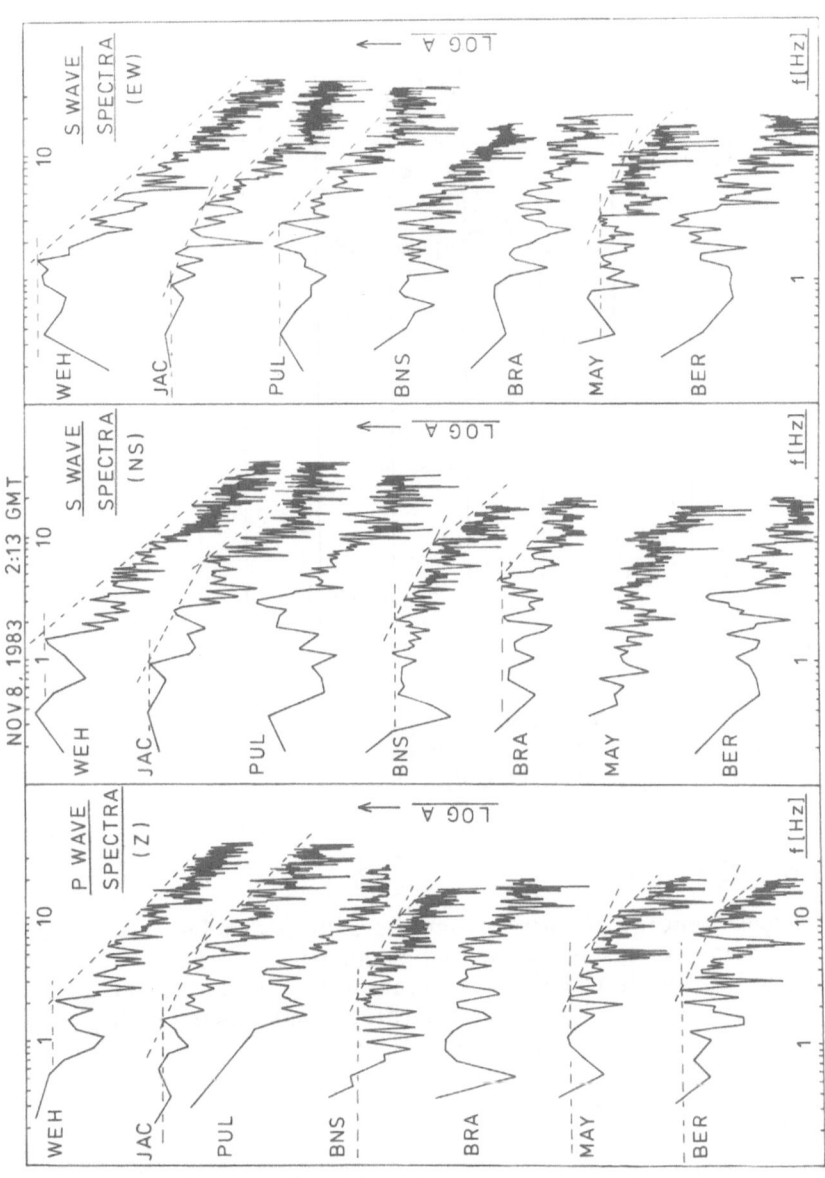

Figure 13. P and S wave displacement spectra for seven stations recording the Liège aftershock at November 8, 2:13 GMT. The amplitude scale (log A) is arbitary in order to show the corner frequencies for all stations. Some stations show two corner frequencies.

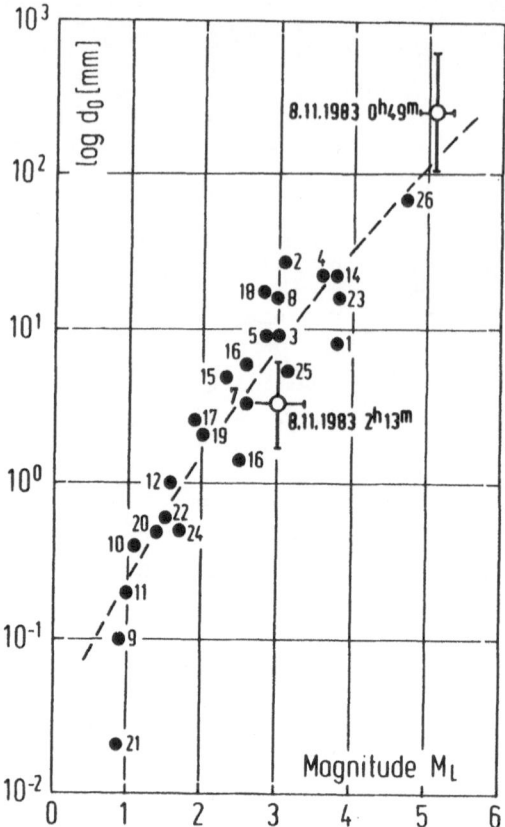

Figure 14. Relationship between average source dislocation d_0 and local magnitude for earthquakes in the Rhenish Massif (Ahorner 1983). The Liège data fit reasonably well.

8.   Source Energy

The seismic energy released from the source can be calculated from ground velocity seismograms. The cumulative energy flow density for a specific station is the time integral over ground velocity squared, resulting from vectorial addition of all three components, multiplied by density ϱ and wave velocity c at the station:

$$E_{sta}(t) = \varrho \, c \int_{t_o}^{t} v^2 \, dt$$

The energy at the source can be calculated from the formula (see e.g. Schneider 1975):

$$E_{so} = 2 \times 0.25 \times 4\pi \, R^2 \, e^{\propto R} \times E_{sta,max}$$

where R = distance to the source, α = absorption coefficient, factor 2 = addition of kinetic and potential energy, factor 0.25 = reduction of free surface amplification.

Typical cumulative energy flow density diagrams for the mainshock recorded at the stations Bensberg (BNS) and Mayen (MAY) are shown in Figure 15. Due to the radiation pattern of the source the ratio of P to S wave energy is smaller for station BNS than for station MAY. The source energy values calculated from the readings of four stations agree reasonably well (Table 7). The mean value for the mainshock is

$$E_{SO} = 1.3 \times 10^{19} \text{ erg}$$

The corresponding value for the largest aftershock at 2:13 GMT is

$$E_{SO} = 2.8 \times 10^{15} \text{ erg}$$

In Figure 16 both values are compared to relationships between seismic energy and local magnitude as found by Richter (1958) for Californian earthquakes (broken line) and by Ahorner (1983) for the Bad Marienberg, Westerwald, earthquake series of June 1982 (solid line and black dots). The values for the two Liège events fit well to both relationships which in fact are almost identical. These results also show that the local mag= nitude as determined with the formula given in section 4 is compatible with the Californian local magnitude scale (original Richter scale).

Table 7. Determination of seismic source energy from integrated velocity seismograms for the Liège mainshock

Station	Distance R(km)	Energy at the Station E_{Sta} (erg/cm^2)	Energy at the Source E_{SO}(erg)
BNS	124	9360	2.2×10^{19}
BER	120	13250	2.8×10^{19}
MAY	129	1922	5.0×10^{18}
BRA	144	2365	8.7×10^{18}

Mean Value from 4 Stations: $\log(E_{SO}) = 19.11 \pm 0.35$
$$E_{SO} = 1.3 \times 10^{19} \text{ erg}$$

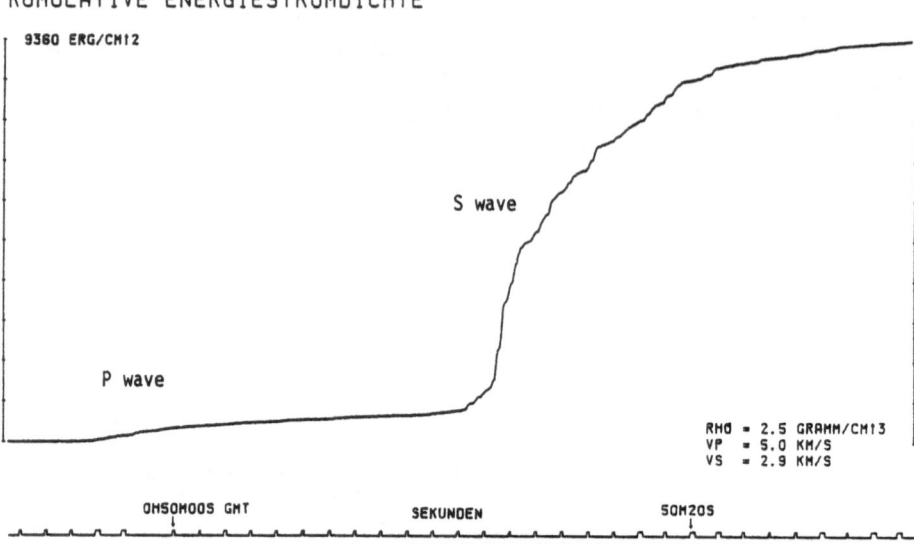

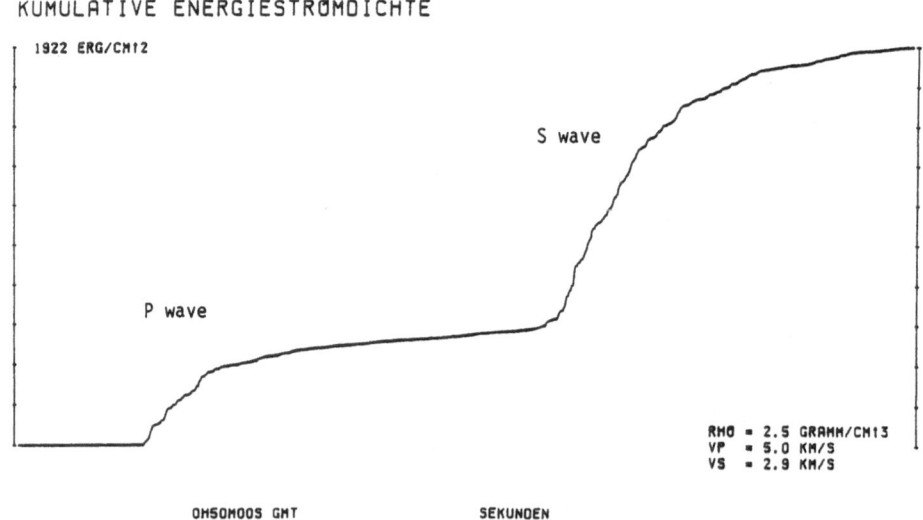

Figure 15.   Cumulative energy flow density diagrams for the Liège
mainshock recorded at station Bensberg (above) and station
Mayen (below). For explanation see text.

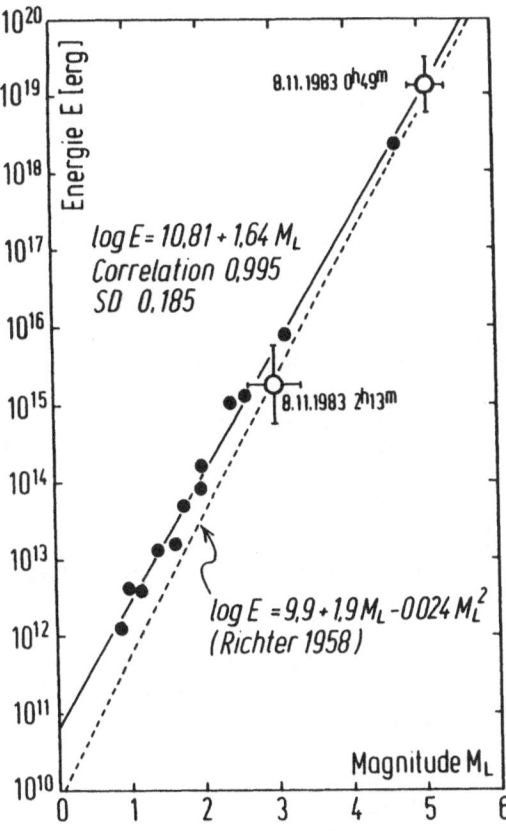

Figure 16.
Relationship between source energy E_{so} and local magnitude M_L for earthquakes in the Rhenish Massif (Ahorner 1983) and in California (Richter 1958). The energy values of the two Liège events agree reasonably with both curves.

9. Concluding Remarks

This contribution shows that the modern network of high-dynamic digitally recording seismic stations which has been installed in the Lo= wer Rhine Embayment and in the Renish Massif during the last years, is capable of determining the hypocenter coordinates and the seismo= tectonic source parameters of earthquakes occurring in the northwestern part of Central Europe with reasonable precision, even for those shocks which are centered outside of the network, like the Liège earthquakes.

For seismic events which are not in a peripheral position to the network, the results would be more accurate. Therefore, in order to improve the seismological research work on future earthquakes in the border region between Belgium, Germany and the Netherlands, it is recommendable to install modern digital seismic stations also in the neighboring countries west and northwest of the West German network.

Acknowledgements

We thank many of our colleagues for supplying us with additional
seismograms from their stations. Our special thank goes to Professor
Wohlenberg from the Department of Applied Geophysics of the Techni=
cal University of Aachen, who kindly made digital recordings of the
stations MAY and BER available. The research work was partly spon=
sored by the German Research Society (DFG) and the Bundesministe=
rium für Forschung und Technologie (BMFT) at Bonn.

References

Ahorner, L. (1975): Present-day stress field and seismotectonic block
    movements along major fault zones in Central Europe. - In:
    Pavoni, N. and Green, R.(eds), Recent Crustal Movements.
    Tectonophysics 29: 233-249; Amsterdam

Ahorner, L. (1983): Historical seismicity and present-day microearth=
    quake activity of the Rhenish Massif, Central Europe. - In:
    K.Fuchs et al.(eds.), Plateau Uplift, p.198-221; Springer-Verlag,
    Heidelberg

Ahorner, L. (1985): The general pattern of seismotectonic dislocations
    in Central Europe as a background for the Liège earthquake
    on November 8, 1983. - (this volume)

Ahorner, L., Baier, B., Bonjer, K.-P. (1983): General pattern of seismo=
    tectonic dislocation and the earthquake generating stress field
    in Central Europe between the Alps and the North Sea. -
    In: K.Fuchs et al.(eds.), Plateau Uplift, p.187-197; Springer-
    Verlag, Heidelberg

Ahorner, L. , Pelzing, R. (1983): Seismotektonische Herdparameter von
    digital registrierten Erdbeben der Jahre 1981 und 1982 in der
    westlichen Niederrheinischen Bucht. - Geol.Jb. E 26: 35-63;
    Hannover

Brune, J.N. (1970): Tectonic stress and spectra of seismic shear waves
    from earthquakes. - J.geophys.Res. 75: 4997-5009; Washington

Crosson, R.S. (1976): Crustal structure modeling of earthquake data.
    1. Simultaneous least squares estimations of hypocenter and
    velocity parameters. - J.geophys.Res. 81: 3036-3046; Washington

Lee, W.H.K., Stewart, S.W. (1981): Principles and applications of micro=
    earthquake networks. - 293 p.; Academic-Press, New York

Mechie, J., Prodehl, C., Fuchs, K. (1983): The long range seismic refrac=
    tion experiment in the Rhenish Massif. - In: K.Fuchs et al.
    (eds.), Plateau Uplift, p.260-275; Springer-Verlag, Heidelberg

Mooney, W.D., Prodehl, C. (1978): Crustal structure of the Rhenish Massif and adjacent areas: a reinterpretation of existing seismic refraction data. - J.Geophys. 44: 573-601, Würzburg

Pelzing, R. (1978): Untersuchungen zur Ortung von Herden seismischer Ereignisse, dargestellt an Beispielen aus einem Stationsnetz im Ruhrbergbaugebiet. - Ber.Inst.Geophys.Ruhr-Universität Bochum, 6: 184 p.; Bochum

Richter, C.F. (1958): Elementary Seismology. - 768 p., W.H.Freeman and Comp., San Francisco

Schneider, G. (1975): Erdbeben. Entstehung-Ausbreitung-Wirkung. - 406 p., Enker-Verlag, Stuttgart

Camelbeeck, T., De Becker, M. (1985): The Liège earthquakes of November 8, 1983, and December 21, 1965. - (this volume)

LE MÉCANISME AU FOYER DU SÉISME DE LIÈGE DU 8 NOVEMBRE 1983

H. Haessler.

Institut de Physique du Globe de Strasbourg.

Abstract.

On November 8, 1983 an earthquake of magnitude 5 and very shallow depth occurred in the western suburb of Liege. This shock, largely felt in Belgium and the surrounding countries reached a maximum intensity VII-VIII at the epicentre. It is the strongest earthquake in this area since the beginning of the century. The focal solution obtained with 37 european stations shows a normal faulting striking E-W with a northward plunge of 45°. An examination of the local geological features shows that this shock is likely to be associated with the Saint-Gilles fault. The focal mechanism is not in good agreement with the general tectonic frame of the Rhenish zone which is dominated by SE-NW compression and SW-NE distension. Nevertheless, recent levelling measurements show a clear plateau uplift in the North-Western part of the Rhenish Massif which may explain the normal faulting observed in this area.

Le séisme du 8 novembre 1983.

Le 8 novembre 1983 à 00 h 49 la région de Liège a été secouée par un séisme important, ressenti non seulement en Belgique mais également dans les régions limitrophes des pays voisins : Pays-Bas, Allemagne Fédérale, France. L'épicentre du séisme était situé dans la banlieue ouest de Liège où il a fait des dégâts correspondant à une intensité VII-VIII de l'échelle MSK, intensité tout à fait compatible avec la magnitude du séisme (5.0) et sa profondeur (3 km environ). Ce séisme est le plus fort survenu depuis le début du siècle dans cette région. Le dernier

P. Melchior (ed.), Seismic Activity in Western Europe, 291–295.
© 1985 by D. Reidel Publishing Company.

événement marquant (de magnitude 4.5) s'était produit le 21
décembre 1965 (Van Gils, 1978) ; nous allons voir qu'il présen-
tait des caractéristiques très voisines de celles du choc récent.
Le réexamen des données historiques permet de penser que certains
séismes survenus au cours des siècles précédents et attribués à
d'autres régions sismiques voisines ont en fait leur épicentre à
proximité de Liège (Vogt, 1983).

La figure 1 présente la répartition des sens des premiers
mouvements verticaux observés dans des stations sismologiques
entourant le foyer et dont la distance épicentrale est inférieure
à 5 degrés pour la plupart d'entre elles. Les observations sont
ramenées sur la demi-sphère focale inférieure représentée en projection stéréographique de Schmidt. Sur les 37 données reportées,
34 ont été déterminées par examen direct des sismogrammes et 3
proviennent d'informations fournies par des stations très proches.
La plupart des données sont relatives à des ondes Pn de même
incidence au foyer et les points correspondants se trouvent en
projection sur un même cercle. Les stations du Fossé Rhénan et du
réseau LDG encadrent ainsi l'épicentre sur près de 180°. Cette
série présente un premier changement de signe entre CDF et GWF
et ensuite entre les sous-réseaux LDG de Normandie et d'Auvergne,
définissant ainsi deux points de passage obligés pour les plans
nodaux. Les données relatives aux ondes Pg sont moins bien con-
traintes en incidence ; cela est dû à la faible profondeur du
foyer et à la connaissance imparfaite du milieu de propagation
des rais sismiques. Elles ont été conventionnellement placées à
une incidence de 90°. La solution focale obtenue correspond au
jeu d'une faille normale de direction E-O et de pendage 45° nord.
Cette solution est assez bien contrainte par les données de type
Pn et peu affectée par l'incertitude des données de type Pg.

L'examen de la sismicité antérieure (Van Gils, 1978) dans
cette région nous indique que le dernier événement majeur
(M = 4.5) du 21 décembre 1965 s'est produit dans la banlieue nord
de Liège (Ans-Vottem). Les paramètres focaux du séisme (hypocentre et mécanisme au foyer) ont amené Ahorner (1972) à conclure au
rejeu en faille normale de la faille de Saint-Gilles. Cette
faille est parallèle, mais de pendage opposé, à la faille eifé-
lienne qui est elle-même le prolongement Est de la faille du
midi. Il est très vraisemblable que ce soit encore cette faille
Saint-Gilles qui ait joué le 8 novembre 1983, suivant le même
mécanisme : l'hypocentre n'est qu'à quelques kilomètres du précédent, dans la banlieue ouest de Liège (Saint-Nicolas), à proximité immédiate de la trace en surface de la faille Saint-
Gilles ; de plus, le mécanisme présenté ici est très voisin du
précédent. Signalons une autre caractéristique relevée par
Ahorner pour les séismes de 1965 (Ahorner, 1972) et que nous
retrouvons ici : le petit nombre des répliques observées pour
un choc principal aussi important. D'une manière plus générale
les observations des vingt dernières années semblent indiquer

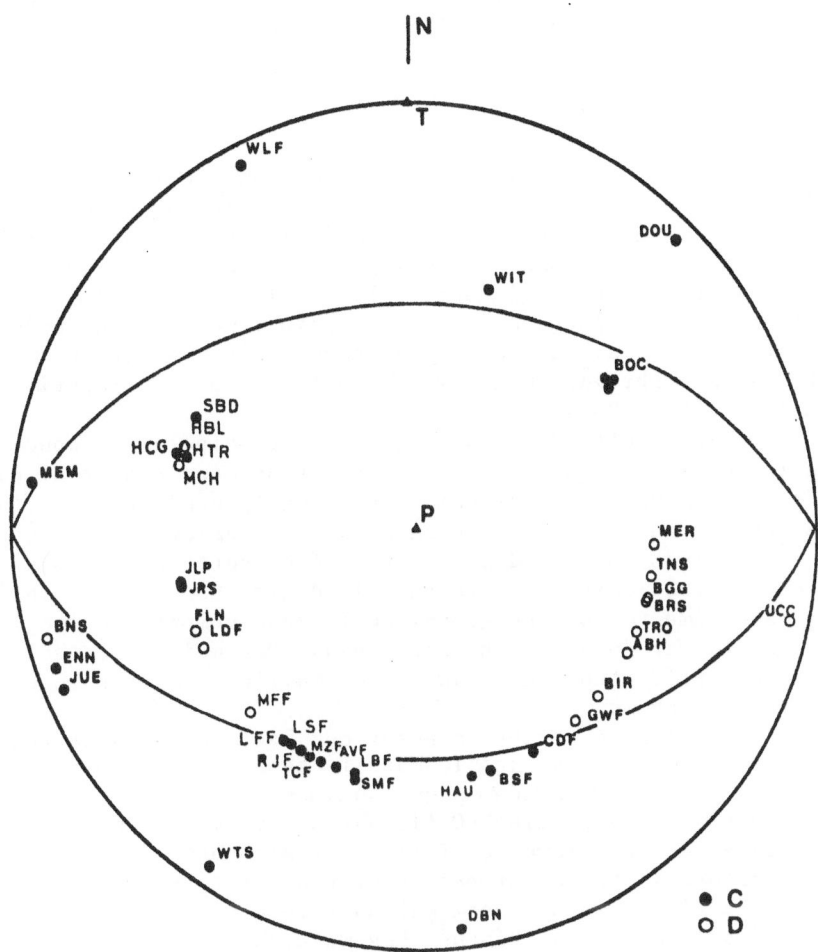

Fig. 1

que les chocs de faible magnitude sont en petite quantité dans
la région de Liège, eu égard à ces deux chocs importants de 1965
et 1983.

Cadre tectonique.

        Ce séisme Liégeois s'inscrit mal dans le schéma global
de la tectonique des Pays Rhénans. Le trait majeur de cette région est en effet une compression de direction SE-NW déduite
aussi bien des mécanismes au foyer des séismes que des mesures
de contrainte "in situ" (Ahorner, 1975 ; Illies, 1981). Sous
l'action de cette poussée, il y a principalement rejeu en décrochement senestre des accidents du fossé du Rhin supérieur et pour-
suite de la taphrogenèse au niveau du fossé du Rhin inférieur
(Ahorner, 1975, 1983).
        La sismicité de la Belgique est principalement concentrée
dans le Hainaut et la région de Liège, tous deux traversés par
la "faille du midi". L'alignement E-O des épicentres ainsi que
le mécanisme au foyer (décrochement dextre) de certains chocs
(Bruxelles 1938) ont conduit à interpréter cette zone comme une
région de cisaillement dextre qui viendrait assez bien s'insérer
dans le schéma tectonique global de la zone comprise entre les
Alpes et la Mer du Nord (Ahorner, 1975). Néanmoins, les séismes
belges qui présentent un mécanisme de faille normale s'insèrent
mal dans ce schéma.
        Il est intéressant de noter que les récentes mesures de
nivellement effectuées dans le Massif Rhénan et ses abords
(Mälzer, 1983) ont mis en évidence une surrection du massif essentiellement dans sa partie N-O (1,5 mm/an). Ce soulèvement pourrait
être lié à ces mécanismes en faille normale observés dans la
région du Massif Rhénan et qui, sauf pour la taphrogenèse du Rhin
inférieur, s'inscrivent assez mal dans le contexte de la tectonique globale des Pays Rhénans. Les mesures de contrainte "in
situ" effectuées dans le massif sont également en accord avec les
observations sismologiques et géodésiques puisqu'elles présentent
au niveau des composantes horizontales un caractère distensif
pour σ_2 et pour σ_1 et distinguent ainsi ce massif des régions
avoisinantes (Bauman, 1983).
        Le choc du 8 novembre 1983 se présente ainsi comme un
séisme en faille normale à la limite nord du plateau des Ardennes,
dont le taux de soulèvement est un des plus élevés du Massif
Rhénan. La cause de ce soulèvement reste une question ouverte :
il peut s'agir soit d'une poussée locale verticale, soit d'une
flexure de la lithosphère liée à la compression SE-NO.

Références.

Ahorner, L., 1972. Erdbebenchronik für die Rheinlande 1964-1970.
Decheniana, 125, pp. 259-283.

Ahorner, L., 1975. Present-day stress field and seismotectonic block movements along major fault zones in Central Europe. Tectonophysics 29, pp. 233-249.

Ahorner, L., 1983. Historical seismicity and present-day micro-earthquake activity of the Rhenish Massif, Central Europe. In "Plateau Uplift"* .

Ahorner, L., Baier, B., Bonjer, K.-P., 1983. General pattern of seismotectonic dislocation and the earthquake-generating stress field in Central Europe between the Alps and the North Sea. In "Plateau Uplift"* .

Baumann, H. and Illies, H.J., 1983. Stress field and strain release in the Rhenish Massif. In "Plateau Uplift"* .

Illies, J.H., 1981. Stress pattern and strain release in the Alpine foreland. Tectonophysics, 71 (1981) pp. 157-172.

Mälzer, H., Hein, H., and Zippelt, K., 1983. Heigt changes in the Rhenish Massif : Determination and analysis. In "Plateau Uplift"* .

Van Gils, J.M., 1978. La séismicité de la Belgique et son application en génie parasismique. Annales des Travaux Publics de Belgique. N° 6-1978.

Vogt, J., 1983. Révision de deux séismes majeurs de la région d'Aix-la-Chapelle - Verviers. - Liège, ressentis en France : 1504, 1692.
Communication aux Journées de Sismicité Historique des Rencontres d'Antibes. Novembre 1983.

*"Plateau Uplift - The Rhenish Massif, A Case History"
Edited by K. Fuchs, K. von Gehlen, H. Mälzer, H. Murawski, and A. Semmel. Springer Verlag, 1983.

MACROSEISMIC MAP OF THE LIÈGE EARTHQUAKE OF NOVEMBER 8, 1983.

L. Ahorner
Abteilung für Erdbebengeologie,
Geologisches Institut, Universität zu Köln,
Federal Republik of Germany.

T. Camelbeeck and M. De Becker
Centre de Géophysique Interne
Observatoire Royal de Belgique
Avenue Circulaire 3, 1180 Bruxelles.

J. Flick
Laboratoire Souterrain de Géodynamique
Walferdange
Grand-Duchy of Luxemburg.

R. Ritsema and G. Houtgast
Royal Netherlands Meteorological Institute
De Bilt
The Netherlands

J. Vogt
Antenne Sismicité
Service Géologique Régional (B.R.G.M.)
Strasbourg 1.

As the earthquake of november 8, 1983 was felt far away from the Belgian borders, a complete macroseismic map is presented, taking into account the observations in the different countries (figure 1).

The scale of intensity used is the one of Medvedev, Sponheuer and Karnik (1).

The difficulty to definish the isoseismal lines, especially those of lower intensities is essentially explained by the fact that the earthquake happened in the middle of the night.

P. Melchior (ed.), Seismic Activity in Western Europe, 297–299.
© *1985 by D. Reidel Publishing Company.*

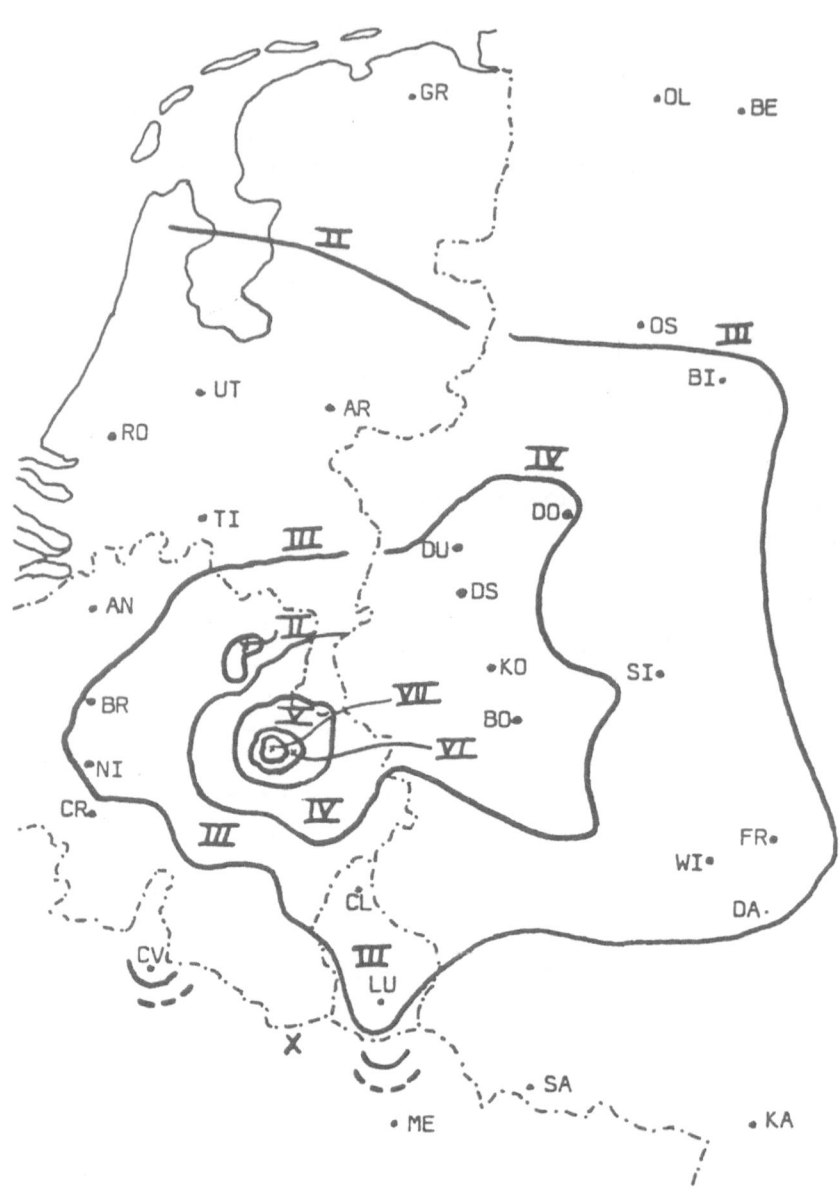

Figure 1.

Caption to figure 1

AN	Antwerpen	LU	Luxembourg
AR	Arnhem		
		ME	Metz
BE	Bremen		
BI	Bielefeld	NI	Nivelles
BO	Bonn		
BR	Bruxelles-Brussel	OL	Oldenburg
		OS	Osnabruck
CL	Clervaux		
CR	Charleroi	RO	Rotterdam
CV	Charleville		
		SA	Saarbrücken
DA	Darmstadt		
DO	Dortmund	TI	Tilburg
DS	Dusseldorf		
DU	Duisbourg	UT	Utrecht
FR	Frankfurt	WI	Wiesbaden
GR	Groningen		
KA	Karlruhe		
KO	Köln		

Data from France :   ‿‿‿figurative isoline of intensity ≤ IV
                      ▬ ▬ ▬limit of the macroseismic area
                   X   isolated indication.

BIBLIOGRAPHY

Medvedev S., Sponheuer W., Karnik V. : Neue Seismische Skala.
    Veröff. Inst. Jena - Heft 77 - Akademie Verlag - Berlin 1964.

MACROSEISMIC INQUIRY OF THE LIÈGE EARTHQUAKE OF NOVEMBER 8, 1983.

M. De Becker, T. Camelbeeck

Centre de Géophysique Interne
Observatoire Royal de Belgique
Avenue Circulaire·3,   1180 Bruxelles.

The earthquake of november 8 in Liège was not really an exceptional event in consideration to his magnitude and to historical
seismicity in this region but it caused exceptional damages. The
total cost of the damages is actually estimated as being in excess
of 3 milliards belgian francs : this means about 60 millions $.

Liège is a typical popular and industrial city of Western
Europe. A precise evaluation of the damages after this moderate
earthquake will be interesting as a good example of what can hap-
pen in many other parts of Western Europe.

HISTORICAL AND GEOGRAPHIC CONTEXT

Liège is one of the most important cities of Belgium in consideration to his amount of population and his economical impact as
a large European industrial centre.

His actual aspect is the result of the industrial boom of the
nineteenth century when the coal deposits began to be extensively
exploited and when the large metallurgies were founded.

The town is built on alluvial deposits in the bottom of the
valley of the Meuse river and on filling up soil along the hills.

This fact can partially explain the high vulnerability of the
buildings to any kind of ground motion.

P. Melchior (ed.), Seismic Activity in Western Europe, 301–309.
© *1985 by D. Reidel Publishing Company.*

MACROSEISMIC INQUIRY

The macroseismic inquiry of the november 8 earthquake was performed using the formularies sent by the Royal Observatory of
Belgium to the administrations of all the belgian towns. These
formularies are filled in under the responsability of the politi-
cal authorities of each district {figure 1}.

On the basis of this information, a macroseismic map was drawn
{figure 2} : **See separate map.**
The macroseismic epicentre is localised some 3 km
East of the microseismic epicentre. The most damaged districts,
for which we estimated that the intensity was VII on the MSK scale,
are Liège, St. Nicolas, Grâce-Hollogne, Ans and Seraing. The earthquake was felt everywhere in Belgium so that the isoline II extents
out of our country to the Netherlands, Germany and North of France.
The complete macroseismic map is also published in this book.

Concerning the most important aftershock of november 8, 1983
at 2h 13m 22.2s (UT), a macroseismic inquiry was also performed in
the neighbourhood of Liège. The answers obtained from the people
already shaken one hour before and the difficulties to distinguish
between the damages caused by the aftershock or by the main shock
didn't permit us to draw a detailed map : very roughly, the area
were the aftershock was felt was defined {figure 3}.

The same thing applies for the later aftershocks for which an
instrumental localisation of the hypocenter is not even available.
The aftershock of november 20 at 8h 03m 27s (UT) was felt in
Sclessin and Ans, whereas the aftershock of december 1st at 21h
52m 49s (UT) was felt in Seraing, Grâce-Hollogne and Flemalle.

MACROSEISMIC COMPUTATIONS

Using the macroseismic map established by the Royal Observatory
of Belgium, an estimation of the depth and the magnitude was obtained.

The empirical relationship of Kövesligethy was used, between
the maximum intensity I_o, the macroseismic radius of intensity I_p,
the hypocentral depth h and the attenuation factor k :

$$I_o - I_p = 3 \log_{10} \frac{\sqrt{h^2 + R_p^2}}{h} + 0{,}4343 \, k \, (\sqrt{h^2 + R_p^2} - h)$$

With the averaged radii R_7, R_6, R_5 measured on the map

$$R_7 = 3.87 \text{ km}$$
$$R_6 = 10.63 \text{ km}$$
$$R_5 = 21.47 \text{ km}$$

OBSERVATOIRE ROYAL DE BELGIQUE

COMMUNE : *Aiseau-Presles*   Tremblement de Terre du :
PROVINCE : *de Hainaut —*

QUESTIONNAIRE   *4*   ⊦ 8 -11- 1983

1h 49m

1.- <u>Quelques</u> habitants ont-ils perçu :
    des vibrations légères   *oui, avec un craquement*
    ~~un bruit caractéristique : grondement ou roulement~~

2.- De <u>nombreux</u> habitants ont-ils perçu :
    un ébranlement net de l'immeuble
    un fort grondement

3.- A-t-on observé :
    l'oscillation de lustres ou d'objets suspendus librement *}la nuit —*
    l'oscillation de chaises ou de lits
    le craquement des plafonds
    le réveil de dormeurs   ⟶ *de quelques uns*

4.- A-t-on observé ou constaté :
    le frémissement de vitres et de portes   ⟶ *oui —*
    le tremblement d'objets mobiliers (tables, vases, bibelots..)

5.-   le choc de verres, de vaisselle dans les buffets
    l'oscillation de cadres suspendus au mur   *} non*
    le craquement des planchers
    la chute de <u>petits fragments</u> de crépis des plafonds
    un émoi des habitants (des personnes sortent des habitations)

6.- A-t-on constaté :
    le déplacement de bibelots
    le déplacement d'objets légers, lesquels ?
    la chute de cadres pendus au mur   *} non —*
    la chute d'objets légers posés sur des meubles polis ou un marbre
    des vitres fêlées ou brisées
    des objets de vaisselle brisés.

7.- Des objets légers (vases, pots de fleurs,..) se sont-ils renversés
    Des meubles légers se sont-ils déplacés
    Des meubles lourds (bahuts, poêles,..) ont-ils vibré   *} non —*
    Des cloches ont-elles tinté

8.- Des meubles lourds se sont-ils déplacés
    Des meubles légers se sont-ils renversés
    Des portes ont-elles été déplacées sur leurs gonds

9.- A-t-on constaté :
    la chute de <u>morceaux de platras</u> des plafonds
    des fissures dans les platras des murs   *} non*
    une chute exceptionnelle d'une cheminée en mauvais état
    une chute exceptionnelle de quelques briques   *} non*

10.- Y a-t-il eu plusieurs cheminées endommagées, combien ?
    plusieurs cheminées renversées, combien ? % pour la commune ?
    des chutes d'ornements de façades, de quelle importance ?
    des lézardes profondes dans les murs (brique ? béton ?)
    des glissements le long des berges ou des talus, Importance ?

OBSERVATIONS COMPLÉMENTAIRES :

<u>N.B.</u>- Si l'on n'a rien observé dans la commune, veuillez renvoyer ce
questionnaire avec la mention "NEANT"; ces renseignements sont indispen-
sables pour délimiter les zones de silence et la limite d'extinction.

3

le Bourgmestre
MORADX

<u>Figure 1</u>

Figure 3. Area where the aftershock of november 8, 1983 at
2h 13m (UT) has been felt.

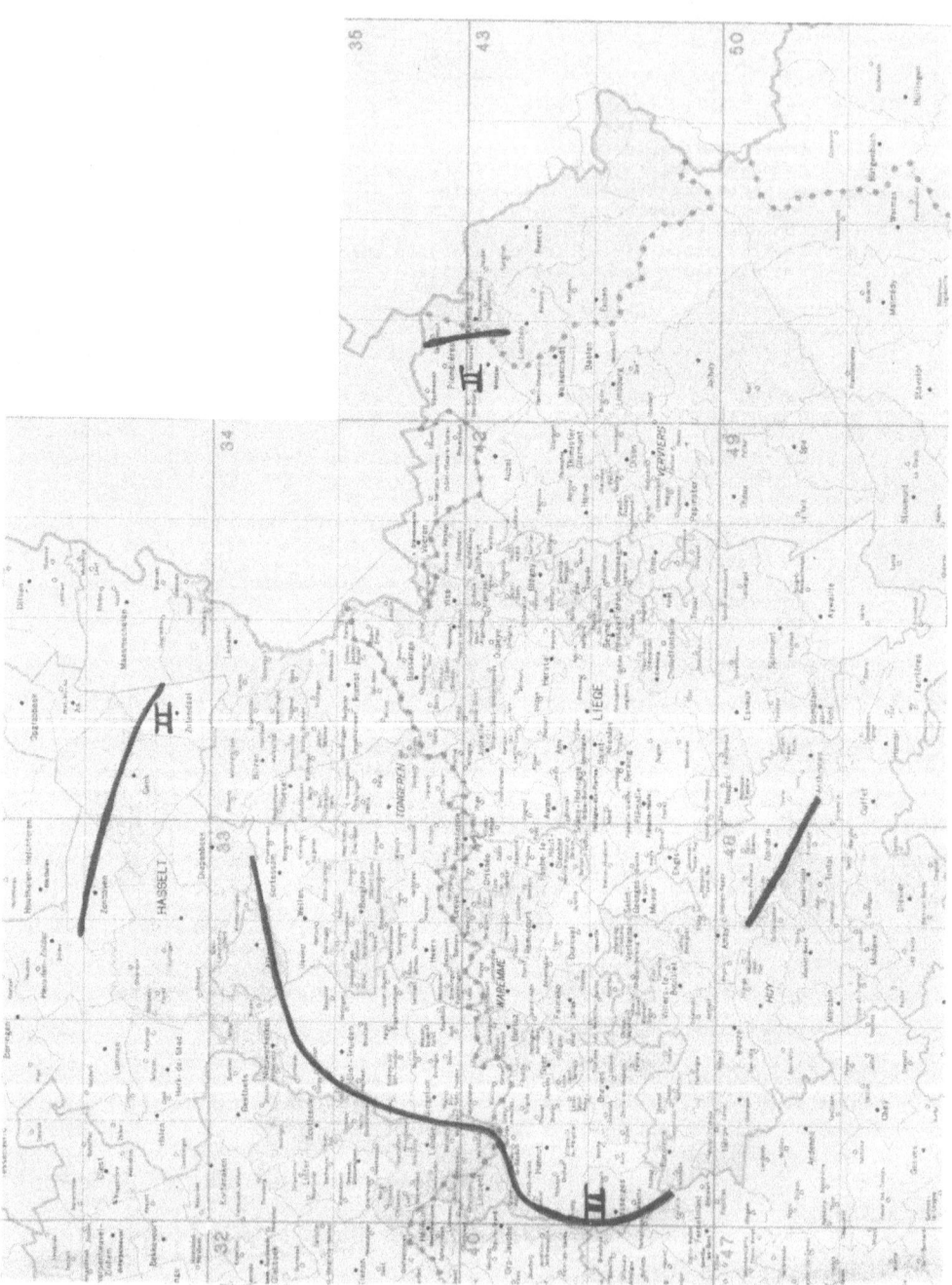

this equation was solved for
$$\begin{array}{c} I_7 - I_6 \\ I_6 - I_5 \\ I_7 - I_5 \end{array}$$

With values of h ranging from 2 to 5 km.

The solution was chosen in such a way that at the depth h corresponds a same value of k for the three intensity differences {figure 4}.

The following results were obtained : $k = 0.033 \text{ km}^{-1}$
$$h = 4.1 \text{ km}$$
$$I_o = 7.44,$$

which are in good agreement with the microseismic values.

The magnitude was also computed using the relation :

$$M = 0.52 \ I_o + 1.56 \ \log_{10} h + 0.7 \ k.h$$

The macroseismic magnitude is M = 4.92.

On the basis of the macroseismic map, the attenuation curve has been drawn for the main shock {figure 5}.

DETAILED MACROSEISMIC INVESTIGATIONS

In the most damaged area, a relatively high vertical downward acceleration combined with an horizontal component was observed in many places.

This fact, together with local site conditions like the topography (more damages were observed on the top of the hills than in the valley), local soil, presence of underground coalmine galleries explain the heavy damages in domestic property.

On the other side, much of the more severe damage suffered by the buildings was mainly due to their high vulnerability, most of them pre-First-World brick constructions already weakened through neglect, inproper repairs and alterations mainly due to the air pollution in this industrial area and to collapse of underground coal mines.

There was no significant damage to the more modern and better built houses in spite of the relatively high ground accelerations that must have prevailed locally.

It must be noted that in some places, the damages caused by the earthquake of november 8 have been growing even some weeks after, probably due to the collapse of underground mine workings.

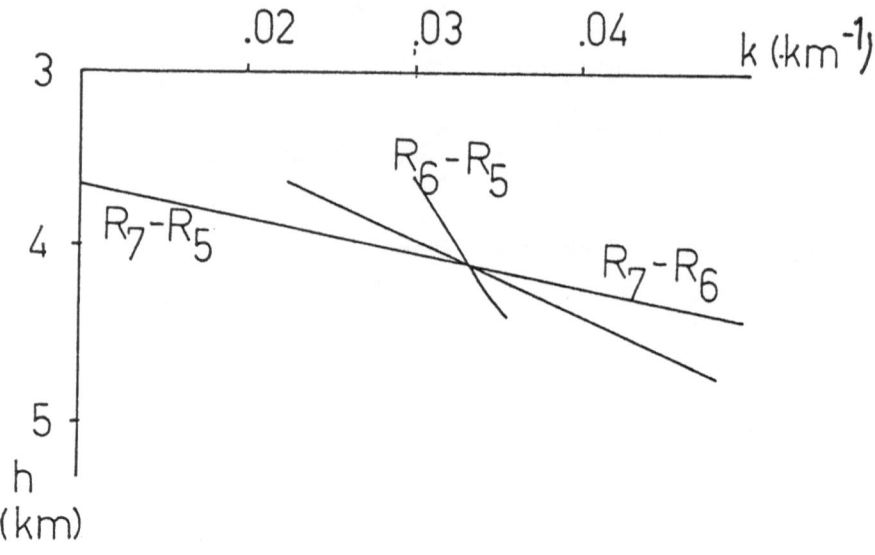

Figure 4. Relation between depth (h) and attenuation factor (k) for the intensity differences R_7-R_5, R_6-R_5, R_7-R_6.

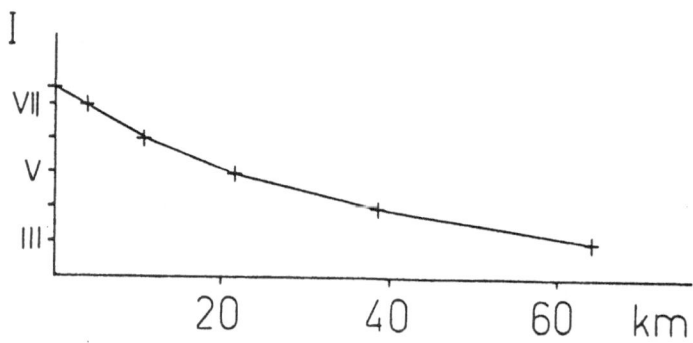

Figure 5. Intensity attenuation curve for the main shock.

TABLE 1: District of Liège : situation at March 14th, 1984

1. Intoxication by CO (as a result of damages to chimneys)
   4 persons killed
   20 persons transfered to hospital

2. 22 decrees of demolition

3. 38 decrees of inhabitability for 36 private houses and two
   churches

4. 235 families had to be rehoused by the administration.
   (Many families have temporarily moved to the houses of a
   member of the family).

Detailed informations concerning the damages to constructions are
available for all the district.

TABLE 2. Suburb of St.Nicolas (=13 km^2): Informations available

on december 29th, 1983.

CHIMNEYS	162 requests of intervention for a chimney
	813 interventions made by the firemen
	4 interventions made by the owner of the house
FACADES	211 requests for stanchion of the facade
	356 stanchions made by official services
ROOFCOVER	2 demands
	100 furnished by the official services
HABITABILITY	55 requests for a visit of an official expert
	451 visits made by an official expert
	93 conclusions for inhabitability by an official expert
	155 conclusions for habitability by an official expert
DEMOLITION OF HOUSES	15 houses had to be demolished completely.

Within an area of more than 5 km^2, thousands of buildings were affected by the earthquake. Hundreds of these were in some danger to collapse due to structural damage while many of them had to be knocked down or subjected to major structural repairs.

The weakest points in the constructions were, of course, the brick chimneys and thousands of them were affected. Two people died immediately after the earthquake and a few were wounded.

Most of the industrial installations lay outside the epicentral region and a little of them suffered damages.

In the installations of Cockerill in Seraing, a gazometer as well as the beams in ferro-concrete under the coke ovens were slightly damaged.

QUANTITATIVE ESTIMATES OF THE DAMAGES

In cooperation with the administrations of the most affected districts a detailed study of the damages is actually pursued. Definitive results cannot be presented already but at least a better evaluation of the total cost of the earthquake of Liège will be available within a few months.

The data already available are presented here in a few diagrams
and maps   for the district of Liège   :   table 1
             for the district of St. Nicolas   :   table 2
             for the district of Flémalle   :   table 3

CONCLUSIONS

Actually, it seems to be difficult to give really a precise idea of the cost of the earthquake of november 8, 1983.

The situation was so catastrophic that the Belgian Government decided to declare the neighbourhood of Liège "disaster zone". On the basis of the work done by the administrations for the "Fonds des Calamités" in order to allocate some money for repairs, a better estimation will be made.

The most urgent problem seems to be actually the immediate repairs which how to be made.

Evidently, this earthquake has damaged the weakest points in the structures, like chimneys and lintels but has also created

TABLE 3 : District of Flémalle : situation at 15th November 1983.

CHIMNEYS	339 chimneys were declared to be in danger to collapse and had to be repared within a few days after the earthquake. Damages to chimneys were evidently accompanied by damages, sometimes important, to the roofs.
	34 chimneys were completely destroyed by the earthquake.
FACADES	163 facades were declared to have important cracks.
INTERIOR OF THE HOUSES	247 houses were declared to have important damages inside : ceilings which felt in, cracks in chimneys and walls, ruptures of delivery pipes for water or gas, etc ...

new weaknesses. The buildings should have to be restored not only to their original strength but to a strength adequate for the proper stability of the structure as a whole. In particular, it must be taken care of strengthening the structures of the buildings in order to resist futur earthquakes or other loads with as little damage as possible.

BIBLIOGRAPHY

Medvedev S., Sponheuer W., Karnik V. : Neue Seismische Skala. Veröff. Inst. Jena - Heft 77 - Akademie Verlag - Berlin 1964.

Kövesligethy R. : Seismicher Stärkegrad und Intensität der Beben - Gerl. Beitr. Geophys., Vol VIII, Leipzig 1907.

MACROSEISMIC OBSERVATIONS OF THE LIEGE EARTHQUAKE OF
NOVEMBER 8, 1983 IN THE NETHERLANDS

Gerhard Houtgast

Royal Netherlands Meteorological Institute
De Bilt, The Netherlands

The hundreds of messages received in the Netherlands
both by telephone and in writing concerning the macroseismic data of the Liège earthquake have been collected
and finally been compiled into a map (see figure).

In the Maastricht region of Southern Limburg
the earthquake was felt by many people according to
intensity V on the Modified Mercalli Scale: buildings
trembled throughout, many people woke up. No reports
of damage were received however. Farther North a gradual
decrease of the macroseismic effects have been noticed,
without local variations in the observations.
Notably, many cases of felt vibrations observed by
individual people indoors and at rest, especially on
the upper floors of high structures, have been reported
from large cities as Amsterdam, Utrecht and Arnhem.
In the most favourable circumstances (such as people
at rest, awake in bed or sitting in a chair at upper
floors) the earthquake was felt at a maximum distance
of 230 kilometers from the epicenter (Alkmaar).

P. Melchior (ed.), Seismic Activity in Western Europe, 311–312.
© 1985 by D. Reidel Publishing Company.

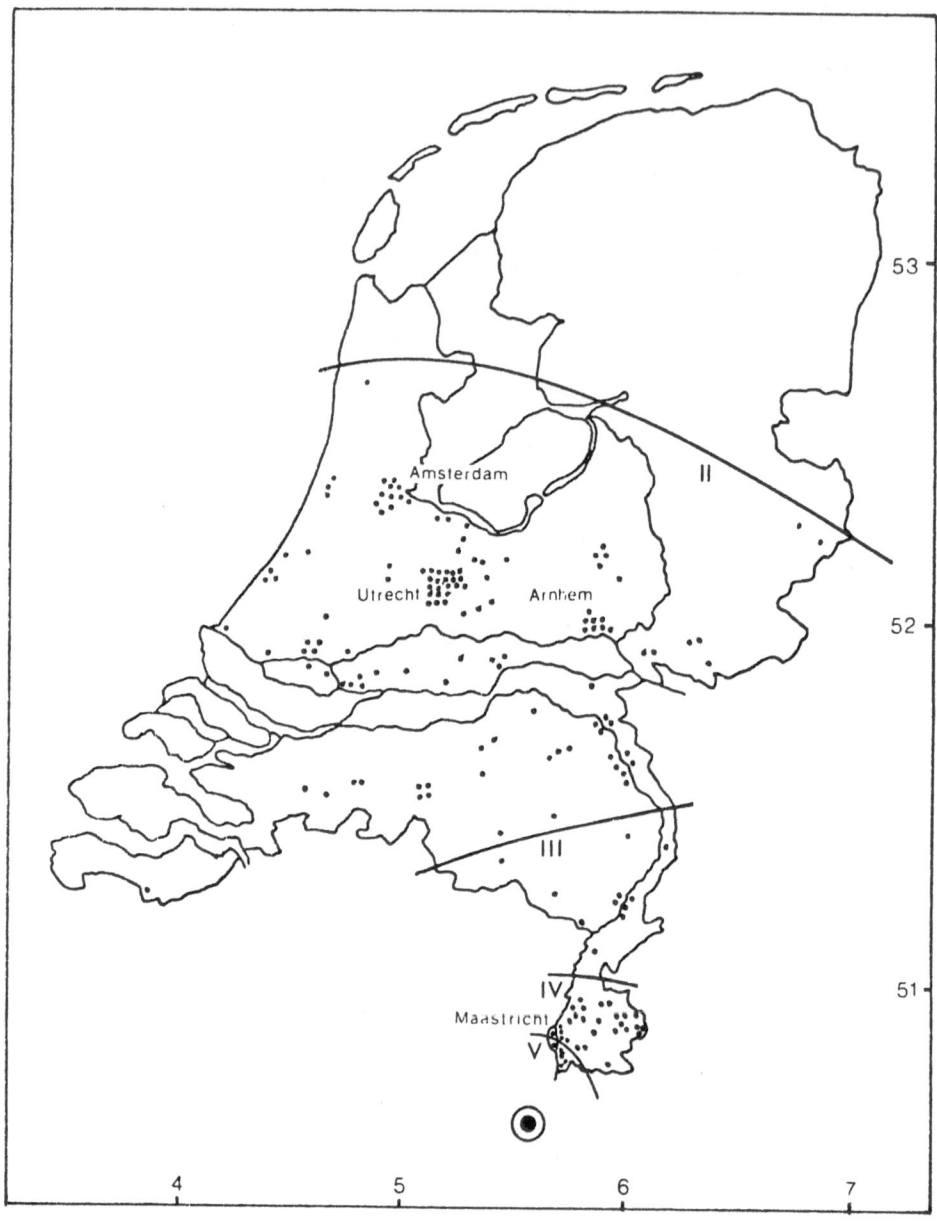

Figure: Location in the Netherlands from where messages by
telephone and in writing have been reported.
The zones of intensity indicated are according to the
Modified Mercalli scale.

MACROSEISMIC OBSERVATIONS OF THE LIÈGE EARTHQUAKE ON NOVEMBER 8, 1983, IN WEST GERMANY

L. Ahorner

Abteilung für Erdbebengeologie, Geologisches Institut, Universität zu Köln, Federal Republik of Germany

Abstract

The damaging Liège earthquake on November 8, 1983, at 0:49 GMT with epicentral intensity VII MSK and local magnitude 5.1 has been felt in West Germany as far as Detmold (270 km), Marburg (230 km) and Darmstadt (240 km). The Rhine valley between Koblenz and Duis= burg and the Lower Rhine Embayment were shaken with intensity IV. In the Belgium-Germany border region near Aachen (45 km east of the epicenter) a maximum intensity IV-V has been observed. No sig= nificant case of structural damage was reported from the West German territory. From the comparison of recorded ground motion with macro= seismic intensity observed at the sites of the seismograph stations it is evident that intensity IV of the MSK-scale can be correlated with peak values of ground velocity in the order of 1 mm/sec.

Introduction

The Liège earthquake on November 8, 1983, at 0:49 GMT, causing slight to moderate damage in the Liege area of eastern Belgium, is the most severe seismic event in the northwestern part of Central Europe since more than thirty years. From the recordings of the seismic station network in West Germany a local magnitude of 5.1 ± 0.2 of the Richter scale has been calculated (see Ahorner and Pelzing, this volume). The maximum observed intensity in the epicentral region of St. Nicolas, a western suburb of the town of Liège, was about VII of the MSK-scale. One person was killed immediately by the quake, another one by a heart-attack, thirty persons were injured. The struc= tural damage was concentrated to a rather small region around the

313

P. Melchior (ed.), Seismic Activity in Western Europe, 313–318.
© *1985 by D. Reidel Publishing Company.*

epicenter with a radius of about 10 km. This points to a shallow-focus event. The focal depth determined from the macroseismic intensity distribution is about 4 km (see De Becker and Camelbeeck, this volume). The Liège earthquake has been felt, in spite of the nightly time, over a large area which covers the total Belgium territory and also wide parts of the Netherlands, West Germany, Luxembourg and northeastern France. The macroseismic observations in West Germany are described in more detail in the following text.

Macroseismic Observations in West Germany

The Department of Earthquake Geology of the Geological Institute of the University of Cologne started immediately after the Liège earth= quake a macroseismic inquiry about the observed effects on the terri= tory of West Germany. People were asked with the help of newspapers, radio- and television stations to report their observations concerning the earthquake. More than 200 reports from 70 different localities reached our institute as a response to this inquiry. Some additional reports have been made available by Dr. Baier from the Geophysical Institute of the University of Frankfurt.

The macroseismic intensity at each locality was estimated on the basis of the twelve-degree MSK-scale (a modified Mercalli-scale). The resulting macroseismic map is shown in Figure 1. The radius of the felt area (with respect to the macroseismic epicenter near Liège) is about 250 km in West Germany. The most distant localities reporting the shock are Detmold (270 km) in eastern Westphalia, Marburg (230 km) at the eastern border of the Rhenish Massif, and Darmstadt (240 km) in the northernmost part of the Upper Rhinegraben. Only a few persons, living in a quiet environment, have observed the earthquake at these far distant places (intensity III).

More distinct effects with intensity IV were reported from the Rhine valley between Koblenz and Duisburg and from the central and southern parts of the Lower Rhine Embayment, where sleepers were wakened, and windows, dishes and doors rattled. In some cases unstable small objects like bottles and books were overthrown.

The maximum intensity IV-V has been observed on the German territory in the border region near Aachen (situated about 45 km east of the epicenter of the Liège earthquake). No significant case of structural damage was reported from the German area.

The general picture of the macroseismic intensity distribution in West Germany exhibits a clear focussing of seismic energy in a north= eastern direction. This is evident mainly from the elongation of the isoseismal IV and the positive intensity anomaly at the northeastern border of the Lower Rhine Embayment between Cologne (Köln) and Duisburg.

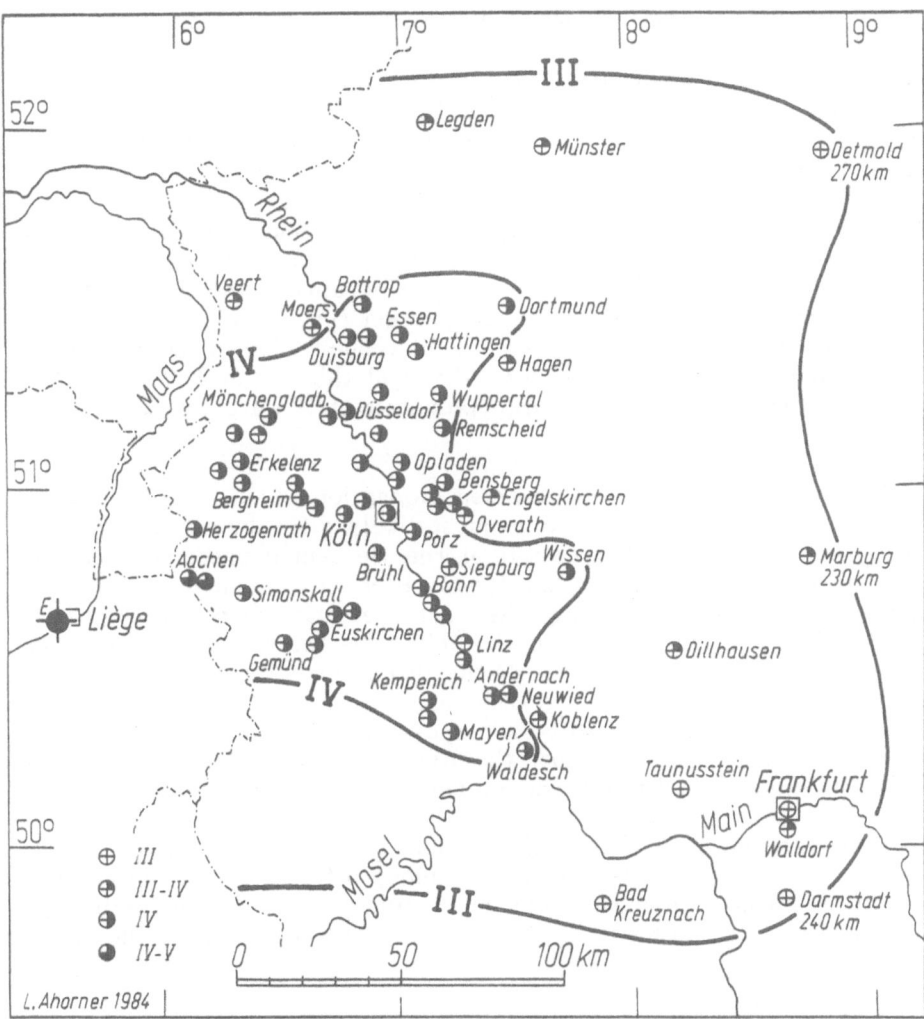

Figure 1.   Macroseismic observations of the Liège earthquake on November 8, 1983, in West Germany. Intensities are based on the MSK-scale. The epicenter near Liège is marked by a black dot with the letter E.

The elongation of macroseismic isoseismals in a southwest-northeast direction is a common effect which has been observed in case of many earthquakes occurring in the past in the northwestern foreland of the Rhenish Massif between Aachen and Brussels. Isoseismal maps showing this effect are published for instance in the earthquake catalogues of Sieberg (1940) and Sponheuer (1952). More recent examples are given by the macroseismic fields of the earthquakes on June 25, 1960 (Ahorner and Van Gils 1963) and February 18, 1971 (Ahorner and Pelzing 1983).

The focussing of seismic energy towards northeast might be partly due to the radiation pattern of the earthquake source which is control= led by the focal mechanism. From focal mechanism studies it is known that the prevailing seismotectonic dislocation type of the Belgian earth= quake zone is a left-lateral strike slip movement along a nearly west-east trending fault plane (Ahorner 1975, and this volume). This mecha= nism type gives a prevailing energy radiation of S waves in eastern direction.

The main reason for the observed elongation of the macroseismic field to the northeast, however, is most likely given by the structural heterogeneity of the rock material of the Hercynian basement complex in the deeper underground of the Lower Rhine Embayment, which is characterized by strong folding with fold axes striking southwest-north= east. In the direction parallel to the fold axes the propagation of seis= mic energy is much better than in the direction perpendicular to the fold axes. Regional soil amplification effects, connected with the wedge-like deposition of Tertiary soft-soil layers at the eastern border of the Lower Rhine Embayment against the Rhenish Massif, might be additional factors in explaining the positive intensity anomaly observed.

The dense network of seismic stations in the Lower Rhine Embay= ment (see Ahorner and Pelzing, this volume) offers the possibility of correlating seismic ground motion, as recorded by the seismographs, with macroseismic intensity observed at or near the sites of the seismic stations.

Table 1 gives the recorded peak values of ground velocities and the corresponding frequencies of the Liège earthquake for the stations Wassenberg (WAS), Jülich (JUE), Bergheim (BEG), and Bensberg (BNS). The stations are equipped with high-dynamic digital seismographs. The seismometers are of the short period type with natural periods between 0.8 and 1.0 seconds and have velocity transducers. Seismograms of the Liège earthquake recorded by these stations are published in the paper of Ahorner and Pelzing (this volume) . The station underground con= sists of thick soft-soil layers (Tertiary and Quaternary sediments with a thickness of 300 m to 1000 m) at the stations WAS, JUE and BEG, whereas the station BNS has firm rocks (Devonian shales and sandstones) as underground. The stations are located in the distance range between 70 km and 124 km northeast of the epicenter of the Liège earthquake.

At the sites or in the close vicinity of all these stations the macro=
seismic intensity IV has been observed. From the data of Table 1 it
is evident that the intensity IV of the MSK-scale can be correlated
in our region with recorded peak values of ground velocity in the
order of 1 mm/sec. The prevailing frequency of strong ground motion
is about 1.4 Hz to 2.6 Hz in the distance range of 70 km to 124 km
from the source of a magnitude 5.1 shock.

The recorded peak values of ground velocity at the four stations
show a rather low attenuation with distance. This confirms the idea
of a favoured energy propagation in northeastern direction underneath
the Lower Rhine Embayment as deduced from the elongation of macro=
seismic isoseismals in Figure 1.

Table 1.   Macroseismic intensities and recorded peak ground velocities
           at seismic stations in the Lower Rhine Embayment

Station		Distance km	Intensity MSK	Peak Ground Velocity NS(mm/s)	EW(mm/s)	Frequency Hz
WAS	Wassenberg	70	IV	0.760	0.988	1.4
JUE	Jülich	74	IV	0.757	0.999	1.9
BEG	Bergheim	90	IV	1.118	0.938	2.6
BNS	Bensberg	124	IV	0.688	1.000	2.6

References

Ahorner, L. (1975): Present-day stress field and seismotectonic block
    movements along major fault zones in Central Europe. -
    Tectonophysics, 29: 233-249; Amsterdam

Ahorner, L., Van Gils, J.-M. (1963): Das Erdbeben vom 25. Juni 1960
    im belgisch-niederländischen Grenzgebiet. - Sonderveröff.
    Geol.Inst.Univ.Köln, 9: 1-27; Köln

Ahorner, L., Pelzing, R. (1983): Seismotektonische Herdparameter von
    digital registrierten Erdbeben der Jahre 1981 und 1982 in
    der westlichen Niederrheinischen Bucht. - Geol.Jb., E26:
    35-63; Hannover

Ahorner, L., Pelzing, R. (1985): The source characteristics of the Liège
    earthquake on November 8, 1983, from digital recordings
    in West Germany . - (this volume)

De Becker, M., Camelbeeck, T. (1985): Macroseismic inquiry of the
        Liège earthquakes of November 8, 1983. - (this volume)

Sieberg, A. (1940): Beiträge zum Erdbebenkatalog Deutschlands und
        der angrenzenden Gebiete für die Jahre 58 bis 1799. -
        Mitt.deutsch.Reichserdbebendienst 2: 1-111; Berlin

Sponheuer, W. (1952): Erdbebenkatalog Deutschlands und der angren=
        zenden Gebiete für die Jahre 1800-1899. - Mitt.deutsch.
        Erdbebendienst 3: 1-195; Berlin

A TEMPORARY ARRAY SEARCH FOR AFTERSHOCKS OF THE 1983 NOVEMBER 8, LIÈGE, BELGIUM, EARTHQUAKE

W.P. Aspinall[1] and G.C.P. King[2]

[1]Principia Mechanica Ltd., London U.K.
[2]Dept. of Earth Sciences, Cambridge University, U.K.

An earthquake of Mb4.9 caused damage of intensity VI to VII in Liège, Belgium on 8 November 1983. A temporary array of eight seismographs deployed between 10 and 16 November detected seven probable after-shocks, seven mine-related events and seven near events. The aftershocks imply a vertical source plane striking NE-SW and a hypocentral location directly below the area of maximum damage.

1. INTRODUCTION

At 01.49 a.m. (local time, 00.49 UTC) on 8 November 1983 the town of Liège in Eastern Belgium was struck by an earthquake of approximate body-wave magnitude $M_b 4.9$.

Eight temporary microseismic stations were deployed in and around Liège between 10th and 16th November 1983 using Sprengnether MEQ-800 smoked-paper recorders with Mark Products L4C $1H_z$ vertical seismometers. The selection of sites was severely constrained by having to work in a built-up area and individual station gains had to be adjusted to suit the conditions. At some sites wind noise was troublesome (Well, Chapel) and in the centre of the town, House and Hotel were affected by traffic noise although this was less severe over a holiday weekend than it might have been otherwise. The locations of the stations are given in Table 1.

P. Melchior (ed.), Seismic Activity in Western Europe, 319–329.
© *1985 by D. Reidel Publishing Company.*

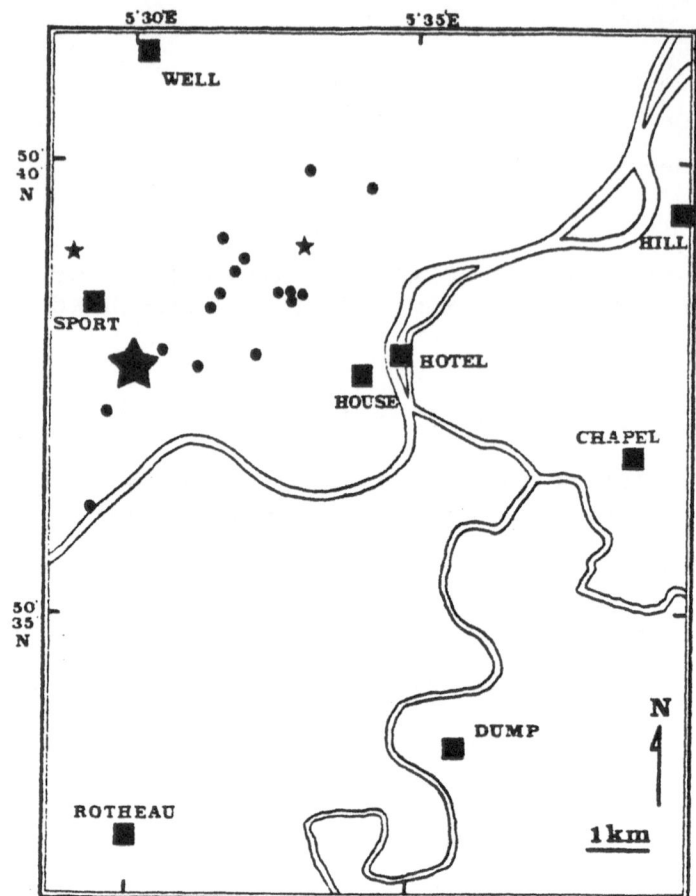

FIG. 1   Epicentres of fourteen events detected by the
          temporary array, 10-16 November 1983.
          Stations are marked as squares and the stars
          mark the preliminary epicentres of the main
          8 November earthquake and two aftershocks as
          determined by Uccle Observatory.

Table 1 Liège temporary array-station locations

Station	Lat	Long	Elevation
Well	$50^{\circ}41.28$N	$5^{\circ}30.10$E	160m
Sport	$50^{\circ}38.53$N	$5^{\circ}29.36$E	168m
Hill	$50^{\circ}39.51$N	$5^{\circ}39.43$E	120m
Hotel	$50^{\circ}37.80$N	$5^{\circ}34.52$E	60m
House	$50^{\circ}37.57$N	$5^{\circ}33.73$E	115m
Dump	$50^{\circ}33.52$N	$5^{\circ}35.71$E	185m
Rotheau	$50^{\circ}32.52$N	$5^{\circ}29.88$E	270m
Chapel	$50^{\circ}37.01$N	$5^{\circ}38.67$E	222m

2. OBSERVATIONS AND LOCATIONS OF AFTERSHOCKS, MICROEARTHQUAKES AND MINE EVENTS

Twenty-one events were detected in the five-day period of instrumental operation (Table 2). Of these, three were recorded by 7 stations, two by 6 stations, seven by 5 stations and two by 4 stations. Locations of the events were computed using the HYPOELLIPSE program (Lahr, 1980) using P-phase arrivals only and a simple half-space crustal velocity model with V_P = 6.0 km/s. A preliminary run , using Ahorner's (1983) values of V_P from the nearby Rhenish Massif area, exhibited solution clustering in alternate layers of the model and was discarded as being probably artefactual. In Fig. 1 are given the epicentral locations of 14 of the events located by the array together with preliminary solutions by Uccle Observatory (Camelbeeck and de Becker, 1983) for the main shock and two aftershocks using fixed European stations. Seven other microearthquakes picked up by the temporary array fell outside the boundaries of Fig. 1.

The microearthquakes fall close to the Uccle solutions and suggest a NE-SW trend. The epicentral centroid of seven events (i.e. those with non-surface hypocentres) is at $50°30.5'$ N $05°$ $32.0'$E and is consistent with the area of maximum epicentral intensity. This is about 3km north-east of Uccle's main shock location. When the microearthquake hypocentres are plotted onto a vertical plane orthogonal to the NE-SW trend it can be seen (Fig. 2) that half the events cluster close to the surface and the remainder fall tentatively onto a near vertical plane. The shallow events are assumed to be mine-related as that part of Liège has been extensively worked with shafts and galleries. The deeper aftershocks range in depth from 1km to 7kms and are not inconsistent with the depth of 4km given by Uccle for the mainshock.

The few aftershocks recorded thus imply a source of about 6 kms diameter or an area A of about $3 \times 10^9 \mathrm{cm}^2$. Assuming a value of $3 \times 10''$ dyne.cm^{-2} for the modulus of rigidity μ and a preliminary seismic moment of 1.2 $\times 10^{23}$ dyne.cm (Camelbeeck and de Becker, 1983), the average slip D in the earthquake is given by:

$$D = \frac{M_o}{\mu A} \sim 15cm$$

If this is taken to occur at a depth of 4km then surface displacement from dislocation theory would be of the order of a few millimetres. This is consistent with the damage observed.

3. FOCAL PARAMETERS OF THE AFTERSHOCKS

The hypocentral solutions obtained for the fourteen events are listed in Table 2. The quality of the locations are indicated by the RMS quality (as defined for HYPOELLIPSE; Lahr, 1980) and by the condition number γ of the matrix of linear equations for each. γ is the ratio of the largest to smallest singular values of the matrix and a condition number of order 10^3 or greater implies on error in the solution comparable to the solution itself. (see Lee and Stewart, 1981; p121-22).

The Sprengnether MEQ-800 - Mark Products L4C combination has a peak response at about $20H_z$; Bakun and Lindh (1977) give a relationship for total signal duration to local magnitude for this system for close microearthquakes in the range $0.5 < M_L < 1.5$ as follows:

$$M_L = 0.71 \log T + 0.28$$

where T is total signal duration in seconds. This relationship has been used in Table 2 and gives local magnitudes ranging from $M_L 0.6$ to $M_L 1.1$ for the aftershocks and from $M_L 0.5$ to $M_L 0.9$ for the surficial mining events. A comparison with estimates of local magnitude using the calibration of the Belgian stations (Camelbeeck and de Becker,1983) has not been made so far.
The polarities of a few clear first-motions were read for the seven aftershocks and plotted in composite form. However, no coherent pattern emerged and in fact there were big gaps in station coverage on the critical NE-SW axis (see Fig. 1). Thus no additional insight was afforded into the focal mechanism of the principal event.

Table 2 Liège temporary array – Hypocentral solutions

Origin	Time	Lat	Long	Depth	N	M_L	Q	γ	Comment
83 11	12 02 53 56.1	50N38.80	5E31.73	0.1	5	0.6	B	6×10^3	Mine event
83 11	12 03 05 04.0	50N38.96	5E38.96	0.3	5	0.5	B	1×10^3	Mine event
83 11	12 10 24 22.1	50N38.54	5E32.62	1.1	6	1.1	B	$3 \times 10'$	Aftershock
83 11	12 15 39 09.5	50N38.57	5E32.46	0.2	6	0.7	B	9×10^2	Mine event
83 11	12 18 51 56.1	50N39.90	5E33.06	7.1	5	0.8	D	$2 \times 10'$	Aftershock
83 11	12 21 20 47.6	50N38.40	5E31.30	0.5	7	0.9	B	2×10^2	Mine event
83 11	12 23 05 02.5	50N38.52	5E32.79	4.1	7	0.8	D	7×10^0	Aftershock
83 11	13 21 54 35.6	50N37.24	5E29.46	0.6	5	0.8	D	4×10^2	Mine event
83 11	13 23 30 50.5	50N37.91	5E30.76	0.2	5	0.8	D	2×10^3	Mine event
83 11	14 03 48 11.6	50N38.57	5E31.45	0.2	5	0.7	B	2×10^3	Mine event
83 11	14 09 41 34.0	50N39.19	5E31.46	1.2	4	0.8	C	$5 \times 10'$	Aftershock
83 11	15 06 41 08.1	50N38.59	5E32.90	5.1	4	0.6	D	$7 \times 10'$	Aftershock
83 11	15 21 03 13.0	50N37.71	5E31.32	4.0	5	0.8	D	$1 \times 10'$	Aftershock
83 11	15 22 38 37.4	50N37.88	5E32.09	2.0	7	0.8	B	$1 \times 10'$	Aftershock

Origin time is local time; N is number of stations recording event; M_L is local (total duration) magnitude; Q is RMS quality (HYPOELLIPSE); γ is the condition number of the matrix of partial derivatives of the linear equations

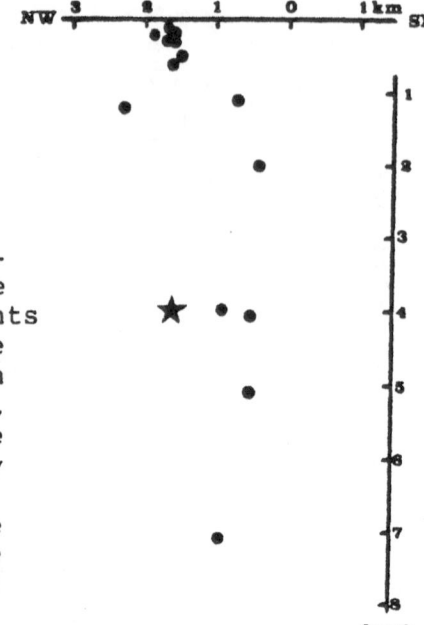

FIG. 2   Cross-
section of the
hypocentres of events
detected by the
array, projected on
to a NW-SE plane.
The star marks the
Uccle Observatory
main event hypo-
centre, and the
cluster near the
surface are assumed
to be mine-related
events.

4.   ARRAY SENSITIVITY

The temporary stations were sited to afford an optimum
capability for event location; in particular two
central stations were used for good depth control
following Duschenes et al. (1983).   The five most
commonly used stations, Well, Hill, House, Dump and
Rotheau (Fig. 1) were used as a sub-net and analysed
for their precision of location in the area in which
microearthquakes were found.   This was done by
deriving correct travel times for a given grid point
on the map, perturbing the arrival times with random
errors and then rerunning the location program.   The
standard error of changes to P-arrival times was 0.1
second to reflect the accuracy with which these phases
could be timed in practice.   For each of the 64 points
on the grid, five dummy hypocentres were computed,
giving an estimate of the variance of epicentre
locations and depth determinations, for earthquakes of
assumed depth 4km.   The one standard deviation
contours of precision for epicentral location are
shown in Fig. 3a and range from ±40m near the centre of

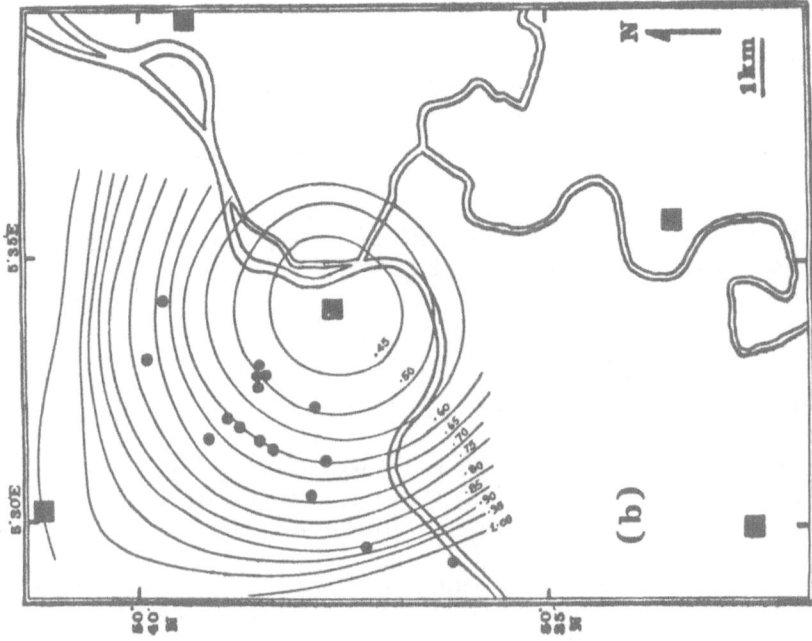

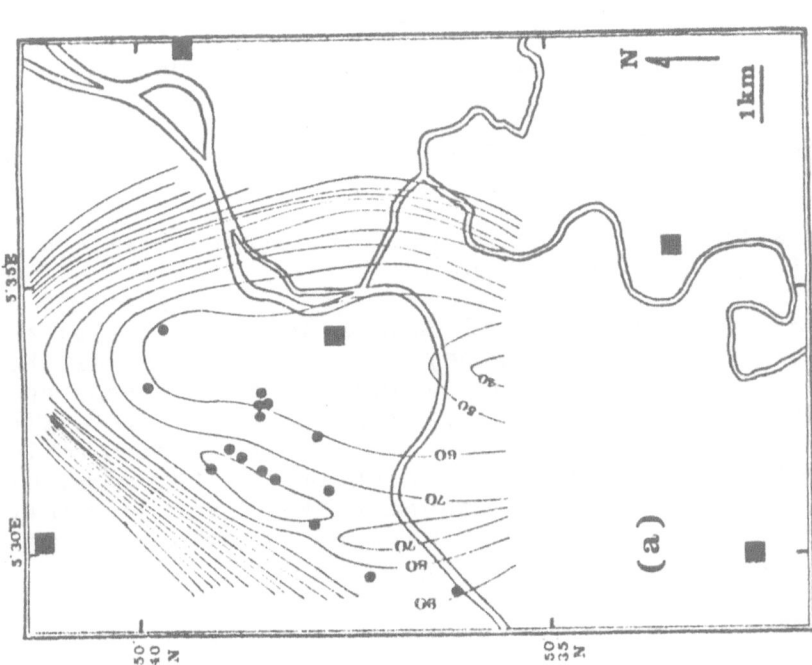

FIG. 3   Analysis of the sensitivity of location of a five-
station subset of the temporary array. a) ±1 standard deviation
contours for epicentral precision in meters, and b) ±1 standard
deviation contours for depth precision in kms.

the array to about ±200m at the edges. The contours
are reasonably flat and uniform, and do not suggest
any potential for serious mislocation of events within
the array. The one-standard deviation contours for
depth are given in Fig. 3b and these range from ±0.45km
to ±1.0km within the array. The controlling influence
of the central station, House, is clearly seen.

It must be stressed that these analyses only treat
errors in seismogram timing and reading and do not
account for uncertainties in the crustal structure
model.

5.   DAMAGE EFFECTS IN LIEGE

Damage was concentrated in the district of St.
Nicholas and its suburbs of Glain, Montagnée and St.
Gilles and the principal types observed were:

a)   failure of domestic chimney stacks, most often by
     crumbling and loss of individual bricks rather than
     failure as whole units

b)   failure of arches and lintels over door and window
     openings

c)   a few examples of diagonal cracking of walls in
     shear

d)   loss of non-structural facings, stonework parapets
     and decorative pinnacles

e)   the extension of cracks in walls already damaged
     by foundation settlement

The most prevalent form of damage was that sustained
by the older domestic chimney stacks. A survey of the
percentages damaged in each neighbourhood was made and
smoothed contours of the 1%, 33%, 75% and 100% levels
are plotted in Fig. 4. The outer bound of observable
damage is roughly circular in shape but the area of
most intense damage is elongated NW-SE and probably
reflects the effects of topography as well as local
soil conditions.

In contrast with the older masonry structures, modern buildings in Liège, especially those with reinforced concrete or steel frames, were virtually unscathed. A maximum intensity of MM VI - VII is appropriate for the damage effects observed, although it must be stated that other evidence may exist for locally higher intensities.

Several examples were encountered where displacements and diagonal shear cracks were consistent with strong shaking in a NE-SW direction, and these locations are marked in Fig. 4.

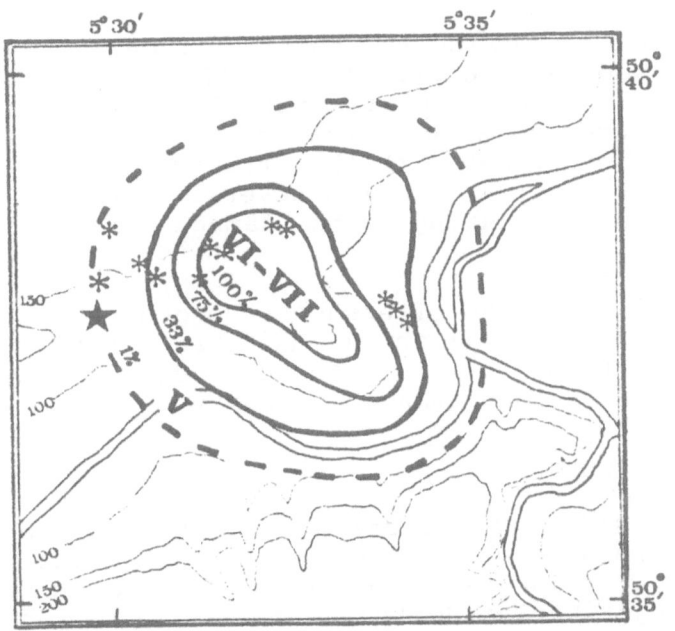

FIG. 4 Contours of the percentages of chimneys damaged on older houses in Liège and the ascribed modified Mercalli intensities. The star marks the Uccle Observatory epicentre of the 8 November 1983 earthquake, and the crosses mark locations where damage consistent with NE-SW strong shaking was observed.

6. CONCLUSIONS

A short field study using an array of 8 temporary
stations in and around Liège recorded 21 local events
of which 7 were conjectured to be aftershocks of the
1983 November 8 earthquake, 7 are probably mine-
induced and the remainder were microearthquakes or
quarry blasts in the vicinity of Liege. Unfortunately
the array was not deployed long enough to record some
aftershocks subsequent to 16 November which were
detected by the permanent Belgian seismograph stations
and thus no direct comparison can be made with the
Uccle solutions.

The aftershocks recorded were all of low magnitude
($<M_L 1.1$) and, allowing for their sparseness, imply a
vertical source area striking NE-SW. This is not
inconsistent with the preliminary focal mechanism for
the main shock given by Uccle Observatory (Camelbeeck
and de Becker, 1983) with the structural trends of
the local tectonics, or with several observations of
NE-SW strongest shaking in the damage effects in
Liege.

The locations of the aftershocks suggest that the
epicentre of the main shock may be in error by about
3km and that the event originated more directly under
the area of maximum damage intensity.

7. REFERENCES

Ahorner,L.,1983: Historical seismicity and present-day
microearthquake activity of the Rhenish Massif,Central
Europe. Chap.6.4 in "Plateau Uplift",eds. K.Fuchs et
al.,Springer-Verlag,Berlin,Heidelberg.

Bakun,S.H. and A.G. Lindh,1977: Local
magnitudes,seismic moments and coda durations for
earthquakes near Oroville,California.
Bull.Seism.Soc.Am. 67, 615-630.

Duschenes,J.,R.C.Lilwall and T.J.G.Francis,1983: The
hypocentral resolution of microearthquake surveys
carried out at sea. Geophys.J.R. astr.Soc. 72, 435-
451.

Lee,W.H.K. and S.W. Stewart,1981: "Principles and
applications of microearthquake networks". Advances
in Geophysics, Supplement 2, Academic Press, New York.

8.   ACKNOWLEDGEMENTS

The authors acknowledge the helpful collaboration of
the staff of the Observatoire Royal de Belgique, Uccle
and the work of Mrs. N. Threlfall in the field. The
study was supported in part by the Central Electricity
Generating Board (U.K.), by British Nuclear Fuels Ltd.
and by Principia Mechanica Ltd.   The loan of an MEQ-
800 by the Department of Geophysics, Imperial College,
London, is also acknowledged.

ASSESSMENT OF SEISMIC HAZARD FOR NPP SITES IN FRANCE - ANALYSIS OF
SEVERAL AFTERSHOCKS OF THE NOVEMBER 8, 1983, LIEGE EARTHQUAKE

B. Mohammadioun, G. Mohammadioun, and A. Bresson

Commissariat à l'Energie Atomique, Institut de Protection
et de Sûreté Nucléaire, Département d'Analyse de Sûreté,
Fontenay-aux-Roses, France

ABSTRACT

Current French practice for assessing seismic hazard on the
sites of nuclear facilities is outlined. The procedure calls for
as rich and varied an assortment of actual earthquake recordings
as can be procured, including earthquakes in France itself and in
nearby countries, recorded by the CEA/IPSN's own staff. Following
the November 8, 1983, Liège earthquake, suitably equipped, temporary recording stations were set up in the epicentral area in order
to record its aftershocks. Ground motion time histories and response spectra were computed for several of these, and a quality
factor Q was derived from these data for the most superficial sedimentary layers of the area. The values obtained show reasonable
agreement with ones found for similar materials in other regions.

1. INTRODUCTION

The evaluation of the seismic risk that should be taken into
account for a critical structure, such as a nuclear power plant,
involves the selection of two earthquakes supposed to be characteristic of the site, known as "reference earthquakes," which can
be considered to quantify this hazard:

1) A less severe level, the "Séisme Maximal Historique Vraisemblable," or maximum credible historical earthquake,
   abbreviated SMHV;

2) The "Séisme Majoré de Sécurité" (safety design earthquake),
   SMS, the intensity of which is one degree greater than
   the preceding (1).

331

P. Melchior (ed.), Seismic Activity in Western Europe, 331–346.
© 1985 by D. Reidel Publishing Company.

These reference earthquakes are derived from a seismotectonic survey of the area in question. They are characterized in terms of
their source parameters - geographic coordinates, magnitude, energy - by their maximum macroseismic intensity, I_0, by the function
$I = f(R)$, and by the on-site intensity I_s. For engineering purposes, these events are most often represented in the form of response spectra.

2. THE DETERMINATION OF SITE-SPECIFIC SPECTRA

Let us examine the different phases of the determination process for site-adapted spectra and research related thereto for NPP
sites in France. The first study concerns the preparation of the
seismotectonic map of France. Its purpose was to compile all data
currently available on geology and seismicity (historical record
and instrumental recordings of more recent earthquakes). This important project entailed the preparation of six thematic maps (1:
250000): a) historical seismicity, b) recent recorded earthquakes,
c) general tectonics, d) deep structures and geophysical data, e)
neotectonics, and f) lineaments. These documents are now available, and the data are in the process of being computerized. Figure 1 shows a sample study concerning southeastern France.

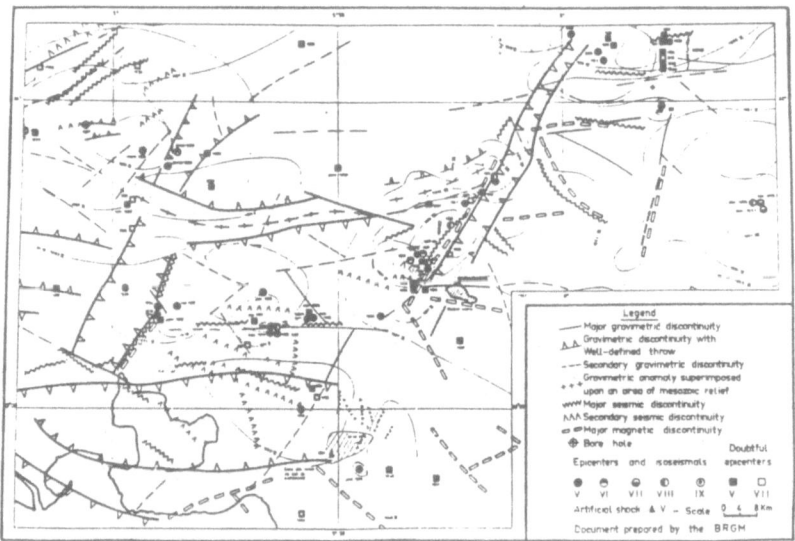

Figure 1. A regional seismotectonic study in southeast
France: interpretive map of the substratum and epicenter
locations.

The reference earthquakes for a site (SMHV, SMS), inferred
from a seismotectonic study of the region involved, are characterized first of all by their expected macroseismic intensity and the
focal distance to the site. Intensity alone (the only characteristic available for historical earthquakes), however, is patently
insufficient when seeking to define vibratory ground motion corr-
sponding to a reference earthquake. The other facet is the collecting of recordings obtained within the epicentral zone and of related information about structural and other damage (intensity).

A considerable effort has been made in various seismically active regions of the world, and notably in California, to install
strong-motion instruments for the purpose of recording severe earthquakes. An initial collection of strong-motion records from the
western United States was compiled and processed by CALTECH in 1976;
information on some characteristics of these events, namely magnitude, focal distance, maximum intensity, on-site intensity, and geological conditions at the site was likewise furnished. Furthermore,
over the past ten years, hundreds of strong-motion instruments have
been installed in certain of the more highly seismic European countries (Italy, Jugoslavia, Greece, among others). Beginning in 1978,
the Institut de Protection et de Sûreté Nucléaire (IPSN), part of
the French Commissariat à l'Energie Atomique, has undertaken to establish a data bank, known as "Sismothèque," in cooperation with
Prof. Ambraseys of the Imperial College of London. Notably, as
much European data has been included as was to be had.

However, data is still lacking in certain areas of particular
interest to safety authorities. This is especially true of the
near field, where severe damage was sustained, and, in general, of
distances to the fault of under 10 km. The Bureau d'Evaluation du
Risque Sismique pour la Sûreté des Installations Nucléaires (BERS-
SIN) of the IPSN, specialized in the measurement and interpretation
of strong ground motion, plays an active rôle, not only in acquir-
ing data from other organizations, but also in obtaining its own
recordings of events selected in line with its individual priori-
ties. Using equipment yielding good approximations of ground motion, this staff has recorded numerous aftershocks of significant
earthquakes in the western portion of Europe and the Mediterranean
basin (Friuli, El Asnam, Swabian Alps, and, most recently, Belgium).

A common practice in engineering seismology consists in choosing a standard shape of spectrum, constructed from a limited number of actual earthquake spectra, and then scaling this to a given
level of ground acceleration. This value is derived from an empirical acceleration-intensity relationship (see Figure 2). As
this intensity is, of course, estimated by comparing the effects
produced in a given locality with those described in a reference
scale, and actually involves quite a number of different parameters
(acceleration, velocity, displacement, duration, etc.), it is a
tenuous affair at best to equate it with a single parameter of motion, acceleration in this instance. A graph prepared by N. Am-

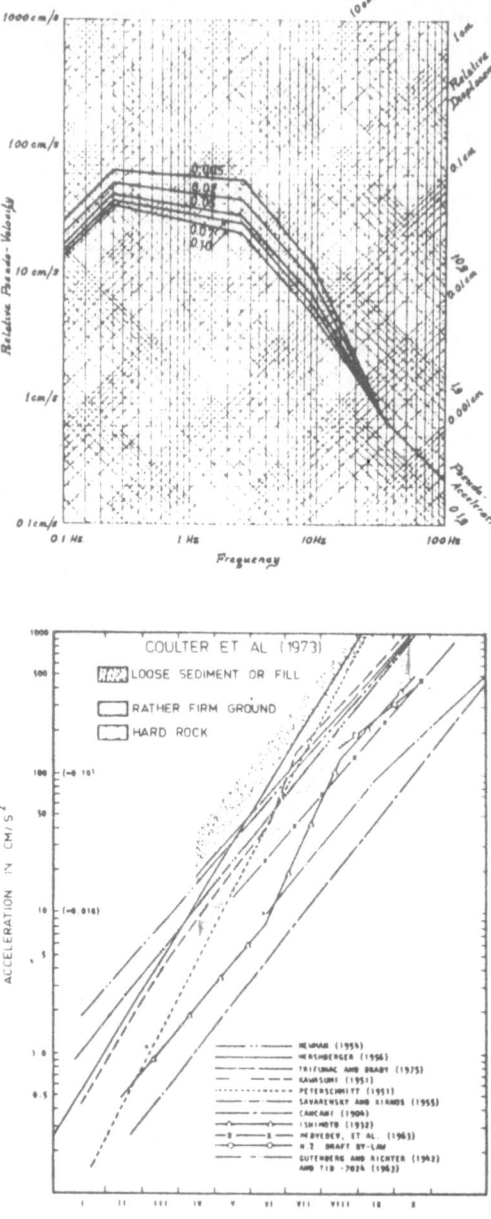

Figure 2.   Elements called upon in choosing standard
spectra: RG-1-60 spectra scaled to 0.15g (above) and
correlations between macroseismic intensity and horizontal acceleration (below).

braseys (2) shows that, as data become more numerous, an ever-increasing degree of scattering results.

The ground-motion spectrum varies, in fact, according to the amount of energy released at the source, the source configuration, the attenuation of seismic waves over the wave path, and local geological conditions. A very sophisticated scientific method for calculating a site-specific reference motion would entail a good knowledge of geological structures through geophysical investigation, together with an exhaustive inventory of all active faults and source zones. It then would proceed to compute a set of possible ground motions using mathematical models to represent the sources and propagating media. Great strides have indeed been made over the past ten years toward the application, at least on an experimental basis, of such an approach (cf. recent work by Haskell (3), Aki, and Bouchon (4), among others). This has come about thanks to advances in seismology theory, which has developed tools for calculating strong earthquake motion in the near field. These studies, however, are still in the stage of fundamental research, and their application is often limited to the comparison of computed results with actual recordings, noting the size and nature of the discrepancies, so as to readjust the starting hypotheses.

An alternative, empirical method is proposed that is currently being called upon in seismic risk analysis. Starting with a collection of strong earthquake motion recordings for different magnitudes, focal depths, and epicentral distances, obtained on a variety of geological formations, an attempt is made to define the influence upon the ground motion of the aforementioned variables. Once the effects of each factor have been isolated, correlations may be established between the spectrum and magnitude and focal distance, for instance. In a study made of the aftershocks of the May 6, 1976, Friuli earthquake (5), the influence of varying magnitude upon the spectra of recordings at relatively constant focal distances was noteworthy: an increase of energy in the low-frequency range was observed when magnitude increased (Figure 3). Wave attenuation with distance can be investigated by comparing, for a given distance, the effects of different earthquakes with the same magnitude.

Lastly, the local transfer function affects the motion recorded on a site. Transfer functions can be derived by comparing recordings on hard rock with those on alluvia, as long as other factors are held constant. If $S_0(f)$ is taken to be the spectrum at the source, $S_R(f)$, the spectrum at focal distance R, with $A(f)$ and $T(f)$ the transfer function along the wave path and that of the local geology, respectively, the following equation may be said to characterize their relationship in linear elasticity:

$$S_R(f) = S_0(f) \cdot A(f) \cdot T(f) \tag{1}$$

Trifunac, in a very detailed study (6), attempted to establish a correlation between the Fourier transform of recorded mo-

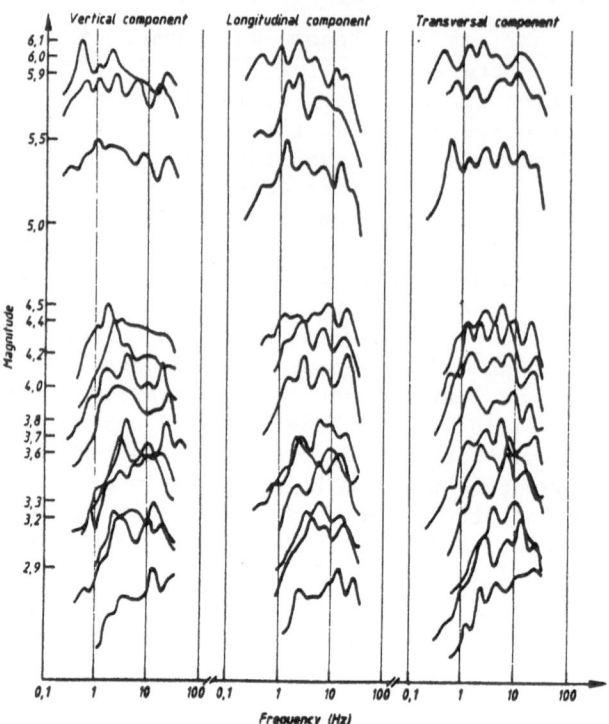

Figure 3. Variation of response spectra of Friuli aftershocks (0% damping) as a function of magnitude (focal distances 8 to 16 km).

tion and different earthquake characteristics (magnitude, focal distance, on-site geological conditions). This sophisticated study, however, shows that the size of the strong-motion recording sample was not sufficient for reliable correlations to be established (especially for site conditions: hard rock, intermediate rock, alluvium). Using a simpler approach, however, promising results were obtained in a study (7) relating all the spectra of the IPSN data bank for a given intensity to magnitude and focal distance:

$$\log S_f = K_f + \alpha_f M + n_f \log R \tag{2}$$

where K_f, α_f, and n_f are frequency-variable regression coefficients. These variables were evaluated for 46 frequencies and for the following simple and composite intensity groups: VI, VII, V-VI, VI-VII, and VII+.

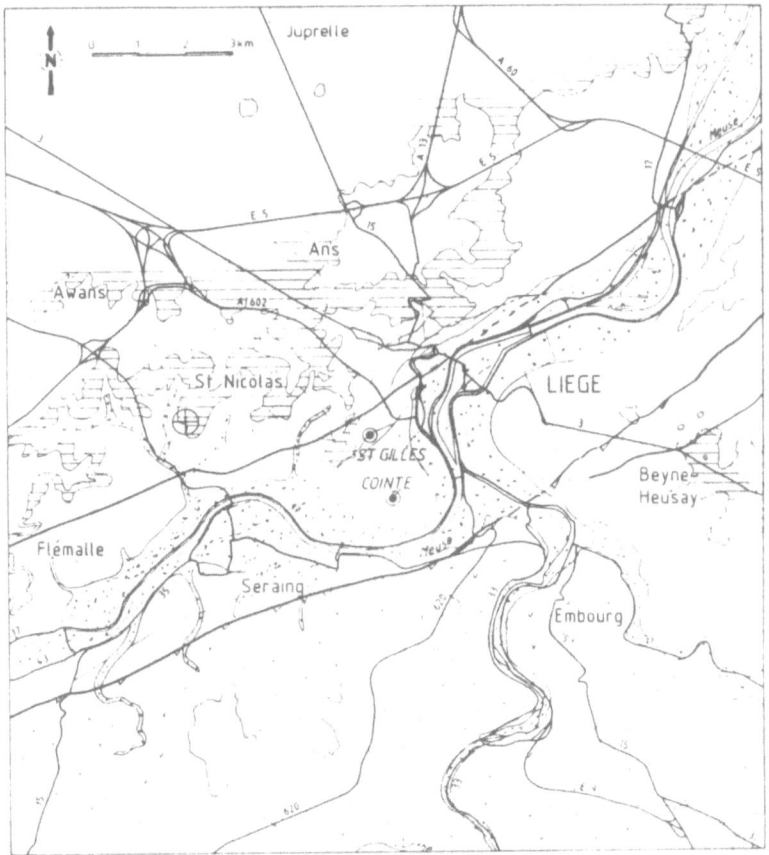

Figure 4. Geological map of Liège and vicinity, showing the locations of the CEA/IPSN recording stations.

3.   THE LIEGE EARTHQUAKE OF NOVEMBER 8, 1983

     Starting two days after the main event, the BERSSIN, with the
help of the Observatoire Royal de Belgique, began installing seismic recording stations in the epicentral area, depicted in Figure
4, in order to record whatever aftershocks might occur. The first
of these stations were equipped with SMA-1 strong-motion accelerographs, of which two were placed, on November 10th, quite near the
main shock epicenter, in the churches of Saint Nicolas and Saint
Gilles, on the off-chance that an aftershock of exceptional intensity should take place. A third instrument of this type was later
installed in the nearby French nuclear power plant at Chooz, where
it is still in operation. The transfer function of these instruments is shown in Figure 5, left. Not surprisingly, none of the

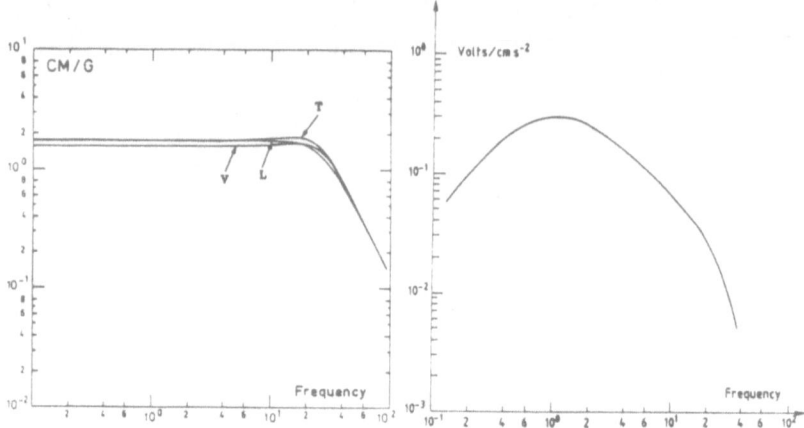

Figure 5.   Transfer functions of equipment installed in
the Liège area: SMA-1 accelerographs (left), PCM/HS-10
stations (right).

Liège aftershocks was sufficiently strong to trigger these instruments, so no recordings of this type were retrieved.
    The BERSSIN likewise installed more sensitive mobile stations
composed of HS-10, 1 Hz., velocity transducers and PCM numerical
recorders.   The transfer function of this combination appears in

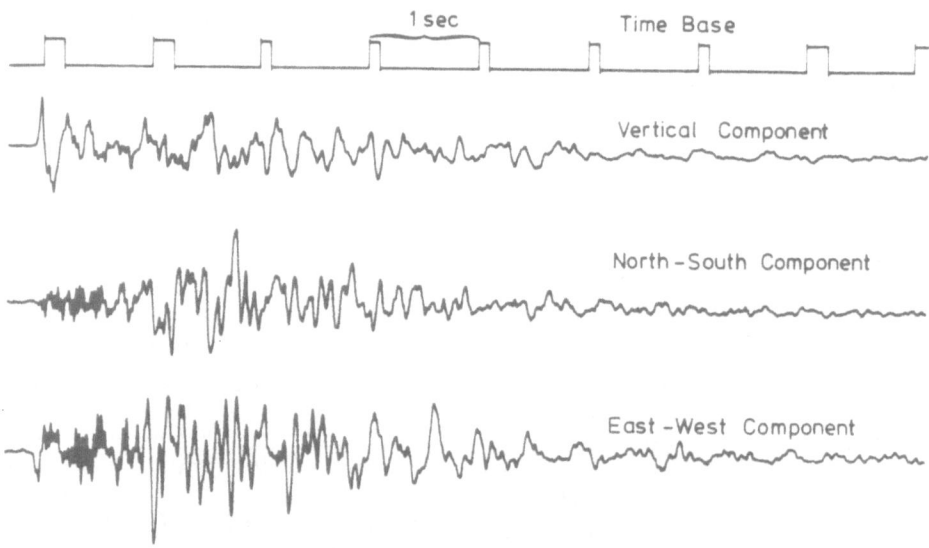

Figure 6.   Uncorrected velocities for the Dèc. 1st event (21h 52m).

Figure 5, right. Initially, they were set up in parallel, and con-
currently, with the first two SMA-1's, but, due to persisent environmental disturbances, the Saint Nicolas station was moved to a
much more favorable site made available in the Cointe Observatory,
where it continued to operate under excellent conditions until January 13, 1984, at which time no further aftershocks were expected.
A third station operated near Tihange, in the La Sarte church, from
November 14th to December 6th, but the conditions were unfavorable,
and no recordings were obtained.

Eight aftershocks in all were recorded during the period of
experiment, the first at Saint Gilles, the others at Cointe, during
December and the first week of January. The main characteristics
of the events are shown on Table 1. Figure 6 shows the three components of uncorrected velocity for the December 1st aftershock.
This, together with the single Saint Gilles record and a December
4th event were selected for processing.

The following procedure was adopted in processing the signals:

1) The recordings were digitized so as to obtain a uniform
    rate of 200 samples per second, which was possible due to
    a satisfactory signal/noise ratio;

2) The time histories were corrected for instrumental dis-
    torsion;

3) Ground acceleration, velocity, and displacement were com-
    puted;

4) Response spectra in pseudo-relative velocity were calcu-
    lated;

5) Displacement spectra were computed for 20% damping.

The ground acceleration, velocity, and displacement time histories
for one component of the Figure 6 example are given in Figure 7,
and its response spectra for five damping values in Figure 8. It
will immediately be seen that the high-frequency content *of seismic origin* in this record is quite considerable, as was the case
with all the components processed, and also presumably with those
that were not, as far as the naked eye can see. This is not unexpected, as the magnitudes were weak (probably less than 3) and
the distances appear to be in the neighborhood of 5 km.

4. THE QUALITY FACTOR OF SUPERFICIAL LAYERS IN THE LIEGE AREA

Knopoff (8,9) conceived of a factor he designated Q, for quality, that would account for all inelastic effects, and that he defined as:

$$\delta E/E = 2\pi/Q \tag{3}$$

where δE is the energy loss per cycle for a lightly damped, sinu-

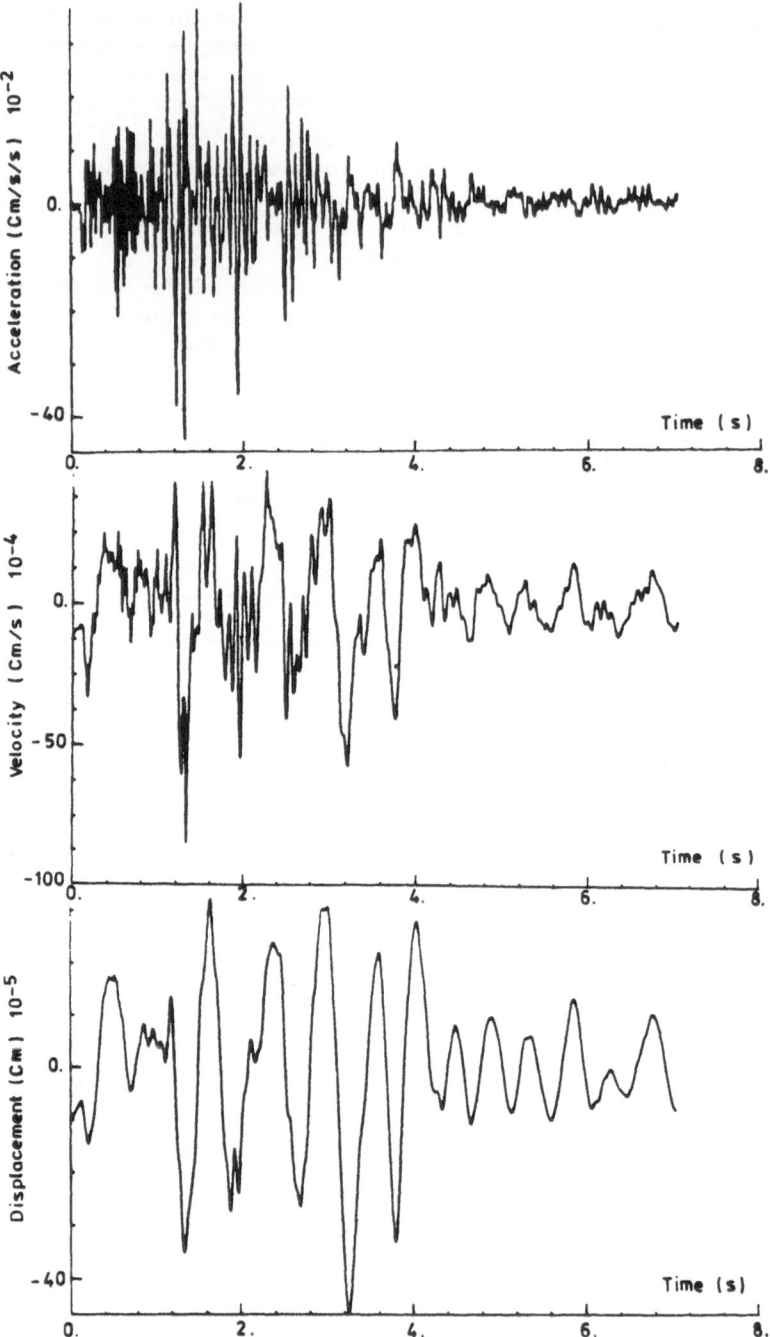

Figure 7. Corrected acceleration, velocity, and displacement time histories for the December 1, 1983, aftershock (East-West component).

Date	Arrival Time (T.U.)	Station	Distance (km)	Component	Peak Value Read		
					Acceleration (cm/s²)	Velocity (cm/s)	Displacement (µ)
Nov. 20	08h 03m 26.017s	St. Gilles	4.5	Z	2.08	0.0255	3.12
Dec. 1	21h 52m 49.886s	Cointe	6.5	Z	0.145	0.00391	1.06
				N	0.237	0.00487	1.00
				E	0.264	0.00611	1.41
Dec. 4	17h 41m 07.634s	Cointe	5.1	Z	0.256	0.00314	0.38
				N	0.343	0.00597	1.04
				E	0.768	0.00989	1.27
Dec. 13	10h 12m 59.273s	Cointe	4.3	Z	0.240	0.00309	0.40
				N	0.214	0.00384	0.69
				E	0.572	0.00735	0.94
Dec. 19	11h 10m 49.020s	Cointe	5.5	Z	0.123	0.00166	0.22
				N	0.149	0.00210	0.30
				E	0.308	0.00435	0.61
Dec. 21	11h 03m 57.479s	Cointe	4.2	Z	0.717	0.00782	0.85
				N	0.729	0.00918	1.80
				E	1.68	0.0265	4.16
Dec. 21	15h 20m 14.644s	Cointe	16	Z	0.120	0.00191	0.96
				N	0.043	0.00066	0.14
				E	0.083	0.00156	0.63
Jan. 5	12h 38m 09.988s	Cointe	6.8	Z	0.287	0.00552	1.06
				N	0.375	0.00669	1.19
				E	0.793	0.0123	2.04

Table 1. Characteristics of aftershocks recorded in the Liège area.

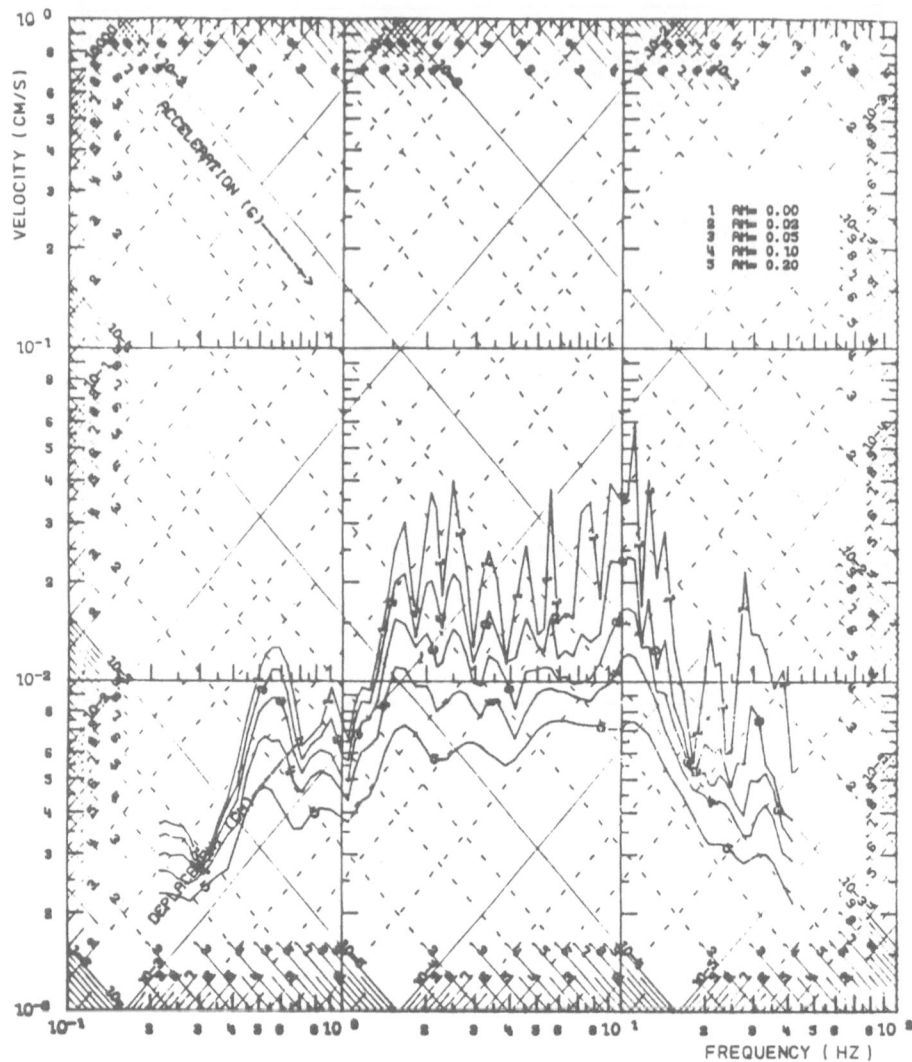

Figure 8.   Response spectra for five values of damping of the December 1, 1983, aftershock (East-West component).

soidal motion.   This quality factor may be evaluated by means of
either time- or frequency-domain analyses.
     In time-domain analyses, a plane wave is considered which pro-
gresses in direction x, with an angular frequency ω and wave number
k.   Energy loss due to inelasticity can be introduced with γ, as
follows:

$$U(x,t) = U_0 \; e^{-\gamma x} \cdot e^{i(kx-\omega t)} \qquad (4)$$

It can be shown that:

$$\gamma = \omega/2VQ \qquad (5)$$

where V is the wave velocity.

When a seismic ray is considered that passes through a heterogeneous medium, in which V and Q vary from one point to another, the damping term, $\exp^{-\gamma x}$, can be generalized as $\exp^{-\gamma s}$, where s is the abscissa of a given point on the ray. For a frequency f, the intrinsic attenuation term becomes:

$$e^{-\pi f} \int_{RAY} (ds/VQ) \qquad (6)$$

or, in simplified form:

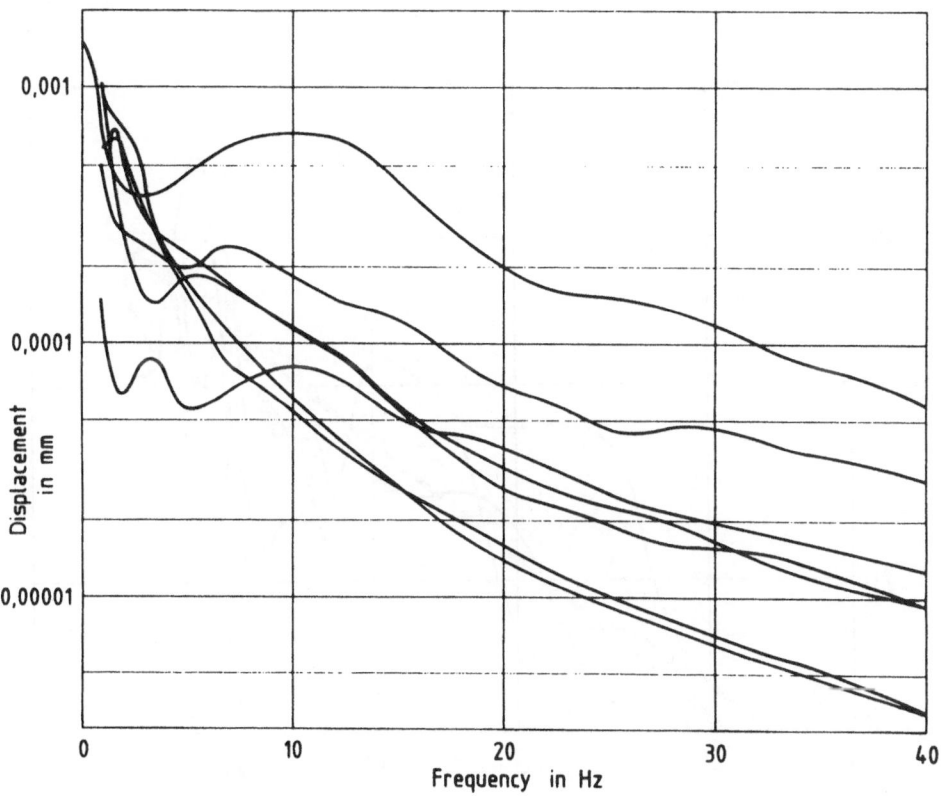

Figure 9. Examples of displacement spectra (20% damping) for the Liège aftershocks.

$$e^{-\pi f t / Q}$$

(7)

where t is the time of propagation.

In frequency-domain analyses, one may make use of the surface wave slope in defining Q, provided the source spectrum is flat over the frequency range in question (10,11). Here, the following equation may be considered to apply:

$$A(\Delta, f) = A_0 \, \Delta^{-\frac{1}{2}} \, e^{-\pi f} \quad (\Delta/VQ)$$

(8)

where Δ is the epicentral distance and V, the average group velocity. For displacements read for f_1 and f_2 at a given distance:

$$Q = \frac{\pi \, (f_1 - f_2) \, \Delta}{V \, \ln \left[A(f_1)/A(f_2) \right]}$$

(9)

An attempt has been made to compute the quality factor from ground displacement spectra, samples of which have been superim-

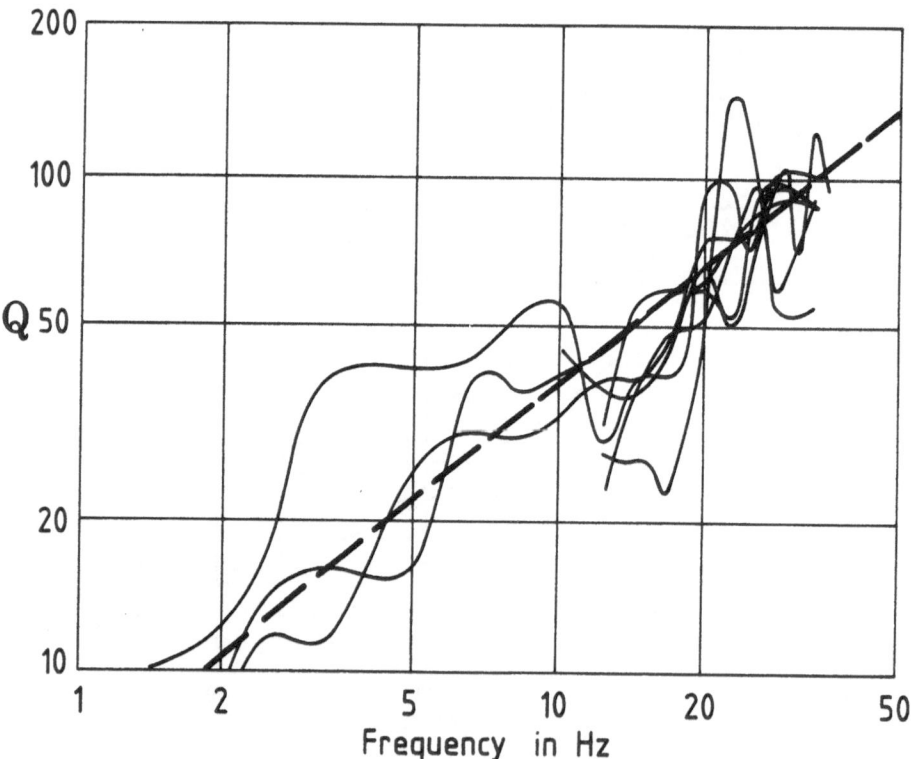

Figure 10.   Estimative Q curves from the Liège aftershocks.

posed on Figure 9, assuming that the maximum amplitude can be ascribed to surface waves. The simplified formula, equation (9), was called upon, and V was taken equal to 3.5 km per second. Figure 10 shows the variation of Q versus frequency for the seven components that were processed, together with a suggested mean slope. Its value is seen to be around 10 for the lowest frequencies and attains 100 for the highest ones. The frequency dependency of Q here is not unlike that observed in California over a quite different frequency range (1 to 4 Hz.) (12).

ACKNOWLEDGEMENTS

We wish to express our appreciation to the staff of the Observatoire Royal de Belgique in Uccle for their valuable aid in solving the material problems encountered by our personnel in the installation of the recording stations, and most particularly for making available to us the Cointe Observatory facility.

REFERENCES

1. *Règles fondamentales de sûreté*, Tome I. Conception générale de la Centrale et principes généraux applicables à l'installation, Chapitre 2. Principes généraux relatifs à la protection contre les agressions externes, Identification c: Détermination des mouvements sismiques à prendre en compte pour la sûreté des installations, Ministère de l'Industrie, Service Central de Sûreté des Installations Nucléaires, France.

2. Ambraseys, N.N. 1973, *Dynamics and Response of Foundation Materials in Epicentral Regions of Strong Earthquakes*, Proc. 5th World Conf. Earthquake Eng., Rome.

3. Haskell, N.A. 1969, *Elastic Displacements in the Near Field of a Propagating Fault*, Bull. Seism. Soc. Am. 59, pp. 865-908.

4. Bouchon, M. and Aki, K. 1977, *Discrete Wave-Number Representation of Seismic-Source Wave Fields*, Bull. Seism. Soc. Am. 67, pp. 259-277.

5. Barbreau, A., Mohammadioun, B., Ferrieux, H., and Mohammadioun, G. 1977, *Etude des répliques du séisme du 6 mai 1976 au Frioul*, Proc. Specialist Meet. on the 1976 Friuli Earthquake and the Antiseismic Design of Nuclear Installations, Rome, pp. 342-374.

6. Trifunac, M.D. 1976, *Preliminary Empirical Model for Scaling Fourier Amplitude Spectra of Strong Ground Acceleration in Terms of Earthquake Magnitude, Source to Station Distance, and Recording Site Conditions*, Bull. Seism. Soc. Am. 66, pp. 1343-1373.

7.   Levret, A. and Mohammadioun, B. 1984, *Determination of Seismic Reference Motion for Nuclear Sites in France*, Eng. Geol. 20, pp. 25-38.

8.   Knopoff, L. 1956, *The Seismic Pulse in Material Possessing Solid Friction*, Bull. Seism. Soc. Am. 46, pp. 175-183.

9.   Knopoff, L. 1964, *Q*, Rev. Geophys. 2, 625-660.

10. Herrmann, R.B. 1980, *Q Estimates Using the Coda of Local Earthquakes*, Bull. Seism. Soc. Am. 70, pp. 447-468.

11. Gupta, I.N., Burnetti, J.A., McElfresh, T.W., von Seggern, D. H., and Wagner, R.A. 1983, *Lateral Variations in Attenuation of Ground Motion in the Eastern United States Based on Propagation of Lg*, NUREG/CR-3555.

12. Singh, S.K., Aspel, R.J., Fried, J., and Brune, J.N. 1982, *Spectral Attenuation of SH Waves along the Imperial Fault*, Bull. Seism. Soc. Am. 72, pp. 2003-2016.

GEOLOGICAL AND MINING ASPECTS OF THE LIÈGE AREA

Robert M.G.L. Liégeois

Chef du Département Mines et Carrières
Iniex, 200, rue du Chéra
4000 Liège

ABSTRACT

After an overview of the major aspects of the geology of Belgium,
in so far as they are essential for the understanding of the local
geology of Liège, we insist on the geology of the Devon and the
Carboniferous geology in the area which was directly touched by
the earthquake. We also draw conclusions from the coal winning in
deep mines and search for a relation between the situation of the
underground constraints and the earthquake of 8 November 1983.

1. LOCALIZATION OF THE EARTHQUAKES IN THE LIEGE/HAUTES-FAGNES AREA

The area affected by the earthquake of 8 November 1983 was limited
in two different ways.
From the administrative point of view, it is a Royal Decree given
at Brussels on 23 November 1983 which has determined the geographical scale of the public calamity which, recording the law,
constitute the damages caused by the earthquake of 8 November 1983
in several localities of the provinces of Liège and Limburg. From
a scientific point of view, the delimitation results from the
tracing of the isoseists by the Royal Observatory of Brussels.
Recently the catalogue of the earthquakes was revised in Belgium,
and the seismicity map of Belgium could be completed. The map from
the "Atlas of Belgium" of 1949 is quite different from the one
which is proposed to us today. This is very important for the
interpretation.
The Liège earthquake of November 1983 and the preceding one in
1965, had their epicentre in the heart of an area which was

347

P. Melchior (ed.), Seismic Activity in Western Europe, 347–368.
© *1985 by D. Reidel Publishing Company.*

supposed to be protected by the existence of the coalfield which
has its outcrop there. In these conditions the hypotheses must be
questioned, formulated by prominent personalities who, at that
time, were involuntarily deceived by informations which later,
were shown to be wrong. It is not impossible that the present
subdivision of Belgium in four seismic areas can be refined on
the basis of a better analysis of the geological structure of
the areas in question.

On the basis of the structural geology of the East of Belgium, I
would be tempted to distinguish, for instance, the earthquakes
of the actual Liège area and those of the so-called Hautes-Fagnes
area.
There are too few earthquakes available to make any statistics,
but it is remarkable to observe that the depth, when known, of
the earthquakes centered on Liège is respectively of 4, 4,1, 3,3
km, whereas for the other quakes it is respectively of 13/5/7,5/
10/5/5/7/5 and 9 km.

With good observation bases and rather precise instruments, the
epicentres of the earthquake of 8 November could be determined as
well as aftershocks 2, 3 and 4. Following the information we received, the distance between these points is of about the possible
estimation error in the localization of the epicentres.

In the list of earthquakes established by the Royal Observatory,
on 31 January 1984, the earthquake and its aftershocks are situated at Liège. One must understand "Liège" in the broad sense,
because the initial shock and aftershock 1/4 are on the territory
of the city of Grâce-Hollogne, whereas aftershock 1/3 is on the
border of the cities of Ans and Glain. As a matter of fact, for
the habitants of Liège, it has always been the earthquake of St-
Nicolas. Undoubtedly because this city was badly tried, on the
one hand, and, on the other hand, the first social dramas of which
one was aware during the night mainly hit the habitants of this
city.

2. GENERAL GEOLOGY OF BELGIUM (lato sensu)

2.1. The large geotectonic entities

The geology of Belgium is characterized by three geotectonic entities which are:
a) A Caledonian tectonic segment in a general East-West direction,
the probable extension of which goes from the Rhine area to Kent
in Great-Britain. The rocks are of the Cambrosilurian and the
foldings are of the Synanticlinal type.

b) The varisc area, which encompasses, in the north, unfolded
rocks deposited on the Caledonian segment, in the South folded
rocks with a general East-West orientation in the central part of
the country. The rocks are of the Devon-Carboniferous. The folding
which affects the southern tectogenic segment is of the Synanticlinal type.
The axial planes take an East-North-East orientation in the East
towards Germany, whereas in the West towards France the West-
North-West direction prevails under Pas-de-Calais.

c) The Mesocenosoic area which consists in a cover in the form of
a dome lowered like the Caledonian segment. According to P.Michot,
the apical zone of this area culminates on the varisc segment. He
concludes from this that the Ardenne dome has, subsequently, a
specific geotectonic evolution between the subsident contemporary
unities which border on it in the South and in the North.

The geology of the Liège area is particularly marked by the varisc
orogen, finished in the higher Westphalian (Asturian phase).

2.2. The geology of the Liège area (lato sensu)

The Liège area, lato sensu, is enclosed in the Herve synclinorium;
the Carboniferous rocks have their outcrop in the town.
The Herve synclinorium joins the Namur synclinorium in a spot
between Namur and Huy. Both the Herve synclinorium and the Namur
synclinorium are characterised by a coal belt with many coal layers which have been worked for a very long time, both in Belgium
and in Northern France.

A cross section indicates that in the coal synclines, the Northern
sides which have Southern inclinations are plane or feebly inclined, whereas the Southern sides are steep. All of this has undergone a pressure from the South-East, pressure which also manifes-
ted itself by a thrust of a few kilometres of the rocks at the
South of a large lengthwise fault called Eifel fault which sepa-
rates the Namur synclinorium from the Dinant synclinorium.

The structure of this Dinant synclinorium is characteristic. West
of a meridian through Namur, the general orientation of the folds
is East-West, whereas East of this meridian the orientation is
East-North-East, West-South-West. The Eastern extremity of the
Dinant synclinorium is observed in the area South of Liège. A
certain incurvation is observed there and a bank by the Vesdre
massif in the area between Liège and Eupen.

The thrust fault itself undergoes directly South of Liège a pronounced incurvation towards the South-East before taking an orientation which is essentially similar to the one it has West of
Liège.

Following a heaving in the thrust fault, the peneplanation, by
its abrasion, has shown a window, called "window of Theux", the
same rocks as those observed a little more to the West in the
western border of the Dinant synclinorium, but with such facies
and thicknesses that one has to admit the deplacement towards the
North-East of the entire mass of the rocks situated South of the
Eifel fault.

On these structures, characteristics of the varisc field in the
Liège area, a network of normal fractures is imposed in a North-
West/South-East direction determining a partitioning in horst
and graben of the entire area East of Liège. From cenozoic to
recent age, this network belongs to the regime of the Rhine rift
valley; according to P. Michot, it might have a varisc antecedent.

2.3. The Caledonian segment; the Bolland ripple

The Belgian Caledonian segment is lithologically characterized by
a monotonous sedimentation with pelitic dominance. In the Brabant
massif, where the thickness is supposed to be more important,
there would be a total of at least 13,500 m. It seems that, from
the beginning of the sedimentation, the Brabant area has had a
more clearly marked geosynclinal nature than the Ardenne area.

On the contrary in the Mesocambrian and Higher Cambrian, the high
Ardenne should have been more active than the Brabant area. From
the Ordovician on, the Caledonian geosynclinal is cut in the
middle by a geoanticlinal slightly subsident, sometimes even stationary: the condrusian body.

Thanks to a deep boring in Bolland, it was possible to understand
the remake of the caledonian folding by a deforming phase dating
from the middle Emsian and called Bolland phase. The result of it
is a stratigraphic break in the series stretching from the Eodevonian to the transgressive Frasnian. This break can be observed
at the South and at the South East of Liège down to the South
slope of the Herve synclinorium (Pepinster, La Reid). It marks
the extension of the emergence era at the North, at the moment
where the subsidence occurred quickly at the South of this zone.

2.4. The varisc field; the Pepinster ripple

The Bolland ripple, born during the Middle Emsian out of a localized folding, as Mr. Michot writes, gradually spread towards the
South in the form of a swell the Southern flank of which was covered with regressive sediments formed by material coming from
already emerged parts and this until the new strata emerged themselves. This sedimentation interruption increased with the erosion
due to this emergence, can be recognized by a sedimentologic break
during the Mesoneodevonian transgression.

TRANSLATION DE LA RIDE DE PEPINSTER

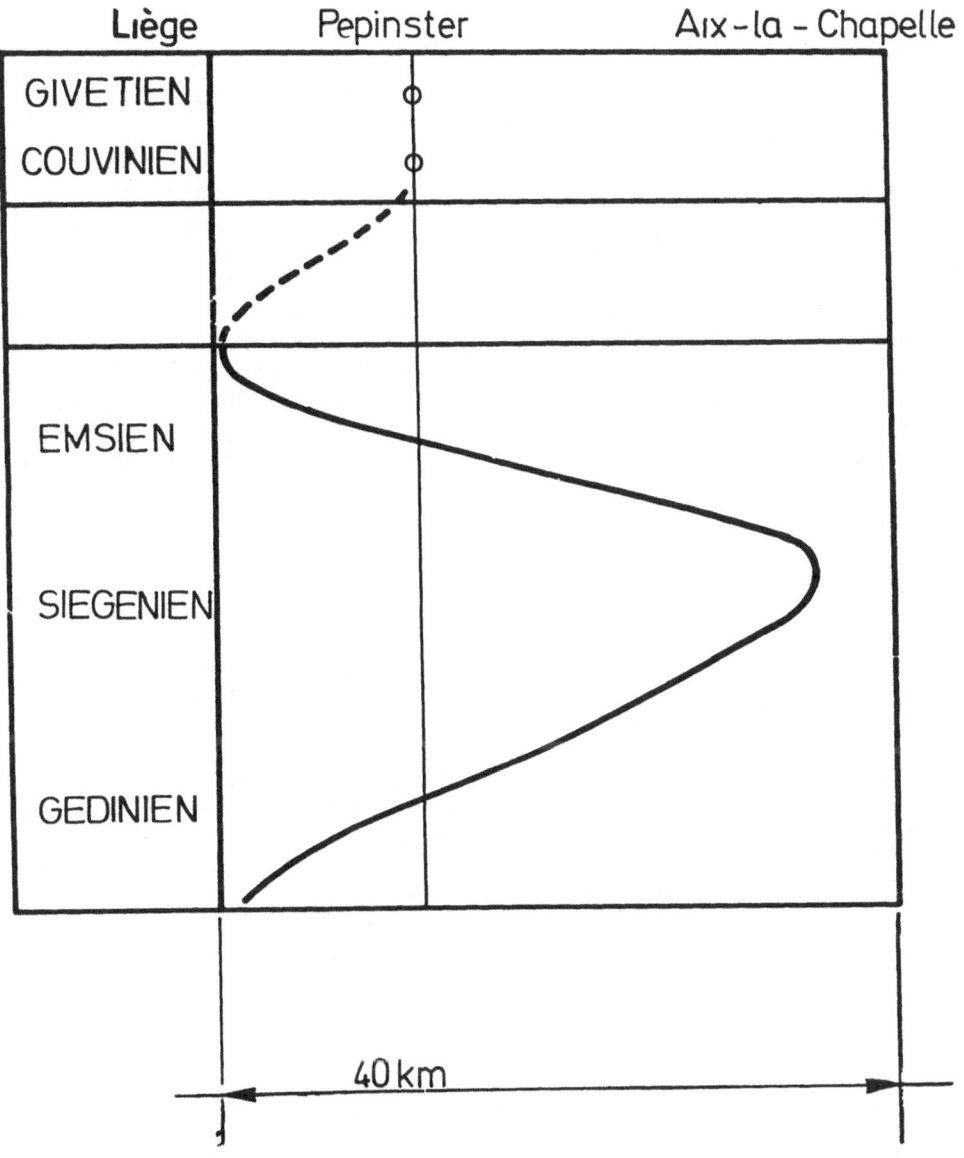

Figure 1.

At the end of the lower Couvinian, the exunded zone stretches
from the Brabant massif towards the South- South-East to the
Stavelot massif. The ripple goes down in this direction, so that
during the Mesodevonian transgression it occurs on an abrasion
surface called the Epibollandian surface by Mr. Michot, comprising a central part formed by the Caledonian base, and surrounded
at the South, South-East and at the East by a discording complex
comprising the Eodevonian and the lower Couvinian.

These phenomena were studied in detail and principally for the
lower Devonian and the Mesodevonian.

In 1952, Mr. Schmit published maps showing the existence of a
transverse ripple at the South-East of Liège at the age of the
lower Devonian and the upper Devonian. A careful study of the
Mesodevonian in the years 51-55 allowed us to modify the maps of
Schmit by specifying the strata attitude in which we observed the
effects of a transversal ripple in the same way as for the lower
Devonian or the upper Devonian. These precise tracings could be
done thanks to very detailed field observations. They are based
for the Couvinian and for the Givetian on the study of isopacks,
isopical lines and the geographical extension of the Mesodevonian
rock horizon in which we found numerous plant deposits.

Our observations completed Schmit's.

We called the transversal axis ripple of Pepinster and we tried
to study its translations. We were led to consider the fact that
this ripple moved laterally in a passage of 40 km length, approximately North 30° West. In this passage you can find the Theux
window. (fig. 1)

We concluded that the most important tectonic characteristics of
the East of Belgium arose at the age of sedimentation. We consider
the transversal axis of Pepinster as a major sedimentation and
tectonic mark. With this we can explain some of the principal tectonic phenomena of the east of Belgium, i. e.:

1. The staggered arrangement of the Dinant synclinorium by the
   one of the Vesdre;
   The deviation of the Eifelian fault;
   The lack of a Northern flank of the Vesdre synclinorium.

2. The axis uprise of the oriental side of the Dinant synclinorium;
   The Xhoris fault.

3. The sedimentologic symmetry observed on both sides of the
   Pepinster axis;
   The accident area immediately around this symmetry axis;

The Theux window.

The Pepinster ripple is almost orthogonal at the axis of the paleobrabantic barrier which followed after the Famenne exundation at the end of the Tournaisian and at the end of the upper Visean.

2.5. The Carboniferous and the transforming faults

Another important parameter of the regional tectonics is the presence of fields of the upper Carboniferous or the Silesian the thickness of which is estimated at at least 3500 m. Such a clastic sediment could only be deposited without the presence of a neighbouring area in process of upheaval which gives the Silesian its "molasse" character.

So, though there is no common mark with the alpine tectonics nor with the Swiss nappes or with the pennine Alps, the varisc tectogenesis may nevertheless be a relatively plastic deformation.

It is the last deformation which affected Belgium, so that the structures we observe today are practically the same as those of the original form at the end of the varisc tectogenesis. Of course there were ulterior deformations, i. e. a generalized swell during the Mesozoicum and the Cenozoicum and a radial fault system depending on the Rhine graben system.

The varisc folding itself dates back from the end of the Westfalian. It is marked by synclinorial formations and by longitudinal shear faults.

Recent deep borings showed the principle of a transforming process explaining how the Silesian strata may have been folded and fractured, as it is the case in the Herve basin, without deforming the underlying strata in the same way. As Prof. Michot writes "so the disharmony between the bottom and the surface can be interpreted as the result of a shingle block by contemporary transforming faults of the major folding phase".

The upper component of the force couple determining the synclinorium pinch is not horizontal but has even a climbing obliquity to the North. Everything happens as if the varisque segment tended to move towards and above its foreland, which seems to go deeper under it, to the South.

2.6. The Condroz thrust, the Eifel fault and the (thrusted) Herve massif

We call Condroz thrust the thrust which explains, by the effect of the eifelian fault, the pressure of the anticlinorial front on the frontal synclinorium Herve-Namur. In this movement the con-

drusian anticlinorium works as a wave pressing the frontal synclinorium Herve-Namur more to the East, the folded Silesian has a
synclinal attitude complicated by an overthrust fault system with
a relatively slight slope to the South (30 to 40°): these are
transforming faults.

Mr. J.M. Graulich published in 1955 a detailed study of the geology
of the Angleur area to which we refer the reader. In this study,
the author demonstrates that the devonian massives of the region
are not overthrust slices, but they form the occidental end of
the substratum of the Herve coal basin plunging to the East.
According to him, there is a hidden autochtonous coal synclinal
of which the axis must be situated between Melen and Pepinster
and on which a tectonic unit, called Herve Massif, is overthrusted,
borded at the North by the fault Aguesses-Asse. The study of the
region is particularly difficult because of the deviation of the
eifelian fault and the intersection of the eifelian fault with
the Aguesses fault. The schematic section passing through the
Colonster boring is practically a radial section in the axial
plane where the third aftershock of the earthquake of 8 November
1983 is found. There we see the relation between the eifelian
fault, the Steppes fault and the Aguesses fault. At a small distance to the east, another section shows the Micheroux fault.
Finally, the author shows in the same manner in a section crossing
two borings and oriented differently that an autochtonous, a Herve
Massif and an overthrust Condroz Massif must be considered.

The detailed study of the Chaudfontaine region, by the same author,
allowed to show the wavy attitude of the eifelian fault in this
place and the existence of other faults, as well those oriented
East-West as the transversal ones. According to Mr. Graulich, the
eifelian fault can be divided into three pieces from the point of
view of its attitude:

A. From Angihoul to Kinkempois (eifelian fault stricto sensu), its
direction is North 70° East; its slope decreases from the South-
West to North-East and passes from 90° at Angihoul to 30° at
Kinkempois.

B. From Kinkempois to La Rochette (fault of the Ourthe and of the
Vesdre); its general direction is North 60° West, but it makes
a series of folds submerging to the South-West. It has a synclinal
attitude in the Woods of Sart-Tilman (Kinkempois), anticlinal at
Streupas, synclinal at Embourg, anticlinal at Henne, synclinal at
Chaudfontaine, anticlinal at La Rochette.

At this point the eifelian fault plunges to the South to reappear
at Theux thanks to a new synclinal attitude.

C. From La Rochette, 3 km to the West of Herve (St-Hadelin fault)

its direction is North 65° East and it slopes about 30° to the
South.

For Mr. Graulich, the Herve Massif is an overthrust massif plucked
from the thrusted Condroz flag. The schematic attitude of the
occidental end of the Herve Massif, the Condroz nappe being carried
away, shows horizontal and longitudinal shifting of the units
forming the Herve Massif.

In a geological section passing along the borings of Chertal, Melen and Pepinster, we can very well see the relative position,
from South to North, the Theux window, the Condroz Massif on the
eifelian fault, the Herve Massif and the Liège syncline. More in
detail, the Herve Massif shows the form of a series of overthrust
wedges.

2.7. The coal fields of the Liège area

The winning of coal in the flat seams of the Campine started after
the first world war. On the other hand the coal mining in the
Southern basins of the country started several centuries ago.
At the West of the country it was concentrated in the three mining
basins of the Borinage, the Centre and Charleroi. At the East it
was concentrated around the City of Liège.

A horizontal section at -200 shows that the coal deposit has two
principal parts. At the North of the Meuse: a basin with the
principal axis oriented to the East-North-East/West-South-West,
and at the South-East of the Meuse, as the rim of a basin with a
similar orientation. For the Northern part there is a real basin
since it straightens up at the west and at the East, as is shown
on the longitudinal sections. A series of transversal sections in
the coal from East to West shows us the marginal fringe totally
at the East, then two productive zones separated by the anticli-
nals of Cointe and La Chartreuse while the Southern part is rather
disturbed and crumpled, the Northern part is specified like a
basin with a flat Northern flank and a steep Southern flank half-
steep and flat when we go on to the West.

The area struck by the earthquake of 8 November 1983 is a coal
zone which was abundantly worked out. The almond-shaped basin
has a Northern flank slightly inclined to the South and a Southern
flank, more straightened and partially truncated. From 1906 on
(O. Ledouble) very numerous fractures can be observed. They were
divided into three categories:

In the first category we find dislocations which are parallel to
the stratification direction and with the same slope as those of
the coal seams. Almost all these dislocations called flat-disloc-
cations, when they are slightly inclined are reverse faults.

They are particularly well known in the part of the basin which is
at the North of the Seraing faults.

In the second category, we place dislocations like the St-Gilles
fault, the Seraing fault, the Marie fault, the six-Bonniers fault,
etc.. These faults have much more inclined dips than the dislocations of the first category. The throw is generally important and
the study of mine workings showed that independently from the
falls or the field straightenings they cause, they also provoked
very important lateral transports. Most faults of this category
have a slope to the South.

The St-Gilles fault is characterized by a general direction North
70° East to North 55° East and a North dip (its slope is irregu-
lar). In some places the fault is affected by a chair fold. The
fault generally seems to straighten the South strata, but at the
extreme West the fault throws down the fields to the South of
about 125 m. We estimate that the lateral displacement is about
1400 m, while the fields situated to the North of the fault are
thrown to the east.

The Seraing fault appears in the Marihaye claim, where it has a
dip of 80° to the South; it makes an angle of 19° to the North
with the West-East line. The fault causes an apparent downthrow
to the South, very variable but always very important. Studies
have demonstrated that the fracture is accompanied by an important
lateral displacement.

The Marie Fault, between the two others, is undulating. It probably merges with the Seraing fault, towards the East. The Marie
Fault is characterised by the fact that the southern part is lower
than the northern part, the vertical displacement varying from
place to place. Lateral movements are probably important.

In a third category, are the faults the direction of which is
around the North-South one. According to O. Ledouble, "those
faults cut and displace the so-called "crains" of the first category. There is no evidence that a crain situated north of the St-
Gilles fault has a continuation south of it. Some of them, even,
among the most important ones, have an end against a fault inclined towards the north, parallel to the St-Gilles fault, north of
it, and having probably the same origin.

Therefore, we can assess that North-South faults were formed
later than the "crains" but maybe initiated at the time of the
St-Gilles fault occurred. However, the St-Gilles fault occurred
more recently, if the North-South faulting observed north of it
corresponds with a similar faulting observed south of it, the
pattern being moved laterally by the important movement of the
St-Gilles fault".

Mr Walgraffe, mining engineer, paid a valuable contribution to the study of the St-Gilles fault in 1942. He observed the diversity of facies of seams and inferred the vertical and horizontal displacement of the fault. The horizontal movement was estimated at 1400 m, as an average, while the vertical movement, rather smaller, could be considered as of the order of magnitude of 50 metres.

The northern part was lifted and this more at the eastern end than at the western one. This indicates that the force inducing the movement was not horizontal but came from the depth.

A number of small boreholes midtown have shown the St-Gilles fault with accuracy.

The tectonic history of the coal basin has been described by Renier in 1919. He said: "The Houiller syncline, acting as a flijsch, very soon was sheared at different depths. Translation movements were stopped in depth, by the Brabant Siluro-Cambrian Massif".

The same comparison was made by F. Kaisin later. Professor Ch. Ancion distinguished 4 tectonic phases:
- a prehercynian phase with early deformations occurring at the time of sedimentation, outlining the future shape of the basin
- a hercynian phase: the Cointe anticline is created and acts as a resistant mole, while folding is happening at its south. The thrust is followed by thrust above the Faille Eifelienne.
  As a consequence, the coal basin is broken in a series of pieces named "écailles" (shells) overthrust wedges.
  In a second Hercynian phase, deformations have occurred in the coal basin, north of the Cointe anticline which would have undergone this deformation.
- A third phase called Neohercynian corresponds to the faulting of the basin through radial faults having the orientation of the main folds: south west/north east. Those faults may be slightly folded at the end of the phase.
- A fourth phase, called post-hercynican, still going on during the Quaternary. Would be characterised by succession and recurring displacements of the main longitudinal faults.

3. ENGINEERING GEOLOGY

3.1. Rock pressure, underground, in collieries

In the mining industry, we call rock pressure, the pressure measured or observed around the cavities created by underground mining.

These rock pressures break the rocks, dislocate strata and provoke ground movements towards the cavities.

Rock pressures dismantle the rocks around the cavities and deform the supports.

Bed separations and rock falls may also occur. Rock pressure finds its origin in constraints due to gravity and to orogenic forces.

The behaviour of a gallery at 900 m depth in a Belgian deep mine is a good example. It is the main gate of a semi-steep seam face.

Plugs were introduced into the walls, roof and floor, in the crown and in the floor of the road (fig.2). All of them are anchored in the same bed; the approach of these strata was measured in time by measuring the convergence of the plugs of the pairs. Considering the pair 3/30 in the centre of the road, one observes that in the hours' time the approach is of about 10 cm, whereas close to the coal solid situated ahead, a telescopic convergence measuring probe has shown a wall approach of about 2 cm. Everything which happens at the working face of the road then indicates the existence of rock pressures which are translated by movements consisting in crushing, flexions, bed separations and the slipping of one bed on another.

When such a road is driven a dozen metres ahead of the front of the mining face, the road already has to be repared at the passage of the front. At 55 m behind the face front (fig.3), the side section of the arches deeply penetrate into the rock of the floor, and the roof, initially inclined in a continuous and regular way, breaks above the crown of the roadway. Repair must be carried out by placing identical profiles in order to have the section equivalent to the initial one. At 270 m from the front face, one rips and dints for the third time and the rocks, initially stratified in a regular way, are completely disturbed (fig. 4).

Considering at present the approach of the surrounding rocks in a longwall face with a 1.25 m opening and supported by mechanical support elements, we observe that, for a working-place advance of about 1.80 m in 24 hours, the approach of the surrounding rocks (or convergence) is of about 10 cm. This approach is inescapable and entails a cleavage of the strata parallel to the front face.

When the support is progressively removed from one end of the face to the other, the roof strata subside at the moment of shifting and everything happens as if the transversal wave had gone through the face from one side to the other. This movement induces fracturings perpendicular to the face front. (fig. 5).

Thus one can observe the behaviour of the rock strata which are

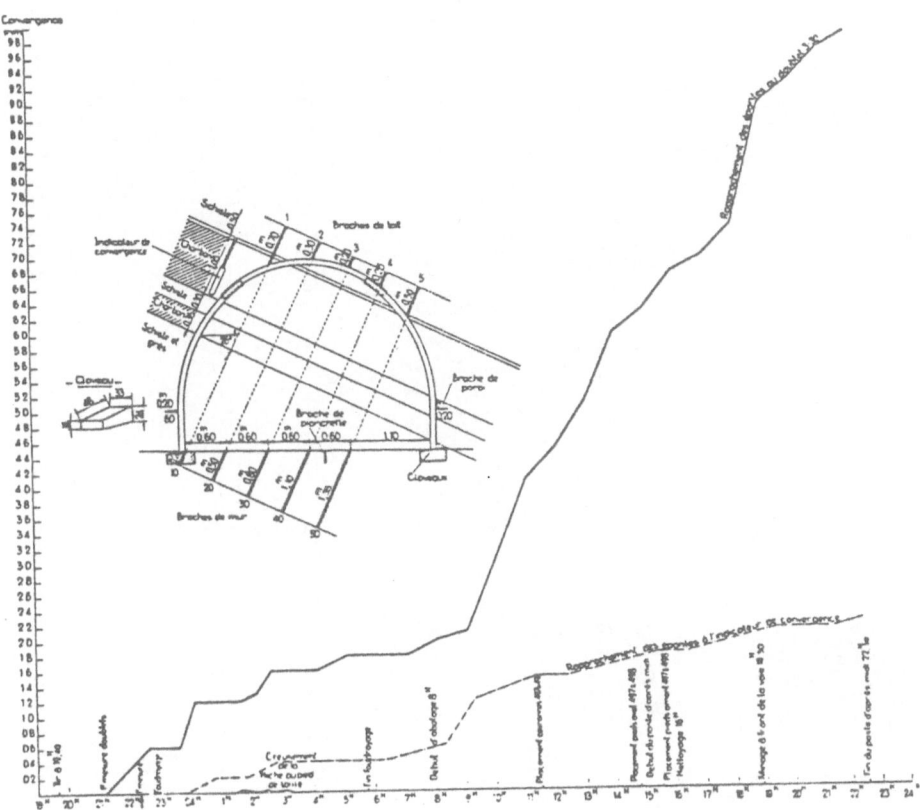

Figure 2

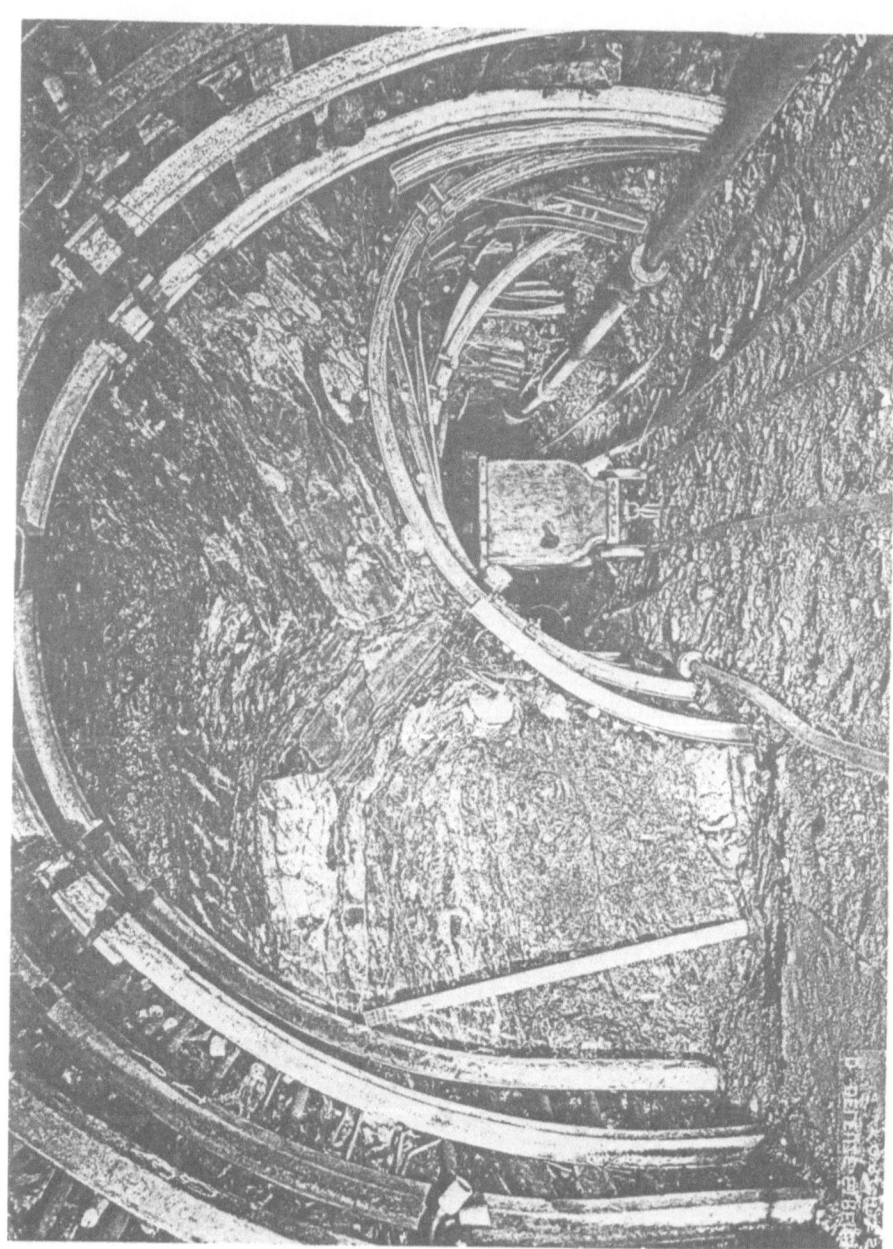

Fig. 3   Geological and mining aspects.

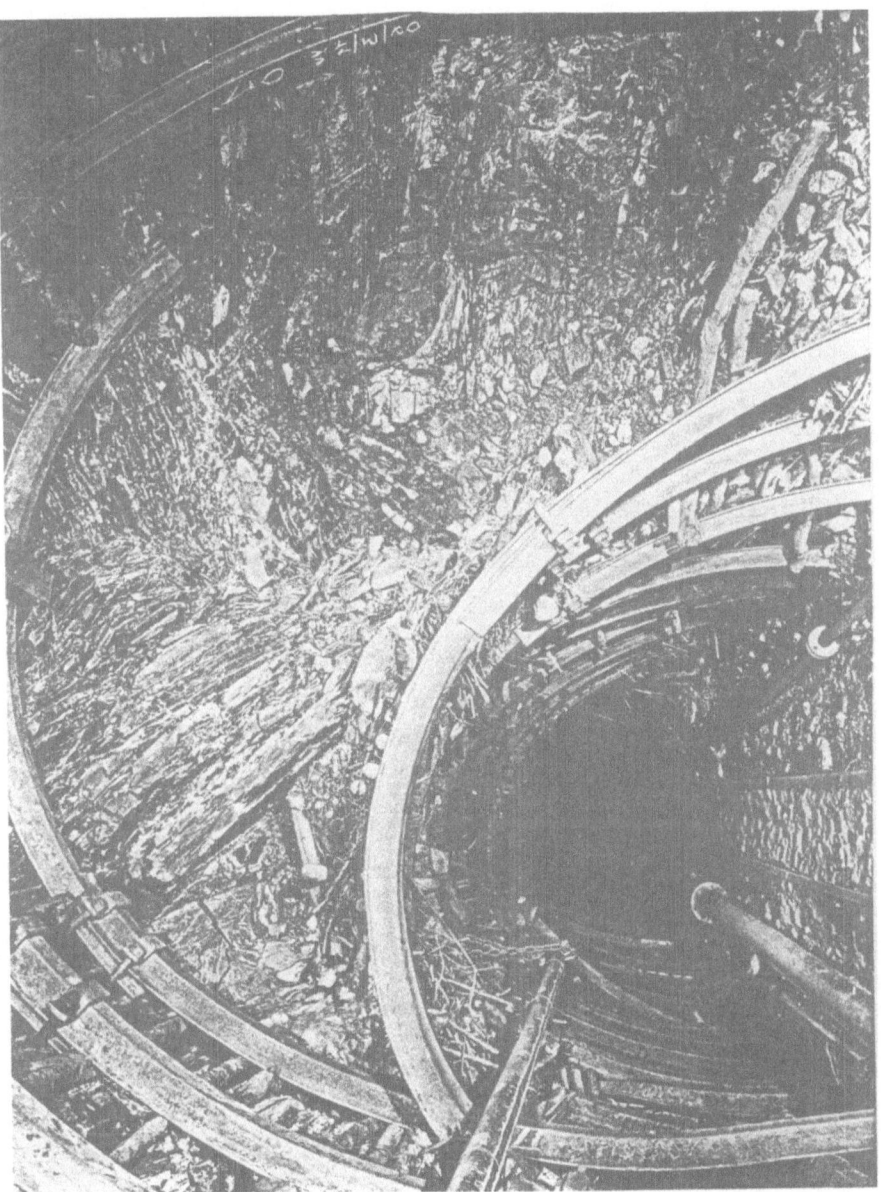

Fig. 4 Geological and mining aspects.

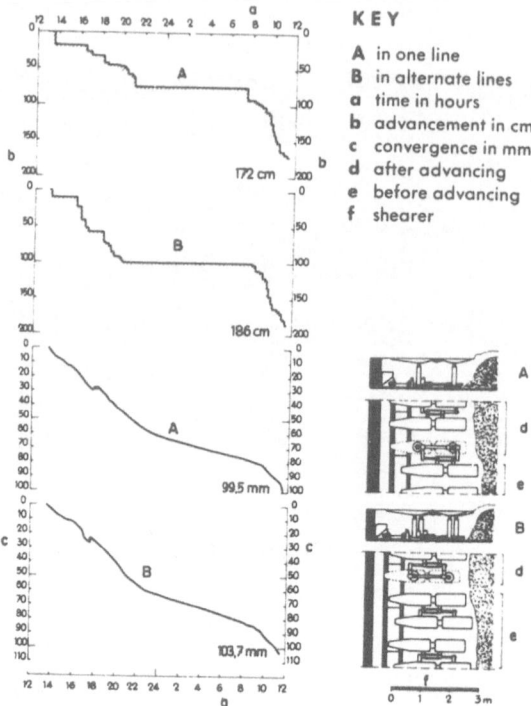

KEY

A   in one line
B   in alternate lines
a   time in hours
b   advancement in cm
c   convergence in mm
d   after advancing
e   before advancing
f   shearer

Figure 5. Convergences at the face in relation to the architecture of the support. Westfalia powered support in the S.A. des Charbonnages de Helchteren-Zolder. {Thickness 1.25 meters, gradient 8 degrees, plough}.

broken following two orthogonal movements, in the way of the earth's crust submitted to the earth tide.

Both of my examples were selected to show the effects of the extraction on the behaviour of the very deep rock strata.

3.2. Tri-dimensional state of stress and relaxation

Is there any possible relationship between the long coal winning in rich rock coal deposits and an earthquake?

In virgin ground, strata are in a tri-dimensional state of stress. The accumulated constraints can attain a limit above which an earthquake may occur.

Mine cavities destress the ground in an area limited to the influence zones of the works.

As shown by the extraction plans of the Liège area coal mines and the statistics on the annual production of the operating works, several hundreds of millions of tons of coal have been extracted from the Liège underground (35.500 hectares of concessions).

Consequently, there has been, on the one hand, a local weight reduction of the earth's crust and, on the other hand, a stress release by the mining works. Raw coal being selected on the surface before sale of the clean coal, heaps of stone (spoil heaps) have been formed, some of which have a considerable volume representing tens of millions of tons of stone. Recovery of the spoil heaps has now started and has been progressing over the last years. A new destressing of the local charge was the result of this.

Considering a cross section of the Liège coal basins, and taking into account the number of working places opened there, we can illustrate the situation by saying that everything happens as though we had made a rubbing of the earth's crust by means of a multitude of small kerfs, we contributed in a certain way to a global destressing of the rocks. Locally, this shearing is relatively important since the mining subsidences can reach and even go beyond ten meters.

3.3. Water in the mine

Coal mining has affected the existence of the underground water tables and their level. As the works went deeper, they had to be unwatered, first by gravity through subhorizontal galleries conveying the water at the level of the river in the valley, and then by pumping systems. The plans of these galleries, called "aerines", which have been found, indicate that the number of pits or shafts in the Liège area was already very important 200 or

300 years ago. The unwatering has in particular as a consequence
the running dry of certain springs and the drying of certain
grounds. With the closing of the coal mines, the unwatering was
stopped and the water rose in the pits again, trying to recover
its initial level. This rising of the water was observed because
it impeded the last active workings. At the closing of the last
works, the water could rise fully and it would seem that springs
which had run dry since several generations became active again.
The water rising in the abandoned mines entailed, in particular,
the filling of the cavities which existed at the end of the winning and the re-establishing of the pressures of several bars on
the deep rocks which had been exposed and brought to atmospheric
pressure by the excavation of galleries and roadway.

Locally, there might be between 375.000 and 500.000 m^3 of water
per km^2 in the Liège mines.

3.4. The puzzle of troughs

Now that the winning has come to an end, there remains at the
ground level some sort of a puzzle of which pieces are the troughs
produced by subsidence. Between the troughs stand rigid dividing
walls. In the troughs, a passive settling occurred that dies away
with time and the influence of which generally disappears after
ten years. On the other hand, with a vibration lasting as long as
4 seconds (the earthquake of 08.11.83), tyxotropy may occur, partly due to the water influw in old workings, and provoke a dynamic
settling.

Coming back to the limits of the mining concessions, we have to
mention that they were not always respected by mine managers.
Historical chronics, reports from courts of justice indicate that
miners were caught in the act of trespassing the limits of mining
concessions, driving galleries under the town walls and working
out the coal below the city itself. The Mine Inspectorate was established during Napoleon's time, i.e. in the beginning of the
19th century. Before that time, no one was bound to show the
drawings of the layout of his mine.

Thanks to survey maps and drawings, we know that coal was extracted since ages in the area affected by the earthquakes of 1965
and 1983.

3.5. Geology and town-planning

Mine damages caused by underground mining are well known.
Before the 8th November 1983 earthquake, one could easily observe
them in the affected area. Now, they are superimposed by the damages caused by the earthquake and it is difficult to distinguish.
Damages to buildings not only result from underground mining.

There may be as well constructional defect as inadaptation to the
site of erection. Moreover, natural substances other than coal
have been extracted in the Liège area, lato sensu.

It is a long time ago since sandstone was worked out. Opencast
mines and deep mines have been discovered in the hills limiting
the alluvial plain north west of Liège City.

A sort of marl has been won on the left bank of the river Meuse,
north of Liège, from quarries covering square kilometers and exploited by rooms and pillars. When abandoned by miners, the left
rooms aren often used for growing mush-rooms. A few years after
the war, hectares of rooms collapsed in one of these undeep
quarries.

The winning of phosphate even though in an anarchic manner, was
conducted in the neighbourhood of Liège, mainly north west of the
town.

In 1955, Professor Calembert made a survey of the area pointing
out parameters that should be taken into consideration when erec-
ting buildings.

After him and Professor Monjoie, who recently developed these
observations, it is possible to mention briefly some natural phenomena that influence the stability of buildings in the area.

The river Meuse was progressively canalised; many branches were
used for the establishing of roads along which houses were built,
and it is normal to observe differential settlings, particularly
favored by the circulation of subjacent water.

The facies of the alluvial deposits varies following the course of
the water and transversally, which may also provoke differential
settlings.

The river Meuse abandoned several terraces which are easily recognizable by the peneplanation levels, and these are, again,
areas where the bed rock is marked by relatively recent alluvial
deposits.

Liège City is characterized by many streets constructed following
the most important slope of the neighbouring hills. More recently,
groupes of houses or isolated houses were built following the
horizontal line on the natural slope. During an earthquake, the
behaviour of the constructions differs according to their situation and their orientation.

Certain recent sediments of the area are relatively soluble in
water. Their natural solution could be accentuated by the town-

planning. Indeed, the water which used to percolate over vast
surfaces is now often conducted by roofs or roads towards impre-
gnation or runoff zones.

The stability of the constructions can also be affected by man
in certain particular cases, such as, e.g., intensive water pumping at the immediate underground level, evacuation of rain water
or domestic water in what are called "lost boreholes", the construction of houses on materials such as ancient spoil heaps.

In an area like the one affected by the earthquake, there are
consequently some precautions to be taken before building. Prudent
owners did so. We shall merely quote the "Maison Liégeoise", a
co-operative society of building construction. It consulted spe-
cialists before constructing 800 appartements in the Burenville
area, in the early 1960's. But this is another story.

4. CONCLUSIONS

The Liège area is a turning-point from the tectonic point of view.
This position is inherited from a distant past.

Important sedimentary and tectonic phenomena bear witness of it:
the Bolland ripple, the Pepinster ripple and the relating phenomena.

The presence of a coal deposit allowed movements in all directions
which we do not necessarily find back at 3 or 4 km depth: flat
dislocations, longitudinal button-hole faults with horizontal
throws, transforming faults.

The underground mining changed the state of stress in the underground by causing a field relaxation in the deposit.

The mining engineers and geologists have by their numerous and
precise observations largely contributed to the knowledge of the
underground of the area covering 355 km^2 and 1000 m depth.

A cooperation between them and seismologists seems indispensable
to us in the interest of science and town-planning.

REFERENCES

Atlas de Belgique

Carte Géologique de Belgique
     3ème état, 1900.

Etudes ...
    Etudes relatives au Bassin Houiller de Liège.
    Soc. Géol. de Belgique. Mémoires (Fasc. I et unique). tome 65,
    1941-1942, 212 p. Liège, Belgique.

Seismicity ...
    Seismicity of Belgium. Catalogue of earthquakes revised on
    1984/01/31. Observatoire Royal de Belgique. Centre de
    géophysique interne. Bruxelles, Belgique.

Calembert (L)
    Géologie, Mines et Urbanisme dans le Pays de Liège.
    Ann. Soc. Géol. Belg., t LXXVIII, juin 1955, p B429-460.

Calembert (L), Monjoie (A) et Lambrecht (L)
    64 sondages au centre de la ville de Liège.
    Professional Paper, 1973, in 12.
    50 sondages dans le secteur nord de la ville de Liège.
    Rive gauche de la Meuse.
    Professional Paper, 1974, n°9.
    75 sondages au sud de Liège. Rive gauche de la Meuse.
    Professional Paper, 1974, n°8.

Calembert (L), Lambrecht (L) et Monjoie (A)
    Géologie du Centre de Liège.
    Ann. Soc. Géol. Belg., t 96, 1973, pp. 157-163, 1973.

Camelbeeck (T) and De Becker (M)
    Report concerning the Liège earthquake of november 8, 1983.
    Observatoire Royal de Belgique, Uccle, Belgique.
    11 pages le 1er déc. 1983; 2 pagés le 13 déc. 1983.

Graulich (J.M.)
    La faille Eifelienne et le Massif de Herve. Ses relations
    avec le Bassin Houiller de Liège.
    Ann. des Mines de Belgique, 1955, pp. 27-41 et pp. 265-281.

Humblet (E) et Massart (G)
    Contribution à l'étude de la faille de Seraing.
    Ann. Soc. Géol. Belg., T 42, 1918-19, p.B109-114.

Ledouble (O)
    Carte des mines du bassin houiller de Liège.
    Ann. des Mines de Belgique, janvier 1906, pp. 3-54.

Liégeois (R)
    Le Mésodévonien du massif de la Vesdre.
    Université de Liège, 15 mars 1955,
    Laboratoire de Géologie Générale, 94 p.

Liégeois (R)
     Analyse des mouvements et des déformations des terrains et
     du soutènement dans une voie de chantier en semi-dressant.
     Ann. des Mines de Belgique. pp. 1284-1311, déc. 1960.

Liégeois (R)
     Powered longwall supports. Fourth International Conference
     on Strata Control and Rock Mechanics.
     N.Y. May 4-8, 1964, pp. 248-265.

Lohest (A)
     La stratigraphie et la tectonique de l'anticlinal Cointe-
     La Chartreuse à l'est de la Meuse.
     Ann. de la Soc. Géol. de Belgique, t LXXXII, n°3, 1958,
     pp. 141-173.

Michot (P)
     Belgique. Introduction à la géologie générale.
     Livret-guide du 26e Congrès Géologique International.
     pp. 486-576, Paris, 7-17 juillet 1980.

Van Gils (J.M.)
     Les séismes des 15 et 21 décembre 1965 et du 16 janvier 1966.
     Obs. Roy. de Belgique, Comm. S.B, n°12, 1966, 27 p.

Walgraffe (Ch), Fourmarier (P), Raucq (P), Ancion (Ch), Humblet (E)
     Diverses études relatives au bassin houiller de Liège.
     Ann. Soc. Géol. Belg., p 65, 1941-1942, Mémoires (fasc.1,
     unique).

MACROSEISMIC EFFECTS OF THE LIEGE EARTHQUAKE WITH PARTICULAR
REFERENCE TO INDUSTRIAL INSTALLATIONS

D W Phillips

Safety and Reliability Directorate,
United Kingdom Atomic Energy Authority

ABSTRACT

This note summarises the principal findings from an assessment
of the Liège earthquake, made as a result of a visit to Liège
and Brussels on 14-16 November 1983. In addition to these
notes, about 100 photographs were taken of buildings and
industrial facilities in or near the epicentre. These photographs complement this summary. Most of the technical
appraisals rely heavily on the significant efforts of other
interested groups, and particularly the Royal Belgian
Observatory, the architect/engineers (Electrobel), operators
(SEMO and INTERCOM) and owner (UEEB) of Tihange power plant site,
and UK seismic survey teams (principally sponsored by CEGB).

1. INTRODUCTION

The Safety and Reliability Directorate of the United Kingdom
Atomic Energy Authority has an interest in the effects of
extreme events like earthquakes on the behaviour of installations. This interest has resulted in the publication of a
number of reports. These reports have dealt with a range of
matters including the probabilistic assessment of seismic
hazard and seismic risk (1), the evaluation of structural
reliability (2) and the consideration of regulatory approaches
to earthquake resistance (3). In the course of these studies it
has clearly emerged that visits to areas affected by damaging
earthquakes can be an invaluable source of information, both
about the hazard and about seismic capability. Similar view-
points have also been expressed elsewhere (4).

P. Melchior (ed.), Seismic Activity in Western Europe, 369–384.
© 1985 by the UKAEA.

Representatives of the Safety and Reliability Directorate, SRD, have made three visits to earthquake epicentres over the last four years. These are:

              November 1980   El Asnam, Algeria
              May 1983   Coalinga, California
              November 1983   Liège, Belgium

Each of these events had special points of interest, both seismologically and as regards the types of structural consequences observed. The early reports of the 1983 Liège earthquake indicated some features which were of particular interest. The main ones were:

Magnitude - at the lower end of the range of magnitudes which are generally held to be risk dominant, at least in the UK (5).

Focal depth - the earthquake was relatively shallow, indicative of the possibility of damaging levels of ground motion in the epicentre.

Location - the epicentre was close to, and possibly in, Liège where buildings and industrial installations might be similar in design and construction to those over a wide area of NW Europe.

These characteristics are simply summarised on Table 1.

Table 1. Characteristics of the Earthquake[*]

Magnitude M_L = 4.9 ± 0.2
        Return Period for Belgium ∿ 40 years
        Return Period for UK   ∿ 80 years
Focal Depth: 2 - 5 km
Epicentral Intensity MM VI to VII
"Equivalent" Peak Ground Acceleration ∿ 10% g
Attenuation: 25 km SW: MM III to IV
              100 km NW: MM II to III
Return Period for UK site exceedance of pga
        ∿ 1000 years
Epicentral duration: 2 - 5 seconds

---

[*]as obtained from reports which appeared in the UK within 6 days of the earthquake

It was for these reasons that SRD felt it would be useful to send an engineer out to Liège in order to carry out a brief assessment of the macroseismic effects of the earthquake, with particular reference to industrial facilities. The resulting

three day visit took place six days after the earthquake
occurred.

2. SEISMOLOGICAL CONTEXT

The seismological context of Liège has been well described and
interpreted elsewhere. However, a brief summary couched in
fairly general terms is included here for completeness.

Liège straddles the rive Meuse in Eastern Belgium some 100 km
from Brussels. The Meuse in this region follows quite closely
an extensively faulted and eroded anticlinal structure which
marks the junction between two distinct tectonic features.
These are the stable Brabant platform on the North and the
exposed Southern end of the Rhenish shield to the South. To the
NW of Liège is the lower Rhine gräben or Rhine gräben embayment.
Liège therefore lies at a recognisable tectonic junction and it
might perhaps be expected that the surrounding region is
associated with a higher than average seismic activity, Fig 1.
Also shown on Fig 1 is a proposed but as yet speculative tectonic feature which has been held to extend from the centre of
the Rhenish shield through Belgium and onwards towards Kent and
the Thames estuary.

The earthquake of 8/11/83 has been assessed as being of Richter
local magnitude 4.9 + 0.3 - 0.1. This size of event has an
approximate 40-50 year return period for Belgium, Table 1.
Table 1 also shows the initial assessment of focal depth,
indicating the event to be relatively shallow, although 3 km is
well below the coal mine workings which pervade the Liege area.

Fig 1 also indicates some of the estimates of the regional
stresses in the crust in the Liège area. The gräben is subsiding in a local tensile field whilst the Brabant-Rhenish
boundary South of Liège is under compression. Early fault
plane solutions from 32 first motion records of a relatively
spatially uniform sample of surrounding seismometer stations
indicate a tensile crustal stress in the Liège region, with the
stress normal to the Meuse "boundary" (6). It is possible that
this tensile stress results from an interaction of the local
gräben spreading with the regional compression.

The earthquake was not recorded on any strong ground motion
instruments, the nearest (at Tihange) being some 25 km from the
epicentre. The macroseismic duration was short, probably
between two and five seconds, which is fairly typical of a
small-medium sized shallow earthquake. The earthquake was not
preceded by any foreshocks, but was followed by a number of
aftershocks. So far, these have been small events, with a

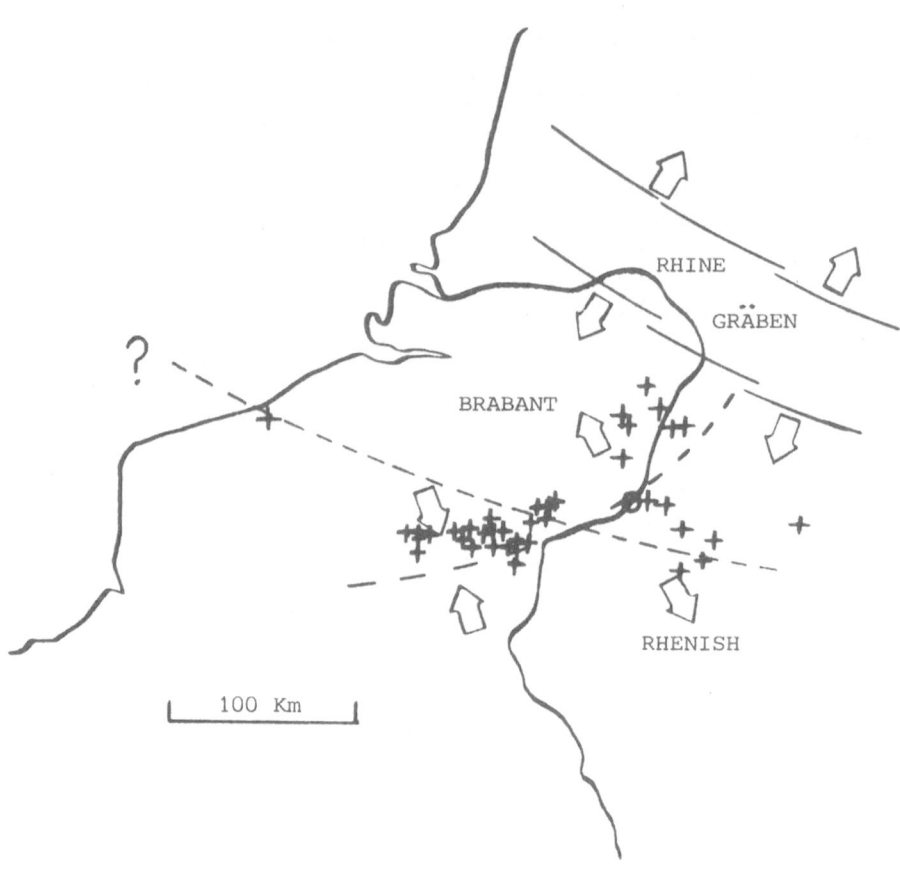

✛ Recorded Earthquake, M < 4.5 [6]

⬆ Estimated Crustal Stress Directions

FIGURE 1   BELGIAN SEISMOTECTONICS

maximum magnitude of about 3.0. The first aftershock occurred 12 minutes after the main shock, and the second some 28 minutes later still. Further small aftershocks have also been reported in some detail elsewhere.

3. DESCRIPTION OF THE EPICENTRAL REGION

Before describing the findings of the visit in detail, it is useful to consider the context of the epicentral region. This serves to illustrate the types of structure prevalent, and their origin and methods of construction. It is worthwhile to note that there is no requirement in Belgium at the moment for domestic, commercial, public and non-nuclear industrial structures and installations to be specifically designed to be earthquake resistant. Thus, the principal lateral loading encountered by designers would be associated with wind effects.

The current estimate of the microseismic epicentre is 50.63N, 5.49E. This is some 3 km West of the macroseismic epicentre which is near the Liège suburb of St Nicolas. Some of the initial reports suggested that within an area of some 5 km² about 16,000 buildings were affected by the earthquake. About 8,000 of these were in some danger of collapse due to structural damage while a further 3,000 would have to be knocked down or subjected to major structural repairs. The local authorities temporarily rehoused 700 families in the first week after the earthquake, and the total cost of the damage was estimated as being in excess of £30M. Most of the large scale operating industrial installations lay outside the epicentral region and, with a few minor exceptions, initial surveys indicated that no industrial installation suffered any damage, Fig 2. In order to appreciate the significance of these findings, it is useful to consider a historical and geographic context.

Liège is an ancient medieval city which rapidly expanded in size and radically altered in character in the late eighteenth/early nineteenth centuries with the extensive exploitation of the coal deposits found in the region. This in turn led to the development of the Meuse into a major waterway capable of being navigated by barges of 1-2 kte and the founding of a major iron, and later steel, industry. The heavy industries tended to be built on the relatively flat alluvial deposits in the bottom of the river valley, whilst the workforce was housed on the hills to the North of the river and along the narrow valley bottoms of the tributaries which joined the Meuse South of the old city centre. In some cases the plentiful mining waste was used as backfill to aid the development of housing sites.

By the twentieth centry, Liège had evolved into an important European industrial centre with a considerable concentration of

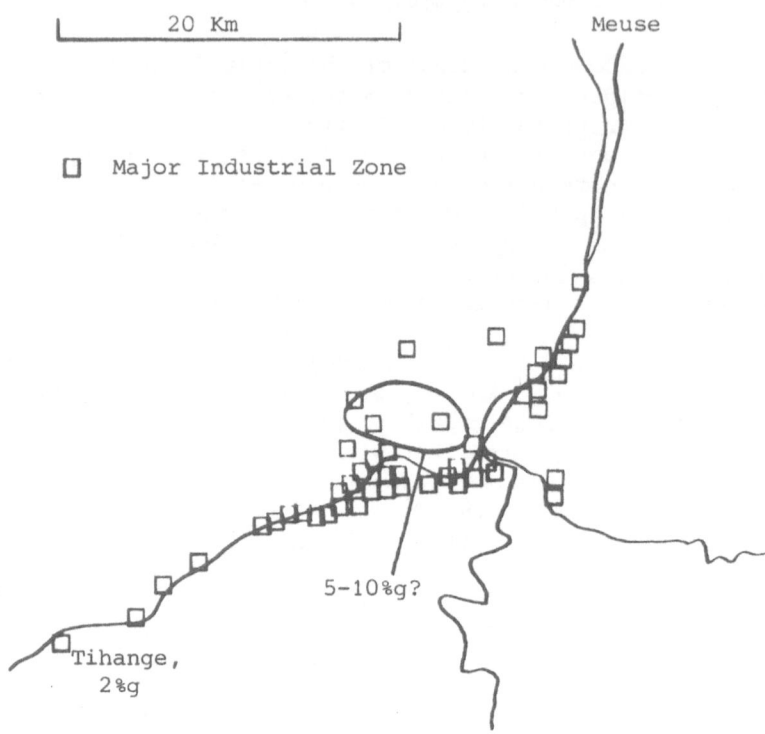

FIGURE 2   DISTRIBUTION OF LIEGE INDUSTRIES

heavy process industries including, in addition to a growing coal and steel industry, sand and gravel extraction, zinc foundries, glass and armaments manufacture, and breweries. The population reached nearly half a million, and the suburbs of Liège extended up to 10 km from the old centre. More recently, the population has declined slightly with the closure of some of the older iron and steel works and manufacturing centres, although the strategic position of Liège in the Western European transport network has attracted new industries. These include a range of light engineering companies, power stations (both nuclear and coal fired) and a major oil/LPG distribution centre.

Liège therefore has some quite striking similarities to other major centres of population in the more heavily industrialised parts of NW Europe. In this sense, Liège could be a valuable indicator of the consequences of a moderate earthquake in such a location. For this reason, it is worthwhile to examine the damage in some detail.

4. VISIT REPORT

It is virtually impossible for one person to gain much more than an impression of the effects of an earthquake when it occurs close to or in a major built-up area, given a visit which lasts only about two days. That some useful findings did emerge from this visit is largely due to the assistance provided by the Royal Belgian Observatory, and in particular to the letter of introduction provided by Professor Melchior and the tour guided by M Camelbeeck. The former made it possible to have constructive interviews with a number of operators of industrial installations in the Liège area while the latter usefully pointed out many of the features of interest.

Although the main intent of the visit was to examine the effects of the earthquake on industrial installations, both with regard to damage and lack of damage, some time was also spent in looking at non-industrial structures. This was useful in pro- viding a basis for making intensity assessments, which could in turn be used to provide a framework for describing and inter- preting the behaviour of industrial installations. The results obtained from the examination of "domestic" structures will therefore be discussed first.

4.1 Domestic Property

Brief tours of the centre and suburbs of Liège provided an indication of the extent of the damaging effects, and thus allowed some initial assessments to be made of felt intensities.

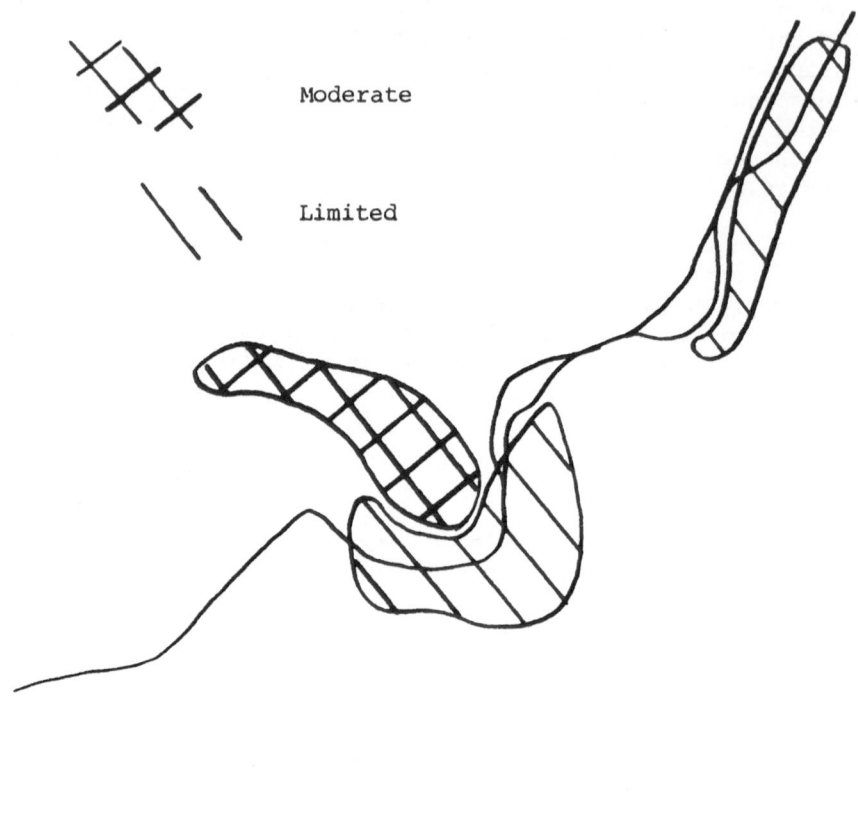

Coverage

Moderate

Limited

5 Km

FIGURE 3   RESIDENTIAL AREAS VISITED

The results of these assessments are indicated on Fig 3. Coverage of the area shown was necessarily incomplete and, particularly in the NE and E, patchy as regards domestic properties. However, some time was spent in epicentral area, and particularly in the suburbs of Glain, Burenville, St Nicolas, En Ster and St Gilles. In addition, a visit was made to the suburbs of Le Perou, and further to the West around Grâce and Wasseige, and out towards the airport. At the time of the visit many houses had received temporary repairs or emergency attention aimed at making them slightly safer. However, many properties were still being repaired and much debris remained on the pavements, and many roads were still closed to through traffic. In some streets in areas such as Glain, a majority of houses had been damaged, the degree of damage being fairly consistent from house to house. However, in other areas such as Haut Douy damage effects were more isolated, this perhaps being a reflection of the importance of design, construction and state of repair of structures for their seismic capability. Also, the use of unconsolidated material in foundations may have played a part in amplifying damage but there was insufficient time during the visit to pursue this possibility.

In general, it was found that the worst structural damage was confined to older and less well built structures. Many of these houses had only a single thickness of brickwork for the main load-bearing walls, but faced with a non-structural skin or cladding which was not tied in to the structural members. In contrast, modern reinforced concrete multi-storey residential properties had survived apparently with no damage, either structural or to shear walls, facing panels, etc. Well constructed brick built dwellings had also fared well, with damage usually confined to chimneys. However, even these properties evidenced shear cracking and other signs of structural distress in places, possibly where foundation conditions were inadequate.

To summarise, domestic property was badly damaged, with houses destroyed by the ground motion or settlement of poor foundation material (and hence the one fatality). Over 10,000 chimneys were affected, and the behaviour of brick chimneys was generally taken as a useful indicator of felt effects. However, it also showed how local site conditions (eg, poor quality backfill, slope failure) had produced local "hotspots" of damage. Also, older chimneys, with clear signs of low quality/chemically attacked mortar, had collapsed when newer chimneys nearby appeared to be intact. For these reasons some of the early estimates of peak felt intensity may be too high and the VI to VII range may be more realistic.

4.2 Industrial Installations (Non-Nuclear)

The distribution of industrial facilities visited, inspected or
simply photographed is shown on Fig 4. Many of the larger
installations are on or close to the Meuse, although there were
some extensive disused factories in Glain and Bolsée, and
modern suburban industrial estates such as at Wasseige. In
addition, smaller or more specific structures were noted in
other areas, such as electricity sub-stations, major road or
rail bridges and viaducts and elevated steel or concrete
storage tanks. The chief difficulty encountered in these visits
and inspections was the limited time available in which to make
an assessment - such an assessment would ideally cover many
aspects including what types of structures, plant and equipment
were there, what was the extent of damage to those items and
what was damaged in the neighbourhood, had the facility been
shut down, etc. In some cases the operators were helpful and
valuable information was obtained, in others a brief visual
inspection from the site boundary had to suffice. The press
reports were sometimes useful in providing general background
material but these cannot always be relied upon - there is no
substitute for primary source material. On Fig 3, the installations visited (and photographed) are classified, very crudely,
into petrochemical (P), electrical (E), steel making (S) and
other (I) activities. There was earthquake damage in only one
industrial facility, a transformer housing near the epicentre
(E5) where a roof collapse led to the electrical supply utility
cutting off power to a small area while clearance work was
carried out. The Cockerill-Sambre steel plant (S2-S4) was partially closed for a brief period soon after the earthquake for a
limited inspection, but resumed normal production with only
minimal delay when no damage could be found. The main oil/
petrol/diesel/LPG distribution centre (P2) was not affected, and
there was no evidence of any damage in the closest petrochemical
depot (P5), a small five tank facility in the suburb of Ougrée.
Further out from the epicentre, at Engis, Cholier and Amay, a
number of major installations were likewise unaffected. These
installations, conventional power plants and cement factories,
had very tall concrete chimneys which are prone to seismic
damage because of the importance of higher dynamic modes and
their susceptibility to collapse due to the formation of a
single plastic hinge.

It would appear then that the bulk of the industrial installations in Liège lie outside the most affected area, and consequently damage to them would not be expected. Exceptions to
this would arise for particularly adverse site conditions, such
as thick alluvial deposits in valley bottoms or on hill-side
sites with inadequate consolidation of backfill. No examples of
such damage were found from the visits, discussions with
operators and other interested personnel, or from press reports.

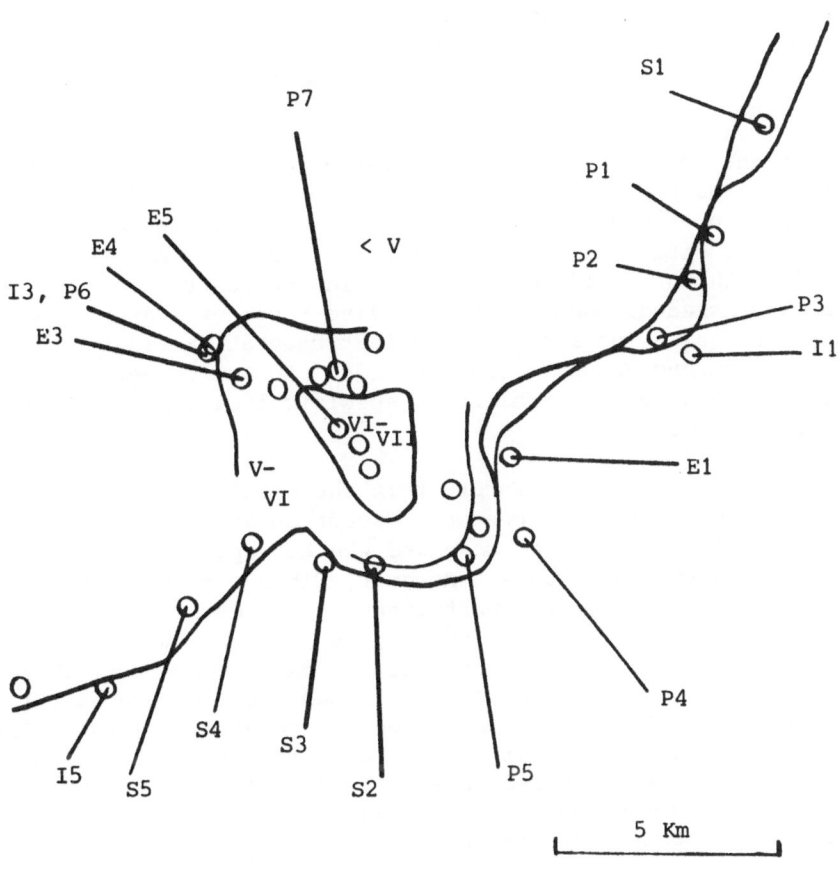

FIGURE 4   MM INTENSITIES AND SITES VISITED

4.3 Industrial Installations (Nuclear)

A principal initial interest was the Tihange triple PWR site
some 5 km upstream of the epicentre, Fig 2. Early press
reports in the UK had implied that Tihange was close to the
epicentre and could therefore have experienced a peak ground
acceleration in excess of the Operating Basis Earthquake, OBE.
Table 2 lists some of the relevant design parameters for
Tihange.

The plant is constructed and operated completely in accordance
with the US NRC philosophy, although only Tihange 2 is seismically instrumented - it is intended, in this respect, to serve
the whole site. None of the stations has a seismic trip as
such, although an alarm is sounded in the control room if the
OBE is exceeded. None of the passive recorders was triggered by
the earthquake (possibly due to instrumentation faults) so there
were no strong ground motion records. An active peak response
recorder near the control room (specific location not stated)
registered 0.02 g.

There was little local interest in the possible effect of the
earthquake on Tihange, the only exception apparently being a
phone caller from Liège who asked whether operations at Tihange
had caused the earthquake. However, Tihange issued a press
statement soon after the earthquake. This detailed some of the
principles of aseismic design used at Tihange, and stated that
the Liège earthquake was too small to be of concern at Tihange.
These press statements received limited local coverage, but
their publication aroused no immediate popular interest. The
directors and staff at Tihange were fully satisfied that there
had been no damage caused on site by the earthquake, and
Tihange 1 and Tihange 2 had continued to operate at full power
during and after the earthquake.

Tihange is an important site for the generation of electricity
in Belgium, since Tihange 1 and 2 together produce about 20% of
Belgium's electricity requirements, Table 2. When Tihange 3
comes on line this will rise to about 30%. It is also of specific interest in that it is the only installation in the area to
be aseismically designed and the only site in the area to have a
strong ground motion recording facility. Therefore, it is of
interest that the operators of Tihange organised a public forum
at Huy one month after the Liège earthquake in order to state
their position in detail and to provide further re-assurance
about the integrity and capability of the installation. Finally,
it should be mentioned that an EEC sponsored study of safety
aspects of Tihange 2, begun in 1981, was reported as concluding
that the site at Tihange is geologically and seismo-tectonically
different to Liège. Thus, a nuclear power plant constructed on

a site with seismo-tectonic characteristics more like those
found at Liège would have to be designed to higher seismic
levels than are felt to be satisfactory for Tihange.

Table 2. Tihange Design Parameters

	Tihange 1	Tihange 2	Tihange 3
Type	PWR	PWR	PWR
MW(e)	870	900	1000
Owner	UEEB	UEEB	UEEB
Operator	SEMO —	Intercom	Intercom
Main Contractor	FRAMACECO	FRAMACECO	ACECOWEN
Architect/Engineer	EdF/Electrobel	Electrobel	Electrobel
Full Power	1975	1982	1985
OBE	0.05 g	0.05 g	0.05 g
SSE	0.10 g	0.17 g	0.17 g
Seismic Trip	NO	NO	NO
Seismic* Instrumentation	NO	YES	NO

*2 types of strong ground motion recording instrument: active
(peak response recorder) and passive (SMA3 and Kinometrics
Engdhal).

4.4 Services

The Liège epicentral area and surroundings has an extensive
network of services, including water, gas and electricity,
telecommunications, and road, rail, air and river transport.
Transport connections were affected to a limited extent, interruptions being confined to some urban roads being temporarily
blocked by fallen debris, restricted in access in order to
minimise further damage, or affected by water main bursts which
led, in at least one case, to significant uplift damage to the
road surface. Rail services were not affected, at least for
main line traffic, with overhead power supplies and signalling
equipment continuing to operate normally. Similarly, sheet
piling and concrete flow control structures appeared to be fully
operable with no sign of any damage, so that river transport was
unaffected. Finally, the small airport at Bierset in the Western
suburbs of Liège had not been affected by the earthquake.

The supply of other services was affected. The gas utility
(L'ALG) reported that they were aware of no significant gas
leaks anywhere in Liège on "their side" of the meters, and
advised the public to be aware of the possibility of gas leaks
on the consumer side, of which there were several (under
Belgian law the consumers are entirely responsible for the
detection and repair of gas leaks on the consumer side of the

meter). The electric utility (L'ALE) reported numerous power
cuts in the epicentral region. These were principally due to
the failure of power line supports where these were attached to
buildings. Controlled power cuts were also used during the
few days following the earthquake as repairs were carried out.
The telephone utility (RTT) reported no failures, but much con-
gestion as relatives and friends attempted to contact those in
the reportedly damaged localities. Public water supplies were
badly affected with numerous leaks. The water supply utility
(CILE) had to deal with many water main breaks, which continued
to appear up to a week after the earthquake. The emergency
services were called in to deal with several severe instances
of local flooding due to such pipe breaks.

5. RESPONSES

5.1 The Local Population

In spite of the broadcasted view of a commentator on Radio
Liège immediately after the earthquake that Liège had been hit
by a Soviet missile (this irresponsibility was later much criti-
cised), many people who were awakened by the earthquake assumed
it to be some large explosion in the steel works. There was
little awareness of the possibility of damaging earthquakes,
which is slightly surprising in view of the seismic activity
near Liège (see Fig 1) and in view of the preponderance of
immigrant workers, particularly in St Nicolas, from the more
earthquake prone countries of Italy and Turkey. In the medium
term, there was concern over the completion of repairs, and re-
housing, before the worst of the winter weather. In the longer
term, the most important issue was that of payment for damage.
The official Belgian "Fonds des Calamities" take some 18-24
months to be allocated and anyway would not pay for the first
£125 of damage (which would be more than the total repair bill
in many cases). Also, few owner-occupiers had earthquake
insurance and Liège itself is financially ill-placed at present
to shoulder a debt burden in excess of £30M, even on a temporary
basis. These difficulties were matched by general lack of
experience in disaster planning and a shortage of damage asses-
sors.

5.2 The Authorities

Belgium has a specialist organisation (CPAS - the Civil
Protection group) for dealing rapidly with emergencies such as
flooding. The CPAS, together with the Red Cross and local
authority social services, police and fire brigades, coped well
with initial aid. In the longer term, repair work seemed to be
progressing mainly due to the activities of individuals and

private contractors, assisted in major tasks by repair teams from the utilities and the CPAS. However, clearing up and demolition work were still far from complete 10 days after the earthquake. This process could well be slowed by the longer term effects of differential settlement which offered a prospect of failures of foundations and essential services for some time to come.

6. CONCLUSIONS

Damage produced by this earthquake seemed to be a reflection of its shallowness (thus leading to an unusual localisation of peak effects) and the local soil and topographic conditions, as well as of its magnitude. Estimated pga levels were below 15% g in the epicentral region, generally too low to be expected to produce damage of any significance in conventional industrial plant. The widespread damage done to domestic property was influenced by the low standards of maintenance and foundation design, leading to an enhancement in the number of chimney failures and many instances of damaging settlement.

After initial panic response, due in part to the propagation of misleading information, the general public evinced little concern over any risk to them from nearby major industrial installations. This lack of concern was quite appropriate in view of the almost complete absence of damage in these facilities. The response of the authorities to the demolition of unsound structures, emergency repair work and re-housing appeared to be adequate in almost all instances, although there remained a potential for longer term funding problems.

There were several lessons for NW Europe from this earthquake. It was clear that the type of brick-built housing with weak chimneys which was so prone to seismic damage is common in NW Europe. An initial assessment of a median seismic fragility in terms of pga for potentially serious (or fatal) chimney failure for such a property would be about 7% g, decreasing with weaker construction to perhaps 4% g and increasing to perhaps 15% g for strong and well maintained construction. It is clear then that chimney/roof damage can be a greater source of seismic risk than damage to a major hazards installation. As other observational evidence suggests, industrial plant is generally robust and can survive pga values in excess of 15% without significant distress.

384

References

1. Alderson, M.A.H.G. 1979, "A method for the estimation of the probability of damage due to earthquakes", UKAEA Report SRD R135.

2. Phillips, D.W. 1980, "The influence of soil behaviour on the aseismic design of nuclear power plants 1. Material properties and modelling choices", UKAEA Report SRD R197.

3. Alderson, M.A.H.G. 1982, "Seismic design criteria and their applicability to major hazard plant within the United Kingdom", UKAEA Report SRD R246.

4. Smith, P.D. and Doug, R.G. 1982, "Correlation of seismic experience data in non-nuclear facilities", Lawrence Livermore National Laboratories Report NUREG/CR-3017.

5. Burton, P.W. 1978, "Perceptible earthquakes in the United Kingdom", Geophys J. R. Astr. Soc. 54, pp. 475-479.

6. Camelbeeck, T. and De Becker, M 1983, private communication.

BASIC CONCEPTS FOR SEISMIC DESIGN

W.J. Ammann

Institute of Structural Engineering
Swiss Federal Institute of Technology Zürich
Switzerland

ABSTRACT

Consequently planning to mitigate earthquake hazards requires a unique engineering approach. The three terms seismic hazard, value at risk of a structure and its vulnerability forming together the seismic risk are briefly discussed. It is shown that the only influence of seismic design to reduce the seismic risk consists in improving the vulnerability of a structure. Basic concepts for seismic design codes are outlined with special emphasis on the role of ductility.

A description of the effects of the Liège earthquake (November 8, 1983) on structures is not subject of this paper. (Reference is made to (31).)

1. INTRODUCTION

About one million deaths, in this century alone, have been caused by earthquakes, and most of these deaths were in damaged or collapsed buildings. Therefore, earthquakes are one of nature's greatest hazards to life on our planet, although the average annual losses due to wind and flood exceed by far those due to earthquakes in many parts of the world (1). Nevertheless, the totally unexpected and nearly instantaneous devastation of a major earthquake has a unique psychological impact to people concerned. This fact makes the big difference to all the other natural hazards.

Hazard imposed by earthquakes are unique in many respects, and

P. Melchior (ed.), Seismic Activity in Western Europe, 385–403.
© 1985 by D. Reidel Publishing Company.

consequently planning to mitigate earthquake hazards requires a
unique engineering approach. An important distinction of the
earthquake problem to other natural hazards is that the hazard
to life is associated almost entirely with man-made structures.
Except for earthquake triggered landslides, the only earthquake
effects that cause extensive loss of life are collapses of buildings. Of minor importance in this respect are damages to bridges,
dams and other work of man.

By the way, it was this fact that led to the great emphasis put
on earthquake prediction. However, it has to be said that even a
successful prediction cannot eliminate the earthquake hazard;
even if all the people are evacuated safely, the structures which
largely determine the standard of living of the community remain,
and their destruction could be a disastrous loss to the regional
economy. This aspect of earthquake hazard can be countered only
by the design and construction of earthquake resistant structures,
and therefore a completely successful earthquake prediction programme could not eliminate the need for effective earthquake
engineering. On the other hand, with efficient application of
earthquake engineering knowledge, the collapse of structures and
the resulting hazard to life can be reduced to a small extent, a
fact which considerably reduces the value of any earthquake prediction programme.

Another unique feature of the earthquake excitation is its essential difference to wind loading. Wind loads are externally applied pressure and suction forces due to the air moving around
the structure; earthquake loads, on the other hand, are inertia
forces generated all over and inside a structure as a response
to motions of the ground upon which it rests. Despite this fundamental difference, earthquake forces have traditionally been
treated as externally applied loads, since familiar procedures
were available for analysis. In addition, the intensity of the
applied earthquake load depends on the properties of the structure itself. Allowing the structure to undergo extensive deformations beyond the global elastic stage may result in a markable
reduction in earthquake loading. Therefore, adequate earthquake
resistance may be provided either by the traditional approach of
increasing strength or by the concept of reducing the stiffness
(respectively assuming plastic deformations) and thereby reducing
the forces to be resisted. This latter approach imposes a certain
ductility demand on the structure and is one of the reasons why
understanding the structural behaviour in the context of earthquake engineering is more essential than in other fields of civil
engineering design.

2. PHILOSOPHY OF SEISMIC DESIGN

Seismic design provisions of current building codes generally
prescribe minimum lateral forces, applied statically to the structure, combined with the gravity forces of dead and live loads.
The applied forces are resisted by the structural materials at
some prescribed working stress level or at strength criteria with
load factors. The prescribed lateral forces are generally considered to be substantially smaller than the equivalent forces,
determined elastically and associated with a major earthquake.
However, it is assumed that buildings designed to resist the prescribed forces will be able to withstand a substantial overload
without a catastrophic failure or collapse. This is due to the
ductile behaviour of the structure. Therefore, the seismic design
philosophy for structures tends to cover the following objectives
(see e.g. refs. (2), (3)):

1. No damage to structural and non-structural elements due to
   frequent, minor earthquakes is accepted. The elastic deformations have to be restricted with stiffness criteria.

2. No structural damage and only some damage to non-structural
   elements is accepted for a less frequent, moderate earthquake.
   The structure has to behave elastically which is controlled by
   stress criteria. In the nuclear industry, this type is called
   operational basis earthquake (OBE).

3. Damage also to structural elements is accepted for rare, major
   earthquakes. The requirement that no collapse of the entire
   structure must occur is controlled by a ductility criterion.
   In the nuclear industry, this type of earthquake is called
   safe shutdown earthquake (SSE).

The definitions for frequent, moderate and major earthquakes in
terms of probability of occurrence, and as a consequence, in
terms of ground motion at a specific site depend on the level of
accepted damage. In code provisions often only one level has to
be explicitly taken into account, the two others are considered
to be covered automatically or by some additional design criteria
like restriction of interstory-drift, etc.

Obviously, the main topic of earthquake engineering is protection of life by minimizing the hazard to life. Therefore, avoid-
ing total or even partial collapse of a structure in a major
earthquake must be the primary objective of earthquake resistant
design. As far as the global integrity of the structure is as-
sured, structural damage is admitted to a certain extent. It may
be less expensive to repair or replace a certain number of structures, hit by a major earthquake, than to build all the structures
resistant enough to avoid damage.

This may not be true for highly industrialized areas like countries in Western Europe with only a moderate seismicity. In these
countries restriction to collapse prevention only would lead to
extensive damage to non-structural elements. In this case, reduction of non-structural damage might even become of primary importance. Figure 1 from (4) shows the damage quantities for different earthquakes as well as a maximum value (curve A) and a
minimum value (curve B) for damage to the Swiss building configuration in function of an increasing intensity. Curve C represents
the expected damage in percent of the total building cost. The
Liège earthquake with intensity VII and a provisional damage
estimation of about 120 Mio SFr. lies even beyond curve A, the
maximum expected value for the Swiss Middlands.

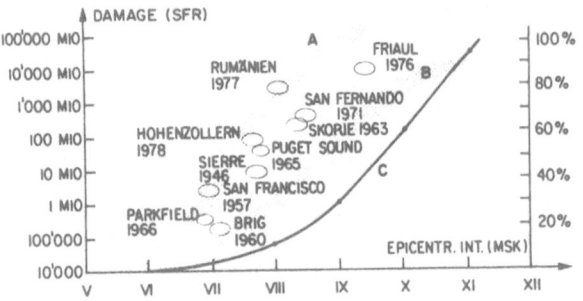

Figure 1. Damage due to earthquakes with different epicentral
          intensities (in SFr., basis: Dec. 1978, from (4).

          A: expected maximum value for the Swiss Middlands.
          B: expected minimum value for the Swiss Alps.
          C: expected damage in percent of initial building cost
             for the Swiss Middlands.

Besides the two criteria of collapse prevention and damage limitation a third goal of earthquake provisions has to be mentioned.
That is the need to improve the capability of essential structures
like hospitals, watersupply stations, bridges, highways, etc. to
remain functionable during and especially after an earthquake.
These structures are called to have lifeline character.

3. SEISMIC RISK

To get more insight into the current state-of-the-art of seismic
design and the various influences affecting this procedure, a
short discussion on the definition of seismic risk might be helpful. Following the UNESCO definition (5) for seismic risk, the

factors entering into the assessment of risk may be expressed by
the following relation:

Seismic Risk = Seismic Hazard * Value at Risk * Vulnerability

The seismic hazard is expressed in terms of probability of occurrence of ground motions within a given period of time and at
a given site or within a given area due to an earthquake of a
particular magnitude. The value at risk is defined by the value
exposed to this hazard, expressed in terms of lives, property,
economic losses, etc. Vulnerability may be defined, following (6),
as degree of damage inflicted by a certain earthquake to man-
made structures and to the ground itself.

For a specific structure at a given site and for a given probability of occurrence of an earthquake both the seismic hazard
and the value at risk are predetermined. Therefore, and under
these considerations, the seismic risk can only be affected by
influencing the vulnerability of the structure. Thus, the limited
influence of all code provision work is obvious.

3.1 Seismic hazard and seismic zoning

The main information on seismic hazard is normally expressed by
sets of curves of a physical parameter which is significant for
assessing the vulnerability of a structure, depicting the probability of occurrence of ground motions within a given period of
time. The distribution of these parameters (e.g. intensities, accelerations, velocities, displacements) is characterized by the
probability of being or not being exceeded within a given period
of time, or in terms of probability density functions. Although
seismic hazard is beyond human control, the seismotectonic study
of the region and the evaluation of its long-term seismicity provides some insight. To estimate the seismic hazard at a specific
site or for a given area within this probabilistic concept the
following steps have to be performed (7), (8):

1. Putting together of information on the probability of occurrence of earthquakes with various magnitudes or intensities at
   various focal distances from the site or within a certain area.
   Determination of maximum probable magnitudes or intensities
   (9).

2. Identification of probable locations of focal regions and
   mechanism of event. Characterization of regions with similar
   seismic properties.

3. Determination of attenuation laws correlating the chosen seismic parameter (e.g. max. acceleration at the site) with magnitude or intensity and epicentral distance.

4. Superposition of the contributions of the different seismic regions to the seismic hazard at a given site or a given area. Representation of the results in an adequate manner, e.g. by a set of curves.

5. If only a specific site is considered, the influence of local soil conditions has also to be taken into account. This cannot be done if a whole area is concerned. The corresponding seismic parameters are then referred to the bedrock.

As an example of such investigations (see Sägesser, Mayer-Rosa (8)), figure 2 shows isoseismals for Switzerland with a given probability of occurrence of $p = 10^{-3}$ p.a.. For these isoseismals 2800 historic seismic events between 200 and 1975 had been taken into account. Switzerland and the surroundings were divided into 22 different seismic source regions, each of them contributing to the seismic hazard throughout Switzerland according to its specific seismicity and the correlated attenuation laws.

For convenient use in seismic design codes and despite of loosing a lot of detailed information, the isoseismals are normally replaced and summarized by some seismic zones. In general, not

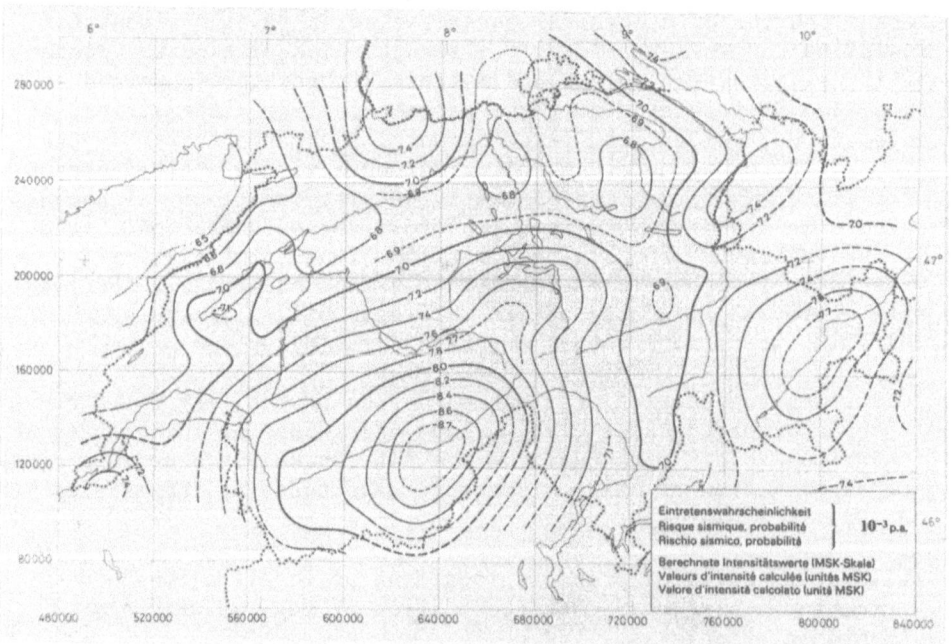

Figure 2. Isoseismals for Switzerland for a probability of occurrence of $p = 10^{-3}$ p.a. (from (8))

more than three or four zones are separated. Figure 3 shows a zoning proposal for Switzerland with three different zones (from (10)) based on (11). For the sake of consiseness and to limit discussions on the authoritative value of seismic hazard within the transition of two zones, the borders may follow jurisdictional or pronounced geographical boundaries. In addition, Switzerland has the advantage that nearly for all probabilities of occurrence the band-width between minimum and maximum value of observed intensity contains two intensity steps (8). Therefore, it was possible to establish the seismic zones regardless of a fixed probability of occurrence of the earthquakes.

Table 1 shows the probability of a structure (in %) to be exposed to an earthquake of intensity I_{MSK} = VII or VIII, for at least one time during the varying life time of the structure. The indicated seismic zones and cities, respectively, are related to figure 3. The corresponding probabilities of occurrence of an earthquake with intensities VII and VIII are also indicated (earthquake with I_{MSK} = VIII in zone 3 or with I_{MSK} = VII in zone 2 show the same probabilities).

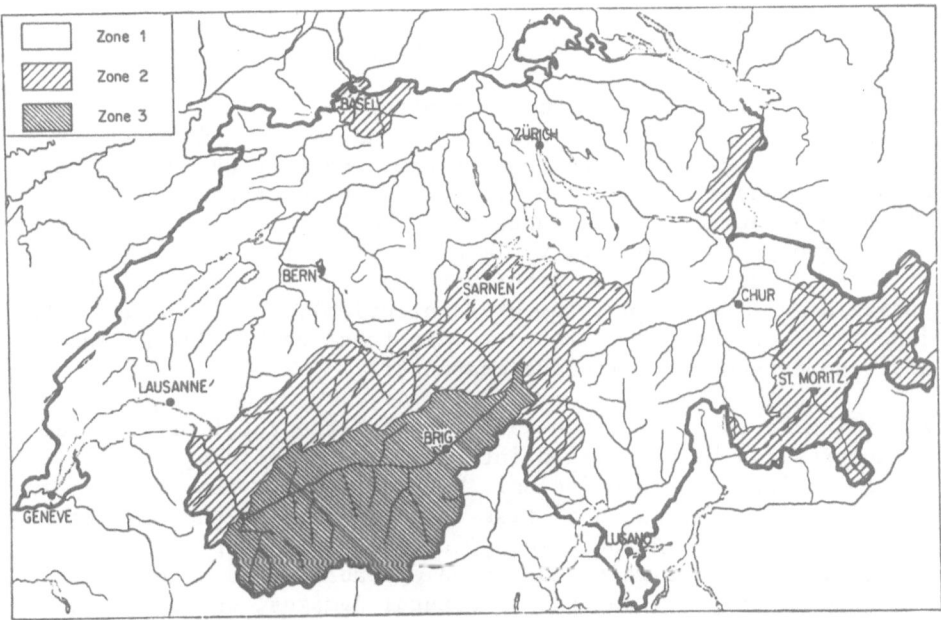

Figure 3. Seismic zones for Switzerland, independent of probability of occurrence (from (10) based on (11))

I_{MSK}	Life-time (years)		Zone 1		Zone 2			Zone 3
			Zürich	Lausanne	Basel	Sarnen	St.Moritz	Brig
VII		Probability of occ.	$5\cdot10^{-4}$	$15\cdot10^{-4}$	$25\cdot10^{-4}$	$30\cdot10^{-4}$	$40\cdot10^{-4}$	$180\cdot10^{-4}$
	10		0.5	1.5	2.5	3.0	3.9	16.5
	50		2.5	7.2	11.8	14.0	18.2	59.4
	100		4.9	14.0	22.1	26.0	33.0	83.5
	200		9.5	26.0	39.4	45.2	55.1	97.2
	500		22.1	52.8	71.4	77.8	86.5	99.9
VIII		Probability of occ.	$0.5\cdot10^{-4}$	$2\cdot10^{-4}$	$5\cdot10^{-4}$	$4\cdot10^{-4}$	$5\cdot10^{-4}$	$40\cdot10^{-4}$
	10		0.05	0.2	0.5	0.4	0.5	3.9
	50		0.2	1.0	2.5	2.0	2.5	18.2
	100		0.5	2.0	4.9	4.1	4.9	33.0
	200		1.0	3.9	9.5	7.7	9.5	55.1
	500		2.5	9.5	22.1	18.1	22.1	86.5

Table 1. Probability (in %) for a structure to be exposed at least one time during its varying life time to an earthquake of intensity I_{MSK} = VII and VIII (probability of occurrence also indicated)

3.2 Value at risk and building categories

The value at risk of a man-made structure as second contribution factor to the seismic risk depends on the type of intended occupancy and its related magnitude of expected consequences in case of an earthquake. This value at risk may be expressed in terms of intended number of occupants or probable maximum loss of lives, respectively the monetary value of the structure and its economic loss due to production interruptions, the potential for consecutive damage (e.g. environmental pollution), and especially the importance of the structure to remain functionable during and after an earthquake (lifeline character).

According to their expected value at risk the total of man-made structures can be classified into different "building categories", or, following (3), into "seismic hazard exposure groups" or just "groups" (12). The value at risk is correlated to the objectives of the seismic design philosophy introduced in chapter 2. For example, a hospital which has to be operable after an earthquake may undergo only limited non-structural damages, whereas a dwelling-house just may not collapse. To visualize this correlation, specific damage patterns, which are a kind of synonym to the vulnerability term to be introduced in the next section, are associated to each building category.

Assuming e.g. three building categories (3), (10), category I is characterized by extensive non-structural damage but only moderate structural damage. No collapse may occur. The structure must no longer remain functionable but people to be to evacuate. Damage may be considerable and probably out of economic repairing. To this category I belong e.g. appartments and office buildings of limited height, industrial facilities of minor importance, parkings, etc.

Structures of category II may suffer some non-structural damage. Minor structural damage is allowed. A reduced serviceability after the earthquake is desired. Repairing is possible with considerable effort. Structures in this group may have a large number of occupants, precious facilities inside, etc. To this category II belong appartments and office buildings exceeding a certain height, schools, churches, open-air stands, shopping centers, hotels, public buildings, chimneys, etc.

Structures of category III may not suffer any structural damage and only very limited non-structural damage. They all have lifeline character as mentioned in chapter 2. To this category III belong all structures which are of importance after an earthquake like e.g. hospitals, fire-suppression facilities, emergency-preparedness centres, important bridges, etc.

3.3 Vulnerability and seismic performance categories

The vulnerability of a man-made structure is characterized by the degree of damage caused by an earthquake of a given magnitude (6) and is therefore directly correlated with the damage patterns introduced before. It can be determined in a deterministic or a probabilistic manner for individual structures or for a group of structures. The vulnerability is associated with the building categories and as stated earlier is the only term in the seismic risk equation which can be controlled by an appropriate engineering technique and thus by code provisions. Assuming that for a given earthquake the vulnerability of a certain structure of one of the three building categories remains constant over all the seismic zones an adequate engineering technique has to ensure that the structure may resist in the same extent to the increasing load levels. This fact can be represented by Figure 4, a matrix formed by the seismic zones and the building categories, respectively their vulnerability. In analogy to ATC-3 (3), the resulting matrix positions or groups of them are called seismic performance categories. The corresponding design criteria, ensuring a given vulnerability, e.g. at position 3/III (zone 3, building category III), have to be much more restrictive than for position 1/I, where the more extended vulnerability may probably be attained without any other design provisions than just originally asked for dead loads, wind loading, etc.

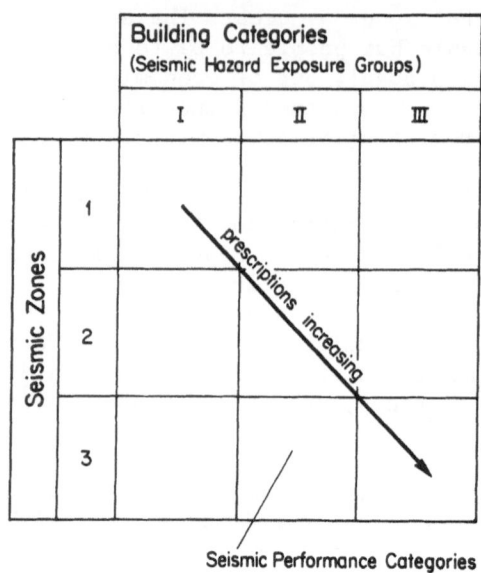

Figure 4. Matrix of seismic performance categories
          formed by the seismic zones and the building
          categories

The extent of these additional seismic design provisions is closely related to the assumed probability of occurrence of the earthquake. To select an appropriate probability fixing this design
level the well known question "how safe is safe engough?" (Fig.5)
has to be answered in the context of the three factors seismic
hazard, value at risk and vulnerability, contributing to seismic risk. The expected damage and the acceptable seismic risk
have to be in an equilibrium, ensured by the seismic performance
categories. The main influence of seismic codes is restricted to
these categories and contains provisions on the layout, the general seismic design, the type of construction, materials, foundation, analysis, equipment, non-structural elements, etc.

The most common methods for evaluating an appropriate seismic
design level are either the "benefit-cost" or the "risk-of-death"
method (6).Performing such an analysis one must never forget that
the seismic risk represents just a small amount of the total
structural risk (comprising all the other loading cases like dead
load, snow, wind, etc.), not to speak about the global accepted
risk as total of all the partial risks (traffic, fire, etc.).
It is one of the most delicate tasks to direct monetary investments and engineering efforts in such a way as to achieve an optimum in the cost-benefit relationship for the total structural
risk (13).

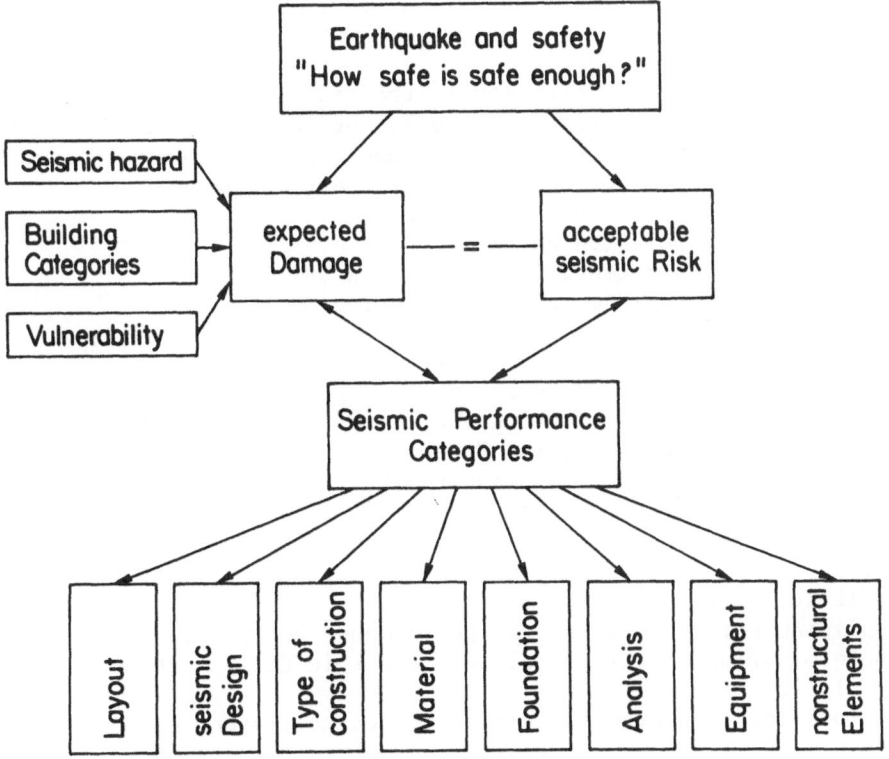

Figure 5. The main influence of seismic codes is restricted
          to design provisions for the seismic performance
          categories

4.   BASIC CONCEPTS FOR SEISMIC DESIGN CODES

4.1   Code evolution

The evolution of the seismic design code of a country is closely
related to its history of recent and historical earthquakes (14).
Very often devasting earthquakes initiated seismic design code
provisions or revisions of existant codes. This is not only true
for recent events but was even the case 200 years before in 1973
when a major earthquake stroke Calabria in Italy causing about
35'000 dead. After the event, a local earthquake commission was
formed to study and to document the consequences of the disaster.
This was the beginning of scientific documentation of earthquakes in the western hemisphere. The era with actual code provisions started after the San Francisco earthquake in 1906. It
was recognized that damage is principally caused by lateral
forces. Therefore, the wind loads were set up to 1.5 kN/m^2. The
earthquake in Messina in 1908 led to a code which introduced for
the first time the concept of horizontally acting inertia loads

prescribing that 10% of the dead load had to be introduced as
lateral force.

Since 1930 a world wide activity in the field of seismic design
provisions can be recognized. But only the last ten years brought
an efficient adaptation of existing knowledge in earthquake engineering to seismic design codes. This fact may be due to
- a pronounced personal need for safety
- an increased population density
- increased hazard potentials of selected buildings
- new building techniques and construction materials
- accumulation of economic values.

The main objective of modern seismic design codes (3), (12), (15),
(16), (17) consists in ensuring the desired vulnerability for a
given seismic hazard scenario. To do so, the codes generally contain information on the following topics (sequence may be changed,
see also refs. (18) and (19)).

1. Scope and field of application of the seismic code provisions
2. Seismic zones
3. Building categories, seismic performance categories
4. Determination of seismic effects (seismic design action,
   analysis procedure)
5. Principles of earthquake resistant design (layout, structural
   system, building configuration, structural design and detail-
   ling, non-structural elements, etc.)

Items 4 and 5 may be vice-versa, indicating that to observe general design principles may be as essential in determining the
vulnerability as performing an extensive analysis. This is mainly true for regions with only moderate seismicity. The first
three items have been covered by the foregoing chapter except the
field of application which is mainly restricted to buildings,
bridges, towers, chimneys, tanks and industrial facilities without secondary hazard potential. Never included in ordinary seismic building codes are NPP-facilities, where much more restrictive design provisions have to be used.

4.2 Determination of seismic effects

Seismic actions result from the vibrations of the soil trans-
mitted to the structure during the earthquake, whereby the intensity of the applied load depends on the properties, especially
the ductility of the structure. Two different analysis procedures
are adopted to calculate the seismic design actions, namely an
equivalent static analysis or a simplified modal analysis procedure (3).

Most of the structures can be treated by equivalent static ana-

lysis procedures. The codes have to define its range of application with restrictions to the fundamental frequency, height,
stiffness redistribution over height, etc. The design lateral
force to be applied to the structure may generally be expressed
by (16)

$$H'_E = f(Z,S,I,R,K,F) \cdot W$$

with

W = total vertical load used for seismic calculation and
    f being a function of the parameters

Z = seismicity at the site taken into account by the seismic
    zoning (Fig. 3)

S = response spectrum value.
    The vibrational response of the structures to the ground
    motions is adequately described by peak ground accelerations
    and elastic response spectra having a form appropriate to the
    seismological conditions (focal mechanism, main frequencies
    of strong ground motion, duration, etc.), the subsoil condi-
    tions and the damping value of the structure (20), (21).
    Ground motions usually are described in terms of strong mo-
    tion recordings. On the other hand, seismic zones (Fig. 3)
    are normally established on the basis of intensities. There-
    fore, a correlation between intensity and peak ground ac-
    celeration must be used, probably leading to questionable
    results (refs. 22, 23). As an example, figure 6 shows elastic
    response spectra for the three seismic zones represented in
    figure 3, assuming a constant value of 5% of critical damp-
    ing for all structures and peak ground accelerations for

Zone 1: $I_{MSK} = VII^-$ $\qquad$ $a_o = 6\%$ g

Zone 2: $I_{MSK} = VII - VIII$ $\qquad$ $a_o = 12\%$ g

Zone 3: $I_{MSK} = VIII^+$ $\qquad$ $a_o = 18\%$ g

These values correspond to a probability of occurrence of
about $p = 2 \cdot 10^{-3}$ p.a.

The factor S is directly dependent on the fundamental frequency f of the structure, which can be evaluated by some
approximation formulas (a list of existing formulas is contained e.g. in ref. 24) or some more sophisticated methods
(e.g. iterative Rayleigh procedures (25)). Assuming a structure in zone 3 with a fundamental frequency between f =
3-10 Hz gives rise to an elastic response of nearly 0.4 g.
This value can then be reduced according to the admitted
ductility of the structure.

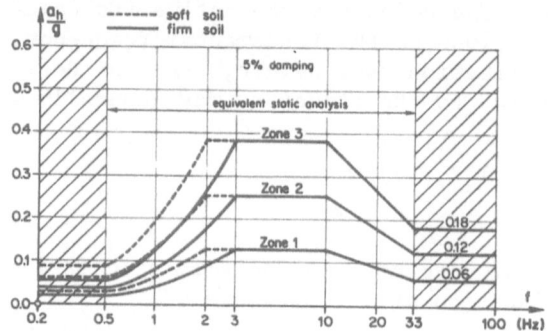

Figure 6. Elastic response spectra for the three seismic
          zones of figure 3 for 5% critical damping and
          two subsoil conditions. The range of applicability
          of the equivalent static analysis is indicated
          (from (10))

I = importance factor.
   The importance factor gives the possibility to take the lifeline character of a structure into account. This can implic-
   itly be achieved by attaching the structure to a certain
   building category or explicitly by multiplying the resulting
   lateral force by a factor I ranging between 1.0 to 1.6 (16).

R = risk factor.
   The risk factor R is often used in combination with I and
   takes e.g. potential risks for environmental pollution into
   account (e.g. oil storage tanks, etc. (16)).

K = construction factor.
   The resulting design load depends very much on the degree of
   tolerated inelastic deformation, respectively ductility, of
   the structure. This ductility demand is strongly correlated
   to the assumed damage pattern, respectively vulnerability of
   the structure (see sections 3.2 and 3.3 and Fig. 4). In Ref.
   (10), a structure of building category III is assumed to behave nearly elastically according to the tolerated damage,
   whereas a structure belonging to category I may undergo extensive inelastic deformation. The capability of the structure to absorb a large amount of energy in the inelastic
   range reduces the original elastic lateral force level according to the ductility. This ductility is essential to
   avoid catastrophic failure. It is defined as the capability
   of a structural element to dissipate energy by non-linear
   deformations without essential decrease of ultimate capacity.
   The resulting reduction factor in the force level can be deduced from figure 7 with a 1-DOF-system. The resulting maximum elastic design load is H_E due to the seismic ground mo-

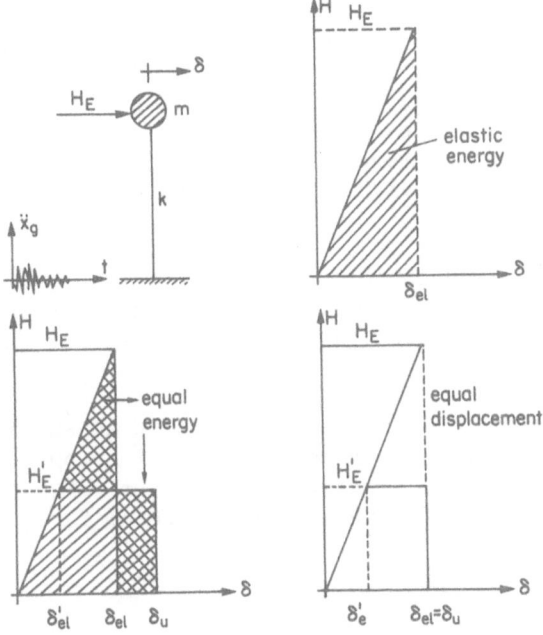

Figure 7. Reduction of seismic forces due to
inelastic deformations

tion $\ddot{x}_g(t)$. Allowing inelastic deformation, the design load
is reduced corresponding to the principles of equal energy or
equal displacements (26). Defining a ductility ratio with

$$\gamma = \frac{\delta_u}{\delta'_{el}}$$

the corresponding load-reduction factor for the equal energy
assumption is

$$K = \frac{1}{\sqrt{2\gamma-1}}$$

and for the equal displacement assumption

$$K = \frac{1}{\gamma}$$

Wether the criteria of equal energy or equal displacement may
be adopted depends on the frequency of the structure. Within
the frequency range, where the equivalent static analysis is
valid, the equal-energy assumption is used. An assumed ductility ratio of $\gamma= 6$ then results in a construction factor of
$K = 0.3$, whereas for $\gamma = 1.1$ (nearly elastic behaviour)

K = 0.9 is attained. The ductility ratio and as a consequence the construction factor K depend very much on the type
of construction and material (steel frames in contrast to
unreinforced masonry, see refs. (3), (27)), but also on the
assumed damage pattern (10), respectively the admitted ductility level (12). These are the main effects for the wide
discrepancies in the resulting lateral force if one performs
comparative analyses with different building codes for the
same structure (28).

F = soil and foundation factor.
    F is either expressed explicitly by a factor or introduced
    in the response spectra (Fig. 6) and stands for the subsoil
    conditions.

Most codes use only a few of these factors in an explicit way,
the rest of them being used in implicit combinations. For example,
in (10) only the response spectrum value and the construction
factor are explicitly introduced in the lateral force, leading
to

$$H'_E = S \cdot K \cdot W$$

This resulting base shear has to be redistributed over the height
of the building in order to take the amplification of the accel-
leration over the height of the structure into account. An appropriate redistribution form is

$$H'_{Ei} = H'_E \cdot \frac{W_i \cdot h_i^k}{\Sigma W_i \cdot h_i^k}$$

where:

H'_{Ei} = horizontal force at floor level i

W_i = weight of floor level i

h_i = height of floor level above ground

k = parameter ranging between 1 and 2

Seismic loads in vertical direction normally do not have to be
considered as the structure provides considerable safety margins
resulting from the conventional design for static loading cases;
otherwise 2/3 of the horizontal seismic acceleration are appropriate to be taken into account. Problems may arise in determining the dominant frequency for vertical vibrations.

Torsional effects due to non-symmetry of the structure and due
to an assumed displacement of the mass out of the shear centre
(in general 5% of the width of the structure) should be considered (16).

To ensure the serviceability of structures after an earthquake
an interstorey-drift limitation is sometimes introduced. The
elastically calculated relative displacements w_{el} multiplied by
the ductility ratio should at any storey satisfy the condition

$$w = w_{el} \cdot \gamma \leq w_{adm} = c \cdot h$$

$c \cdot h$ may be in the order of 2%o of the storey height h.

4.2 Principles of earthquake resistant design

Provisions on the selection of structural systems, building con-
figurations, structural design and detailing represent a very
strong tool in the seismic design of stuctures. Especially in
regions with only moderate seismicity their influence might become even more important than a sophisticated analysis. Depending
on the seismic performance category (Figs. 4, 5) the design provisions are more or less restrictive. They are mostly restrictive
for a structure belonging to the highest seismic zone respectively
building category. Referring to figure 5, provisions are formulated for the type of construction, the building configuration
(with limitations of abrupt changes in lateral stiffness or
resistance), the construction materials (limitations to non-
reinforced brickwalls), the width of joints to prevent mutual
pounding, the adequate fixation of equipment and non-structural
elements, etc. The most extensive provisions are contained in the
ATC-3 (3). One of the most important roles of the design provisions is to ensure an adequate ductility. A lot of research
work has been performed on the design and the adequate reinforcing of shear walls, frames, columns, etc. An extensive amount of
literature on this subject exists (eg. refs. 26, 29, 30).

REFERENCES

(1) Petak W.J, Atkisson A.A., 1982:'Natural Hazard Risk Assess-
     ment and Public Policy. Anticipating the Unexpected'. Ed.
     Springer New York, Heidelberg, Berlin.

(2) Structural Engineers Association of California Seismology
     Committee, 1980: 'Recommended Lateral Force Requirements
     and Commentary'. 4th Edition revised, San Francisco.

(3) Applied Technology Council, 1978: 'Tentative Provisions for
     the Development of Seismic Regulations for Buildings
     (ATC-3-06)', Berkeley, CA, USA.

(4) Sägesser R., 1979: 'Damage Potentials in Switzerland' (in
     German), in: "Tendencies in Earthquake Resistant Design",
     Seminar, ETH Zürich, Aug. 1979, Schweizer Ingenieur and
     Architekt, Vol. 97, No. 49, pp. 1020-1021.

(5) UNESCO, 1979: Final Report of the Intergovernmental Conf.

on the Assessment and Mitigation of Earthquake Risk, Paris,
Feb. 1979. Publication SC/MD/53, UNESCO, Paris.

(6) Ambraseys N.N., 1983: 'Evaluation of Seismic Risk', in
     "Seismicity and Seismic Risk in the Offshore North Sea Area",
     Eds. A.R. Ritsema, A. Gürpinar, D. Reidel Publishing Company,
     pp. 317-345.

(7) Cornell C.A., Merz H.A., 1974: 'Seismic risk analysis of
     Boston', ASCE National Structr. Engineering Meeting, April
     1974, Cincinnati, OH, USA.

(8) Sägesser R., Mayer-Rosa D., 1978: 'Seismic Risk of Switzer-
     land' (in German), Schweiz. Bauzeitung, Vol. 96, pp. 107-123.

(9) Howell G.F. Jr, 1981: 'On the saturation of earthquake magni-
     tudes', BSSA, Vol. 71, No. 5, pp. 1401-1422.

(10) Swiss Association of Engineers and Architects (SIA), 1984:
     'Loads and Design Actions on Structures' (draft, in German,
     to be published), Zürich, December 1984.

(11) Mayer-Rosa D., 1979: 'Earthquake Safety of Structures in
     Switzerland' (in German) in: "Tendencies in Earthquake
     Resistant Design", Seminar ETH Zürich, Schweizer Ingenieur
     und Architekt, Vol. 97, No. 49, pp. 1018-1020.

(12) Comité Européan du Béton (CEB), 1982: 'Basic concepts for
     seismic codes; Seismic design of concrete structures',
     Information Bulletin No. 149, March 1982.

(13) Matusek M., Schneider J., 1983: 'Safety assessment for struc-
     tures. A general concept' (in German), Inst. of Structural
     Engng., ETH Zürich, Report No. 140, Oct. 1983, Birhäuser
     Basel-Boston-Stuttgart.

(14) Leslie St., Biggs J.M., 1972: 'Earthquake Code Evolution and
     the Effect of Seismic Design on the Cost of Buildings',
     MS-Thesis, MIT, Berkeley, USA.

(15) Internat. Standard Organization (ISO), 1982: 'Design for
     seismic actions on structures', TC 98 (5th draft), July 1982.

(16) Internat. Association of Earthquake Engng (IAEE), 1973:
     'Earthquake resistant regulations: A world list', Gakujutsu
     Bunken Fukyu-Kai, Tokyo, with supplements 1976 and 1980.

(17) ACI, 1983: 'Special provisions for seismic design', ACI
     Standard 318-83, Nov. 1983.

(18) Berg G.V., 1983: 'Seismic design codes and procedures',
     Monograph Serie, EERI Berkeley.

(19) Internat. Association of Earthquake Engng. (IAEE), 1980:
     'Basic Concepts of Seismic Design', Vol. 1, IAEE, Tokyo.

(20) Studer J., Ziegler A., 1983: 'Grundlagen zur Berechnung von

Bemessungsbeben', Mitteilungen Nr. 122, Inst. für Grund-
bau und Bodenmechanik, ETH Zürich.

(21) Newmark N.M., Hall W.J., 1982:'Earthquake spectra and de-
      sign', Monograph. Serie, EERI, Berkeley.

(22) Ambraseys N.N., 1975: 'The correlation of intensity with
      ground motion', XIV General Assembly of the European Seis-
      mological Commission, Trieste.

(23) Murphy J.R., O'Brien L.J., 1977: 'The correlation of peak
      ground acceleration amplitude with seismic intensity and
      other physical parameters', BSSA, Vol. 67, No. 3, pp. 877-
      915.

(24) Austrian Association for Earthquake Engng. (OGE), 1981:
      'Evaluation of Friuli Strong Motion Data', OGE Report No.1,
      Vienna.

(25) Clough R.W., Penzien J., 1975: 'Dynamics of Structures',
      McGraw-Hill, Kogakusha Ltd.

(26) Paulay Th., 1982: 'Seismic design of concrete buildings',
      Lecture Notes, ETH Zürich, Inst. of Structr. Engng.,
      Aug. 1982.

(27) Keintzel E., 1981: ' Ductility ratios for reinforced con-
      crete structures in German seismic zones' (in German),
      Dissertation, Techn. Univ. of Karlsruhe.

(28) Ammann W., Bachmann H., 1981: 'Seismic design forces for
      structures due to different seismic codes and analysis
      procedures' (in German), Schweizer Ingenieur + Architekt,
      No. 6, 1981, or Inst. of Structr. Engng., ETH Zürich,
      Report No. 110, Feb. 1981, Birkhäuser Basel-Boston-Stutt-
      gart.

(29) Rosenblueth E. (Ed.), 1980: 'Design of Earthquake Resistant
      Structures', Pentech Press, London.

(30) Dowrick D.J., 1977: 'Earthquake resistant design: A manual
      for engineers and architects', John Wiley & Sons Ltd.

(31) Ammann W., Ziegler A., 1984: 'The Liège Earthquake of
      November 8, 1983' (in German), Schweizer Ingenieur + Archi-
      tekt, Vol. , No. 20, pp. 388-392.

CONCEPTUAL AND METHODOLOGICAL APPROACH TO THE DEFINITION
OF THE "REFERENCE EARTHQUAKE" FOR NUCLEAR POWER PLANTS

Muzzi Francesco, ISMES
Via Taramelli, 14
ROMA, Italy

1.0   INTRODUCTION

The problems connected with the "reference earthquake"
selected for antiseismic design of Nuclear Power Plants
(NPP) present aspects which can be schematically subdivided
into geological-naturalistic aspects and engineering aspects.
There is still much to be clarified concerning the two aspects;
however it is considered that the main knot to be untied
is the point of contact between the two aspects mentioned.

In fact although the geological naturalistic aspect
can enable us to make decisions of great importance, such
as the exclusion of a particular site for the building of
a nuclear plant, or the identification of seismogenetic
geologic structures, most relevant for the "reference
earthquake", neverthless, it is not yet able to supply by
itself numerical parameters which can be used immediately
by the engineer for design purpose.

One of the main difficulties lies in the fact that
when the "reference earthquake" is transformed into engineering
terms, as it must be, it enters into a design process, which
like all engineering activities, proceeds through analytical
or numerical models of the structures in order to foresee
its behaviour.

The process of antiseismic design of a structure is
fairly consolidated in its praxis, or to be more exact,
the meanings of the various hypotheses of schematizing or
of the levels of consevatism innate in each of them and
in the process as a whole are quite well know.

Therefore the transformation of the "reference earthquake"
into engineering terms cannot be made without taking into

P. Melchior (ed.), Seismic Activity in Western Europe, 405–411.
© *1985 by D. Reidel Publishing Company.*

account, on one hand the geological-naturalistic hypotheses which defined it, and on the other hand its adequacy with the schemes and the engineering methods.

Through the discussion of the relevant aspects of the whole process which, beginning with the geological-naturalistic, problems, leads to the definition of the time histories at the base of buildings, the necessity that "the reference earthquake" should be identified with the several phases of such a process and that the entire process should take place with the active and continuous participation of all the elements that take part, will be evident.

2.0   DISCUSSION ON THE MEANING OF THE "REFERENCE EARTHQUAKE" IN ITS VARIOUS FORMS

The definition of the "reference earthquake" requires the introduction of geological, seismological, (soil dynamic) geotechnical and structural models. That is to say, it requires the arrangement in coherent schemes of experimental data in order to supply standards of behaviour and rules for the adoption of suitable measures for the protection of nuclear plants from the foreseeable effects produced by earthquake.

At this moment we should point out that the "reference earthquake", in the course of the process that defines it, takes on different characteristics according to the geological, seismological and geotechnical area it passes through before arriving at the base of the buildings.

It is necessary to verify, however, that it does not loose its initial safety level which defined it according to levels of precaution in proportion to the knowledge (ignorance) of the natural reality. Ignorance is essentially determined both by the extreme variability of the geological environment, and by the entity of natural phenomena in general. This does not mean that it is impossible to find safe technical solutions which are technologically proven.

The conservatism of the antiseismic protection of a nuclear plant must be achieved by harmonizing the safety margins of the whole process, which goes from the definition of the design parameters relative to the movement of the ground, to the determining of the allowable stresses in the structural elements.

As a consequence of what above considered the definition of the "reference earthquake" and its degree of conservatism cannot be represented by a single numerical value which is univocal and absolute, but by a collection of information and evaluations some of which are inevitable subjective, especially when one wishes to give a cautious definition.

2.1 "Reference earthquake" from the geological and seismological point of view

The combination of the geological information concerning the "reference earthquake" are finalized in order to define:
- the seismotectonic character of the area including the site, defined essentially by the tectonic style presently acting;
- the zones related to the tectonic style presently acting, capable of generating a significant vibratory motion at the site for the foundation soil and structural components of the plant;
- the geological, geomorphological and other characteristics of the site.

From a conceptual point of view, the abovementioned geological information and evaluations are in themselves valid for the characterizing of the "reference earthquake" as long as two fundamental postulates are accepted:
a) the tectonic style of the regions under consideration, does not change during the life of a nuclear plant;
b) there is no reason for the random earthquake to exist.

As far as the first postulate is concerned, it emanates from the consideration that the masses in question, the inertia and the viscosity render the hypothesis of sudden variations in the field of stress extremely unlikely.

With reference to the problem of the random earthquake, it is considered that in the natural sciences, both the refusal of the possibility of pure chance on one hand, and the search for the cause-effect relationship on the other hand, correspond to a totally correct attitude from a scientific point of view.

The problem therefore, isn't to establish whether the random earthquake exists, since it has been established that the earthquake like every natural phenomenon, is regulated by the laws of cause and effect. The real problem instead is to recognize in practical cases the cause (seismogenetic zones) which generated the effect (the earthquake).

The sum of the information and evaluations of a seismological nature which constitute the "reference earthquake" can be represented as follows:
- characterization of the seismic, historical and instrumental activity of the tectonic zones which can generate a significant vibratory motion at the site for the foundation soil and the structural components of the plant;
- research on all the historical and instrumental information of the vibratory ground motion at that site (duration, frequency, instrumental registrations of vibratory motion, maps of the isoseisms, multiple quakes during one single seismic event etc.);
- research on all the historical and instrumental information

on the effects produced in the area (superficial cracks in the earth, variations in the levels of underground waters, instability of slops, phenomena of liquefaction, settlements, etc.);
- definition of the attenuation laws of the seismic effects from the seismogenetic zones to the site;
- determining of the maximum macroseismic intensity at the site due to historical earthquakes.

The geological and seismological knowledge which permit the collection of information and the carrying out of the abovementioned evaluations, are generally sufficient to permit also the recognition of the MAXIMA POTENTIAL seismic characteristics with reference to the seismogenetic zones, with the inevitable intervention of professional judgement guided by the knowledge of ,the tectonic style presently acting.

The "reference earthquake" is represented by all the elements indicated before, which can produce the most significant effects at the site, on the foundation soil and on the structural components of the plant.

It is obvious that the information and evaluations carried out in the geological and seismological field, cannot by themselves supply engineering numerical parameters, since the characteristics of the soil and the structures of the plant (liquefaction, dynamic response of the soil, typology of the foundation, etc.) have not yet been taken into consideration.

2.2 "Reference earthquake" from the point of view of the dynamic behaviour of the soil (1) (soil dynamics)

The vibratory motion of the surface of the ground distant from buildings, can be described as the result of the processes of the generation and propagation of the seismic waves, and of the dynamic response of the soil. It is obvious that conditions can exist in which one of the three processes can be prevalent over the others.

As an example, two situations can be described, in which it is not difficult to single out which of the three processes is prevalent.

For a site near to a seismogenetic structure (near-field), the breaking process, (source mechanism) and the position relative to the structure will have a more important influence on the vibratory motion of the site.

For a site in conditions of "far-field", on soft soil, the dynamic response can be prevalently conditioned by the transfer functions of the site, and by the energy content associated with the different ranges of frequency transported by the seismic waves.

These examples serve to indicate that the "reference

earthquake" expressed in terms of response spectrum, or time history, on the surfaces distant from buildings, must be defined taking into account all the processes indicated above.

It is essential at this point, to investigate the sites from the point of view of their dynamic response.

These therefore must be investigated and what is know as a "design soil profile" must be defined. The "design soil profile" is that set of data which represents the average characteristics of the site, and their variability, which are necessary to calculate the dynamic response (measurement of the shear waves velocity, depth of the base-rock, stratification, etc.).

At this point the link with the "reference earthquake" in geological and seismological terms becomes strong.

In fact the information and the evaluations in such areas permit us to characterize the geometrical, mechanical and energy-release elements of the model and of its generation and propagation up to the surface of the site.

In normal practice, by now consolidated, every time correlations are used, for example acceleration, distance, intensity, a model of generation and propagation of the seismic waves is used, and considered applicable to the situation under examination.

The aim of the present contribution is that of pointing out in as much detail as possible the logical concatenation of the steps which lead to the determination of the "reference earthquake" most significant for the foundation soil and the structural elements of the plant, and see if there are presuppositions to improve the correspondence to natural reality of the part of the models used.

In order to make the abovementioned concept more comprehensible, here are some examples.

In the case of site with a problem of potential for liquefaction of the foundation soil, an earthquake characterized by a long duration (a high number of cycles) and a low level of acceleration, can be more significant, than an earthquake of brief duration, with a high peak of acceleration.

In the case of a site on soft soil, if the seismic waves arrive at the site with an energy which can still be felt around the fundamental frequency of the site, the vibrational movement on the surface will be characterized by a prevalent seismic response to that frequency, which could be critical for a pipe or some other structural component.

What has been said above, can only be evaluated, from one side, thanks to a deep knowledge of the dynamic geotechnical characteristics of the site, and on the other from the knowledge of the geometrical, mechanical, energy-release elements of the models of generation and propagation which can be inferred from the information and evaluations relative to

the significant seismogenetic zones and to the attenuation
of their effects with the distance.

2.3 "Reference earthquake" from the point of view of the dynamic behaviour of structures

From the point of view of the dynamic behaviour of
structure the form of the "reference earthquake" is represented
by motion calculated at the base of the foundations. In
fact, we must not forget that it is this last form of the
"reference earthquake" which permits us to calculate the
seismic forces on the structural components of the plant,
and on the foundation soil. Generally the motion at the
base of the formation of buildings of the size and weigth
of NPPs is different from that defined at the surface of
the ground and from that at the foundation depth far from
buildings.
The "reference earthquake", as motion at the foundations,
is always computed by a model which accounts for the so
called soil-structure interaction effects.
From what has been said above, it appears that the
"reference earthquake" in its different aspects, geotectonic,
seismological, soil dynamic, soil-structure interaction
is strictly determined by the seismotectonic characteristics
of the region, and by the local geotechnical characteristics,
as well as by the models of the structures.
Therefore this step, very relevant to the computation
of seismic forces on the structural components, implies
modeling of foundation soil and structures.

3.0 CONCLUSION

The present proven practice characterizes the "reference
earthquake" through empirical correlations and shapes of
design response spectra based on world data, even though
under certain conditions these standards can be unapplicable,
and it is suggested to study the cases one by one.
In the opinion of the author, the growth of seismic
knowledge in general, and the avalaibility of instrumental
data which is increasingly better documented and reliable
has created the presuppositions for an improvement in the
correspondence of the models used to the reality of nature.
In particular for every site the use of standard
correlations can be integrated with the use of data, be
they still empirical, relative to the territory under
examination, or transferable to it from other areas.
The other element which emerges clearly from this
discussion, is the irreplaceable need to determine the
"reference earthquake" in its various phases, through a

strict collaboration of all the experts for the different aspects of the problem.

ACKNOWLEDGEMENT

The writer is grateful to Drs. Iaccarino E., Lojelo L., Magri G., Martinetti S., Pugliese A., Serva L., for their helpful contributions in the course of numerous discussions on the definition of the "reference earthquake" for sites selected for nuclear installations.

NOTE

(1) The term "soil" is used in a general sense. It includes all type of materials (sand, clay, rock, etc.) in different states of aggregation.

FUNDAMENTALS OF SEISMIC MASONRY DESIGN

Jean M.H. MENU

Imperial College of Science and Technology
LONDON

0.0 ABSTRACT

In this paper, an attempt is made to assemble the various updated knowledge concerning the behaviour and design rules of masonry.

Since it is typically a brittle material, masonry is not suited when seismic hazard is not negligible.

In regions of moderate or low seismicity, earthquake shocks usually contain damaging energy precisely in the range of frequencies where masonry constructions are built. Often, even the simple architectural rules of seismic conception which would save buildings, are omitted.

It is also shown how to render masonry more ductile, and then more appropriate in the cases where the potential intensity of shaking is significant.

1.0 INTRODUCTION

Masonry structures are spread in many countries over the world. The reasons for its extensive use are multiple and basically we can say that masonry brings in valuable architectural advantages. This is a material which is almost always available, from basic components easily found in the vicinity of any construction site (mud, clay, gravels).

As a result, it is cheap and fits perfectly with the environment. It also meets several architectural

P. Melchior (ed.), Seismic Activity in Western Europe, 413–429.
© *1985 by D. Reidel Publishing Company.*

constraints, like good protection in either moderate
or hot climate and does so with very satisfying ageing
that only a few of our modern construction materials
are able to provide with the same efficiency.

However, as far as its purely mechanical behaviour is
concerned it has important limitations. Its strength
is certainly not comparable to modern materials like
steel or concrete and this makes its use frustatingly
limited to compression. Its tension and shear
characteristics are low, excluding it from any
significant flexural working condition.

Moreover, its dynamic capabilities are also very
poor, and render it unsuitable for well designed
constructions in seismic areas.
When mentioning dynamic capabilities one should
primarily concentrate on:

a) a small ratio strength/weight as masonry has a
large mass density and is therefore the source of big
inertia forces. These forces are only balanced by a
small intrinsic strength.

b) a very brittle behaviour which means that the
material is not able to sustain or handle large
vibrational energy released during earthquakes.

These two aspects make masonry very prone to early
collapse in any earthquake process. It generates an
important hazard for human life as well as a potential
source of irreparable damages to structures.

2.0 THE BEHAVIOUR OF MASONRY

Masonry is a composite material and as such is very
dependent on the characteristics of its elementary
components: the units (clay blocks, solids or hollow,
concrete blocks, stone units,etc...) and mortar used
to link together the units.

2.1 - Masonry stress-strain law

Figure 1 shows the typical stress-strain law of a
masonry prism tested in pure compression, in a stress
controled experiment.
The main feature of this behaviour is the
brittleness of the material, in particular in the case
of unconfined prisms (column like structures).

2.2 - Role of mortar

In modern constructions, mortar is not only the intermediate product used to smooth the rough contact between the units, but also plays a primary role in the global stress-strain behaviour of masonry prisms.

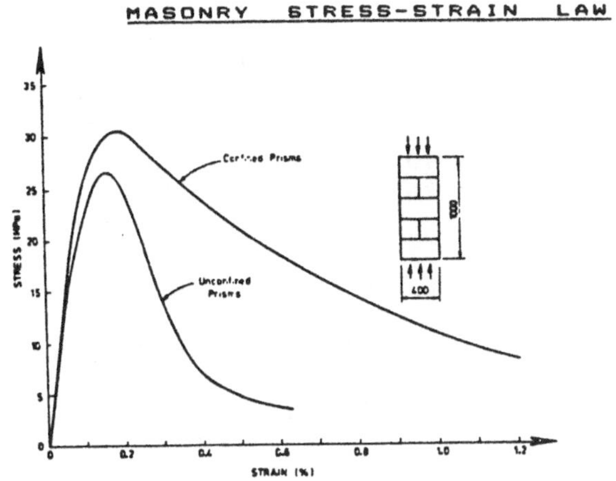

MASONRY STRESS-STRAIN LAW

Fig.1 Stress-strain curve for masonry prisms.

This can be understood as a result of a fundamental difference in Poisson's ratio of the two materials; the value of the ν_{mortar} being several times larger than ν_{unit}.
When compressed, the units generate a lateral "flow" of mortar. The required conditions of compatibility of deformations at the contact mortar/unit cannot be met for large strains. As a result vertical tension fissures are initiated in the units which weaken the prism as a whole.

2.3 - Values of uniaxial compressive strength

In order to assess the characteristics of the prisms, tests have to be carried out. It is agreed that those tests should be realised in pure compression on prisms of about 5 units.
The obtained compressive strength will then serve as a basis to define the various resistances in compression, tension and shear for any subsequent design calculations.

The following table I gives the range of values that can be expected for the uniaxial compressive strength σ'_m (in MPa) of elementary prism.

It can be seen that this basic parameter can vary in a very large proportion, according to the quality of the blocks and the strength of the mortar.

TABLE I — Values of uniaxial compressive strength.
σ'_m (MPa)

	compressive strength of blocks	compressive strength of mortar	
		$\sigma' = 15$ MPa	$\sigma' = 5$ MPa
Special blocks	40 :	16.5	9.3
	28 :	13.8	8.6
Concrete or clay blocks	17 :	10.7	7.6
	13 :	9.3	6.9
Hollow clay blocks	10 :	8.0	6.0
	7 :	6.2	4.3

3.0 DESIGN APPROAHES

The seismic design is governed by the two observations made previously and that the material has a very low strength/weight ratio associated with a very brittle behaviour.

In order to correct these imperfections, it is obvious either to:
    a) ligthen the material
or b) increase its strength
or c) increase its ductility

The first consideration, although theoretically viable, is not the most efficient as the expected gain in units is always accompanied by a reduction in compressive strength (hollow blocks), which diminishes the efficiency of this measure.
On the other hand, points b) and c) can be used to improve the seismic behaviour of masonry structures.

Increase of the strength means the use of good quality material, for the blocks themselves or for the mortar.

However it should be pointed out that highest performance for both components may not be the ideal combination required for the high strength of the composite prism.
We saw that deformation compatibility is an important factor for the global behaviour. Emphasizing the fact that good matching blocks and mortar is at least as fundamental as high compressive strength of elements.

However in terms of seismic response, the most efficient term to play with, is the ductility of the material.
Let us recall that ductility can be defined (among several definitions), as the ratio of plastic to elastic deformation a material can sustain before rupture.
Ductility is favourable in the sense that it permits the material to absorb a large quantity of energy without requiring significant increase in responding stress.

The rest of the paper will describe the design of masonry in the fundamental 3 cases:

3.1 - Non-reinforced masonry

This case corresponds to structures which, for some economical reasons will not be seismically designed.
We shall indicate the minimum basic architectural requirements which should be observed in order to reduce or in most cases surpress any potential damages.

3.2 - Masonry infilled frames

In this case, use is made of steel or concrete frame to provide for the ductility of the ensemble.

3.3 - The reinforced masonry

There are several possibilities of rending a brittle masonry wall ductile by addition of reinforcement in the wall. The technique may be considered as heavy on the site, but is absolutely necessary in areas of moderate to high seismicity when masonry has been chosen as the structural material.

The design methods which will be described are dependent of the contruction techniques and it is not the purpose here to describe them in detail, only give

to architects and engineers some principles of how to
handle the conception.

The methods can be based on the elastic theory, and
the same principles as those dealing with reinforced
concrete, adapted to low resistance material strength.
The elastic approach is necessary when no provision
has been envisaged for increase of ductility. In this
case the peak resistance has to be considered as the
rupture point and a reasonable safety factor should be
considered.

Also, if some ductility can be accounted for, a
pseudo-plastic design can be chosen.
In both cases, it will be sufficient to realise the
design computation starting with a lateral base shear:

$$V = C * S * W$$

where   W = weight of the structure
         C = basic seismic coefficient
         S = structural factor

The basic seismic coefficient C represents the
proportion of vertical gravity that accounts for the
horizontal effect of the design earthquake on a given
structure.

C depends on many factors:
    - The seismicity of the area
    - the soil conditions at the site
    - the natural period of the structure and its
damping
    - the ductility of the structural elements

S depends on the capacity of the structure to
sustain the ductility factor assumed to define C.

The values of the seismic coefficient C given in
various codes of practice assume that structures can
be realised with a ductility factor of 2 to 4.
        C varies typically from 0.05 for regions of low
seismicity to 0.50 for regions of high seismicity; but
it is rarely clear which ductility requirements.

4.0 NON-REINFORCED MASONRY

Non-reinforced masonry or non-infilled frame masonry
are extensively used.

It is not always economically justified that all structures should be strongly designed against destructive earthquakes.

In particular, in low seismicity and non developed areas, it can be sufficient to design masonry structures without the expense of steel or concrete framing or reinforcement.

This can be considered as valid, particularly when low-rise buildings are involved or in higher constructions for the brick partition walls, separated from the main structure and as such can be assimilated with low-rise structures.

These structures should at least meet some fundamental architectural requirements of good conception.

However, it should be kept in mind the BRITTLENESS of the material and whenever some more involved calculations are needed, they must be performed on the basis of the working stress method using a realistic value of elastic response as it will be shown further.

When building non seismically designed masonry structures, the following architectural requirements should at least be satisfied.

4.1 - Global Requirements

4.1.1 Good quality material

Masonry and mortar should be of sufficiently high resistance and have as much as possible deformation compatibility.

Quality testing as suggested by U.S. Building Code Requirements, can advantageously be used.

Composition of mortar is of particular importance for its resistance as well as for its chemical stability to inclemency of the weather or environment.

4.1.2 Adequate position and width of joints

It is well known that buildings should be separated in several elementary blocks, be it for dilatation or differential settlement or else leaving different materials to behave in their own way.

Seismic design also requires to separate the construction in independent blocks. The joints should be up to 5/10 cm wide to allow for free displacement of the structures, preventing any shock between them, since they may not all be in vibration phase.

The position of these joints should aim at cutting
the structures into regular, say rectangular blocks,
having a simple dynamic behaviour. In that sense the
centre of mass should be as close as possible to the
centre of rotation to minimize the torsional mode
effects. Shapes in "T", "U" or "L" are not recommended
and must be cut to form rectangular blocks.

4.2 - Wall Geometry Conditions

Since most masonry structures involve simple walls,
it is important to give simple rules of good
conception (fig. 2).

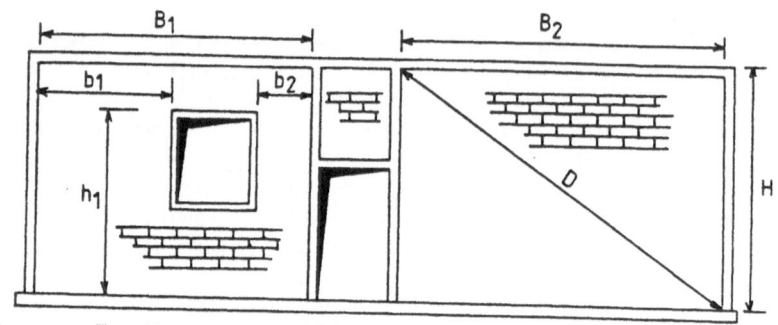

Figure 2. Masonry wall geometry

- thickness   t

- diagonal   d < 50 t

- lengths   B1 or B2 < 5 m.

- areas   H * B < 20 m2.

- pilars   b2 > 0.40 m.

- slenderness h1/b1 < 1 if unreinforced

                      < 3 if reinforced

- opening ratio   < 0.3

4.3 - Horizontal and Vertical Ties.

To provide for minimum tension resistance, all
non-reinforced masonry buildings must have horizontal
and vertical ties. They can be realised in wood, but
preferably in steel or reinforced concrete.

A modern solution consists in simple concrete steel bars passing through special perforated blocks and grouted.
These ties should exist:

- at each level
- at each corner
- around the openings

It is important they ensure a continuity in 3 dimensions. For typical constructions, they must be able to carry a minimum tension of 30 KN.

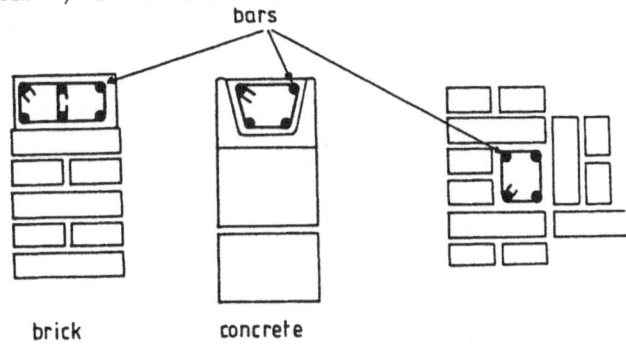

brick   concrete

Horizontal   Vertical

Figure. 3 Examples of concrete ties.

4.4 - Elastic Calculations

In the case where some calculations are needed to insure the stability of the various elements, it must be realised that masonry structures are always fairly sensitive.
Due to low or moderate resistance, contructions are massive and of small height. The natural periods of such "pendulums' are low (of the order of 0.2/0.4 sec), precisely in the range where the dynamic effect is the most damaging.
Moreover. this effect is accentuated in the cases of rock founded structures and also for small or moderate earthquakes.
Figure 4 shows in the case of a moderate shock (a typical aftershock of the FRIULI sequence, May-June 1976) how the acceleration and hence the force acting on structures of natural periods 0.2/0.4 sec are amplified relatively to the recorded peak ground acceleration a_g.

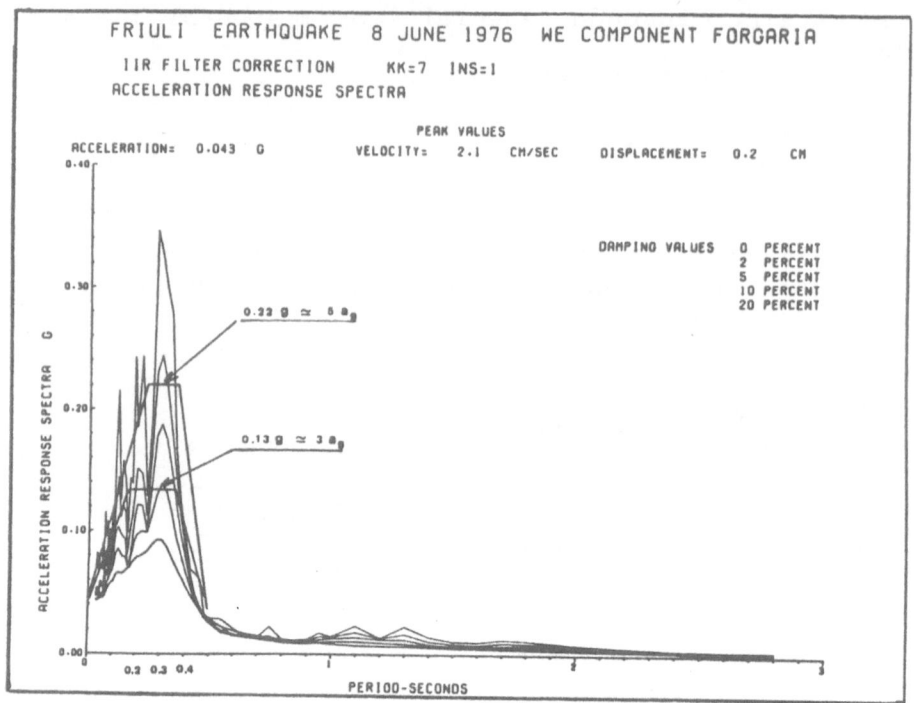

Figure 4. Elastic Response Spectrum of a moderate shock.

Amplifications are of the order of 5 when the equivalent viscous damping of the structure is about 2%, and of 3 when the equivalent viscous damping is about 5%.
The damping of masonry structures lies precisely between those two values.

The equivalent static forces provided by the codes of practice usually rely on some ductility of the structure. This cannot be the case for non-reinforced masonry and it is necessary to use real values of elastic dynamically amplified forces instead.

A linear analysis is required and the obtained stresses should be compared with allowable stresses.
The A.C.I. 531.79 revised 81 has proposed a set of design allowable stresses.

5.0 MASONRY INFILLED FRAMES

Even in low seismicity areas, where no seismic

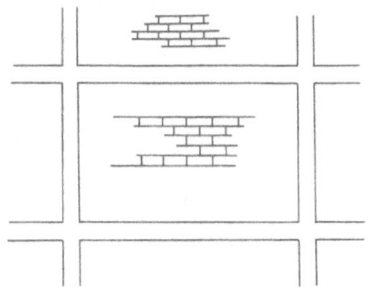

design is considered, masonry walls are often stiffened by steel or concrete frames (fig. 5). This technique can be considered as a first step of good design of masonry in seismic areas provided ductility is carefully analysed.

Figure 5.Masonry infilled wall.

5.1 — Infilled frames behaviour

Calculating this type of wall, taking only the frame into account is greatly in error since it is certainly not on the "safe side".

The presence of masonry inside the frame increases to a large extent the stiffness of the compound wall as is shown in fig.6.

In this test (Leuchars and Scrivener), it is clear that the masonry infilled frame can remain elastic and sustain a load well above the rupture of the frame alone.

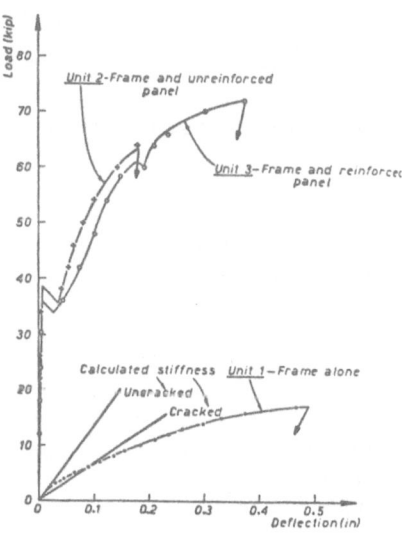

After the elastic limit has been reached, some ductility is available, not only from reinforced panels but also with unreinforced panels. The energy absorption capacity is in fact due mainly to the frame itself. The contribution of the masonry is negligeable before the elastic point.

In the case of earthquakes, the loading is cyclic. Tests show a degradation as the stiffness reduces after each cycle, corresponding to the shear or the crushing of the bricks.

Figure 6. Initial load defection curves.
(after Leuchars and Scrivener)

Parallely, the peak resistance diminishes, which renders the masonry already damaged by previous earthquakes very vulnerable.

It must also be borne in mind that the increase in stiffness is not favourable as far as the dynamic forces are concerned. Any increase in stiffness being a reduction of the fundamental period (down to 0.2 sec or less), the seismic forces will be, at least for most moderate shocks, at its maximum effect. The induced forces will be similar to those sustained by non reinforced walls for the same shaking.
The difference being that here the wall is better equiped to avoid collapse.

5.2 - Elastic Calculations

The calculation in the elastic domain can be inferred from static loading (Stafford-Smith and Carter).
In their work, they consider an equivalent diagonal bracing model (fig 7) and obtain an equivalent relative stiffness:

$$\pi H / 2 \alpha$$

where   H = height of columns betwee centrelines
                  of beams
and α = length of contact between frame and
                  masonry

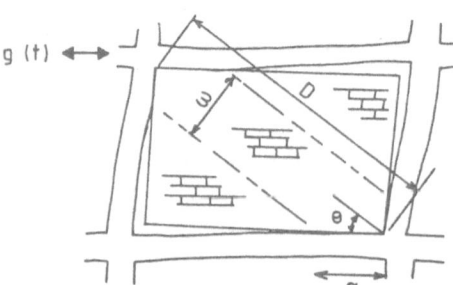

g (t)

w is currently of the order of 0.2 to 0.35 D and is referred to as the equivalent strut.

Figure 7. Equivalent strut model.

5.3 - Ultimate calculations

In order to assess the strength and deformation at failure, the ductile capacity of the infilled masonry wall is easily computed from the "Knee braced model" (fig.8).

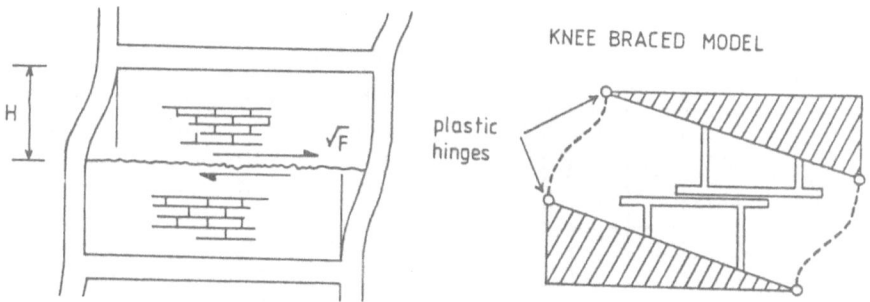

Figure 8. Failure of infilled frames.

This type of rupture, expected when the ratio H/L of the panel is not too big and the masonry of good quality, is reached when the horizontal force is

$$V_u = 2 / H (M_{u1} + M_{u2}) + V_F$$

The first term of the right hand side corresponds to the ultimate moments of the half height columns; the second term being the friction of the two sides of the broken wall.

6.0 REINFORCED MASONRY

There is several techniques for building reinforced

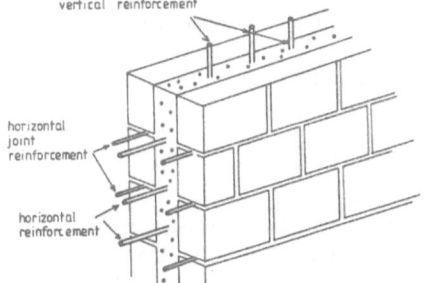

masonry, ranging from the use of special perforated blocks to the sandwich grouted wall. We shall visualise such a wall in the latter case (fig.9) where reinforcing bars can be found either in the grout or in the mortar.

Figure 9. Reinforced masonry wall.

426   J. M. H. MENU

6.1 – General behaviour

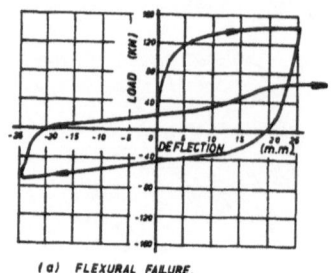

(a) FLEXURAL FAILURE.

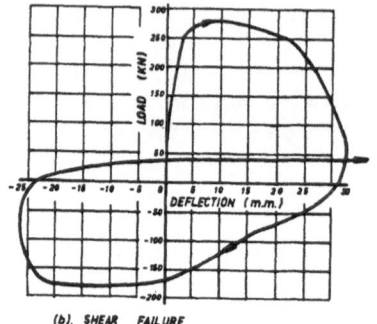

(b). SHEAR FAILURE

Typical load-deflection curve are shown on the fig.10 for the flexural and the shear mode.

We immediately observe the large energy absorption capacity of the wall during the first cycle of its cyclic loading. As soon as the second cycle arises, this capacity is drastically reduced in strength (ratio of 2 in flexion and about 6 in shear).

The stiffness is also reduced and approaches zero in shear mode.

Figure 10. Load-Deflection loop.

6.2 – Ultimate Calculations

The ultimate design of reinforced masonry walls requires the knowledge of the various potential types of failure.
They depend mainly upon the geometry of the wall and the value of the vertically applied load.

6.2.1 Tension failure

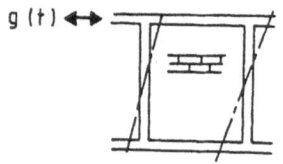

g (t)

This failure (fig.11) occurs when the aspect ratio H/L of the panel is high. The external columns reach first their ultimate resistance in tension or compression.

Figure 11.

6.2.2 Sliding Shear failure

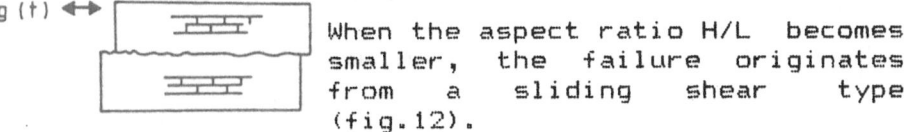

When the aspect ratio H/L becomes smaller, the failure originates from a sliding shear type (fig.12).

Figure 12.

The sliding occurs along a mid-wall horizontal mortar course. It can be estimated to initiate as the horizontal force is:

$$R_s = (0.9 + 0.3 \ L/H) \ t \ H \ T$$

where t = thickness of the wall
      T = shear strees in the mortar

6.2.3 Compression failure

Compression ultimate resistance can be estimated from the relation:

$$R_c = 2/3 \ t \ \sec \theta$$

where α = /2 [4 E_f I_f H/(E_m t sin 2θ)] $^{1/4}$

θ = slope of the diagonal

The coefficient 2/3 was suggested by Leuchars and Scrivener to correspond more accurately to their tests.

6.2.4 Flexural failure

A reinforced masonry wall will reach first a failure in flexion if the reinforcement is low and/or if the vertical load is low.

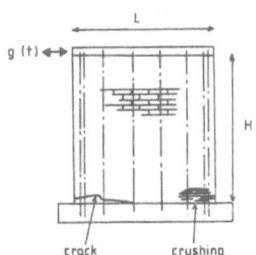

It is also a function of the aspect ratio H/L; a value of 3 can be given as an upper limit.
For this wall, ductility can be such that a structural factor S=1.5 to 2 can be used. Walls become more brittle as the vertical load increases.

Figure 13. Flexural failure.

By inferrence from static loading of concrete, the
ultimate flexural moment can be taken to be:

$$M_u = 0.5\ A_v\ \sigma_v\ L\ [\ 1 + N_u/(A_v\ \sigma_v\ )]\ (1 - c/L\ )$$

(see Cardenas et al)

The flexural ductility is limited by crushing of
masonry at the external corners of the walls (fig.13).

6.2.5 Shear failure

It is encountered when the reinforcement (vertical and
horizontal) is high and/or when the vertical load is
significant.

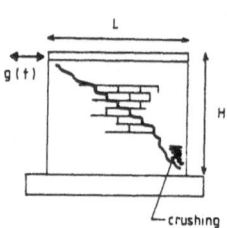

An aspect ratio less than 1
forces this type of rupture
to occur.
The ductility factor can
reach a value of 3, then the
structural factor S can be
taken to up to 3 or 4.

Figure 14.Shear failure.

This valuable ductility in shear is satisfactorily
obtained if the wall is uniformly horizontally
reinforced (fig.14).
Its limitations are again, the quality of the
material and the crushing at the external corner.

7.0 CONCLUSION

The objective of this paper was to to discuss the
main features of Seismic masonry design. It has been
shown that, although masonry is not a well suited
material in seismic areas it is possible, in certain
conditions to improve its performance.
In low seismicity zones, and for low risk
constructions simple architectural rules should be
applied. A significant seismic safety can result at no
obvious extra cost.

REFERENCES

Stafford-Smith B.and Carter C., A method of analysis for infilled frames., Proc.Inst.Civ.Engrs, 1969 Vol 44.

Cardenas A.E.et al., Design provisions fror shear walls, J.Am.Concr.Inst., 1973 Vol 70. No 3.

Meli R., Behaviour of masonry walls under lateral loads., Proc.Fifth W.C.E.E. 1974 (Rome).

Williams D. and Scrivener J.C., Response of reinforced masonry shear walls to static and dynamic cyclic loading.,Proc.Fifth W.C.E.E. 1974 (Rome).

Priestley M.J.N.and Bridgeman D.O. Seismic resistance of brick masonry walls. ,Bull.N.Z.nat. Soc.Earth.Engg.,Vol7, No 4, 1974.

Leuchars J.M. and Scrivener J.C., Masonry infill panels subjected to cyclic in-plane loading., Bull.N.Z.nat.Soc.Earth.Engg., Vol 9, No 2, 1976.

Building Code Requirements for Concrete Masonry Structures, (ACI 531.79 revised 81).

EARTHQUAKES AND TOWN-PLANNING

Robert M.G.L. Liégeois

Chef du Département Mines et Carrières
Iniex, 200, rue du Chéra
4000 Liège

ABSTRACT

An earthquake affected the area of Liège on 8 November 1983. This
is a dwelling area in a mining site. In the beginning of the six-
ties a group of 800 dwelling-places was built in a district which
is subject to mine damage.
Though circumscribed by the isoseismal line 7 of the international
macroscopic intensity scale, this group of dwellings resisted.
It is the result of a preliminary implantation and conception study, based on an geological and mining analysis of the area characteristics.

INTRODUCTION

The earthquake which took place in Liège on 8 November 1983
at 00.50 overtook the population who experienced and suffered it.
There was one dead and material damage, so it was classified in
class VII of the macroseismic scale. According to the observatory
of Uccle it is classified at the magnitude 4.92 of the Richter
scale.

The disaster zone bounded by the Administration with a view
to compensation files, has a surface of 1000 km² and extends
around Liège-City like a triangle of which the base would be Huy-
Verviers and the apex in Genk. The most spectacular damage was
observed in Liège itself and in the neighbourhood of the city, on
the left bank of the Meuse, at St-Nicolas, Ans, Grâce-Hollogne.

431

P. Melchior (ed.), Seismic Activity in Western Europe, 431–439.
© 1985 by D. Reidel Publishing Company.

During an earthquake, all useful information must be collected
as soon as possible on the spot.

If, by chance, dispositions are taken before the earthquake
in order to take into account the strata deconsolidation due to
natural phenomena (faults, dissolution pocket, outcrop bending of
rock banks on steep slopes ...) or due to deep mining, it is very
important to note in how far the effects of the disaster may be
bounded or lessened there where paraseismic precautions were taken.

This is exactly the case in Liège, in the district of Burenville, in the heart of the disaster zone. About 25 years ago, a
housing building company was authorized by the City to renovate a
whole district and to erect dwellings with a capacity of several
thousands of people. The Company took care to entrust a special
team with the study of the implantation, taking into account the
following aspects : geology, past and present mining, geography,
meteorology, sociology, architecture.

Right after the earthquake of 8 November 1983, the geographer
of the team insisted on observing attentively the damage suffered
in the district and the neighbouring country, according to the
localization of the houses and the type of construction. Photogra-
phies were taken and films were made with a view to searching for
a relation between the importance of the damage and the characteristic local parameters like the local tectonic structure, the
allotment of the claims, the situation of the dwelling-places,
their architectural conception, etc ...

The decrepitude of certain dwellings explains the importance
of the damage, but some recent houses were severely damaged.

The pylones resisted well; this is not surprising : the same
phenomenon has been observed elsewhere during more violent earthquakes. As an example, we will observe the contrast between the
damage suffered by roads and houses during the earthquake of
Anchorage in Alaska, and the good state of the pylones in the disaster area.

The water-tower resisted well also. It bore up well against
the mode of vibration induced by the earthquake. Other examples
are known of water-towers destructed because of their own frequency which is itself in relation with the form of the tank and the
degree of filling.

In those parts of the globe where earthquakes are frequent
and ravaging, man has strained his ingenuity in order to face it,
and this for hundreds, even thousands of years. We will illustrate
this with two characteristic examples.

Some very ancient temples resisted to the trepidations thanks to their stream-lined architecture and the flexibility of their wooden joints. With the play of brackets and consoles judiciously assembled, we can put up with vertical, horizontal and sometimes very intricate vibrations, having earthquakes at their origin.

In the region of the Great african Lakes, the natives build flexible and low dwellings with a very reliable resilience. When they are asked to build with hard material they prepare a bed of drift boulders on which they spread out the foundation stone which will serve as a basis for the always small and squat building.

In the coal-mining countries of Western Europe where circumstances led to underground mining under towns, the cavities created by the driving of galleries and the getting in the faces, were not always systematically filled. As a result there are sometimes important collapses.

Very early man found in the Liège region coal outcrops which he first worked in the open and afterwards underground. This occurred first on a small scale, in the family, then industrially. Companies were founded, obtaining claims for the winning of coal all around the City of Liège.

The coal was won out of numerous seams down to hundreds of meters under the ground, over 350 km^2. This explains the subsidence on both sides of the Meuse (10 to 20 m subsidence), so that the river had to be embanked and the sewage from the suburbs had to be pumped away.

Surface subsidence is unevenly distributed because of the natural disposition of the deposit and its administrative division into claims. The claimholder is not authorized to work out the coal to the boundary.

According to the article 4 of the Royal Decree of 20 September 1950 "Along the boundaries of each concession and according to the height of the deposit an unworked massif or barrier of at least 10 m width must be reserved. In case of lease the barrier is brought back to the boundary of the leased part. The division manager of the mine basin will allow the working of a defined part of this barrier". In principle there are vertical partitions of 20 m thickness (or more or less according to time and place) between the subsidence basins.

The concessions have varying shapes, often demarcated by arbitrary boundaries like a meridian, a river, an administrative limit, etc ...

During an earthquake, the accelerations are more important on the palimpsest crests which form the boundaries of the concession.

In order to understand the effects of the mining on a building erected in a mining zone, it is easy to suppose that only one seam in a rectangular panel is worked.

Mr. Kratzsch (1983) demonstrated with simple diagrams the effects on a building depending on its localization and its orientation.

On the other hand, according to Mr. Mebilof (1974), quoted by Kratzsch, it is possible to build a dwelling in a zone submitted to mining subsidence at the rim of a subsidence basin. The building must be stiffened on the exposed side. The reinforcement frame is supported by hollow abutments by interposing a membrane in order to reduce the friction. Should the occasion arise, jacks are mounted in the abutment in order to set the building plumb again after the movements have ceased.

PROTECTION OF THE BUILDINGS AGAINST GROUND MOVEMENTS

Largely before the earthquake, the Liège engineers gave their attention to the protection of buildings against ground movements. For lack of an exhaustive study, we will give two characteristic examples.

The first refers to the construction of a testing station at the place "Grand Bac" at Ougrée (today Seraing) for the Institut National de l'Industrie Charbonnière (Inichar) in 1954.

The station had to be erected right near the boundary of three mining concessions and it was expected to restart the underground working which was temporarily stopped. The precautions taken to face the problems of ground movements have been described at an Information Day organized by Inichar on 2 April 1954, held in Liège. They were published in the "Annales des Mines de Belgique", July 1954, by Messrs. J. Venter, Director of Inichar, L. Pirnay, architect, G. Lesage, consulting engineer.

The second example refers to the renovation of a district of Liège-City, which was particularly underprivileged. In 1959 the cooperative society "La Maison Liégeoise" wanted to create a residential unit with 800 dwellings. The study was carried out by a team of architects. It was not published, but the buildings were erected in the Burenville district, in the heart of the disaster zone. It seemed priority to us to inspect the place after the disaster of 8 November 1983.

First example : the testing station of Grand-Bac.
A first precaution was taken by splitting up the building in an assembling of blocks united as well as separated by flexible

joints of 10 and 15 cm width which allow among others relative
rotations of the blocks under the influence of oblique horizontal
forces (Principle of the partition).

A second precaution was the strutting of the columns there ⌐
where it was compatible with the use of the halls, the offices
and the laboratories (Principle of the belt).

To put up with important horizontal displacements (10 cm)
the following solution was chosen : to superpose roller boxes in
orthogonal sheets (Figure) "These rollers have a small diameter :
3 cm. The intermediate steel plate allows the use of the whole
bearing surface of the contact lines of the rollers with this
plate. Such steel plates were placed above and under the two sheets
in order to have the rollers moving on well faced surfaces". Boxes
with 21 rollers correspond to pillars of 320 tons of vertical
weight. The boxes are filled with a chemically stable plastic
product with a density near to that of water and which does not
emulsify with it. Plumb plates of 10 mm thickness were placed
between the rolling plates and their course in order to put up
with a possible out of parallelism. The base of each pillar was
enlarged and the course was established on the foundation sole-
piece so as to allow the placing of jacks in order to lift the
pillar as a function of the subsidence. The access to the roller
boxes is facilitated by the existence of fluid galleries.

Some years after its construction the building was destroyed
and the ground was struck to construct a plate-mill.

Second example : "La Maison Liégeoise" at Burenville.
    The "Grand bac" testing station was large and heavy. This
was not the case with the Maison Liégeoise at Burenville. To construct 800 new dwellings in a district, one can either built two
towers, like the towers of Liège-Droixhe, or build more modest
dwelling places. This second solution was chosen for two reasons :
1) The experience acquired by the co-operative society La Maison
   Liégeoise showed the absolute necessity to base the implantation on a thorough knowledge of the physical characteristics
   of the ground, here a mine site.
2) The typical neighbourhood unit being more or less 1800 dwelling-places, the group to be constructed counting more or less
   800 dwelling-places could not be organized on its own. The
   district was certainly not unorganized, but the existing residential homes had to be taken into consideration.

The analysis of the physical and social characteristics of
the site of La Maison Liégeoise was considered under the following
aspects :
1. General geographical conditions
2. Geology and mining industry

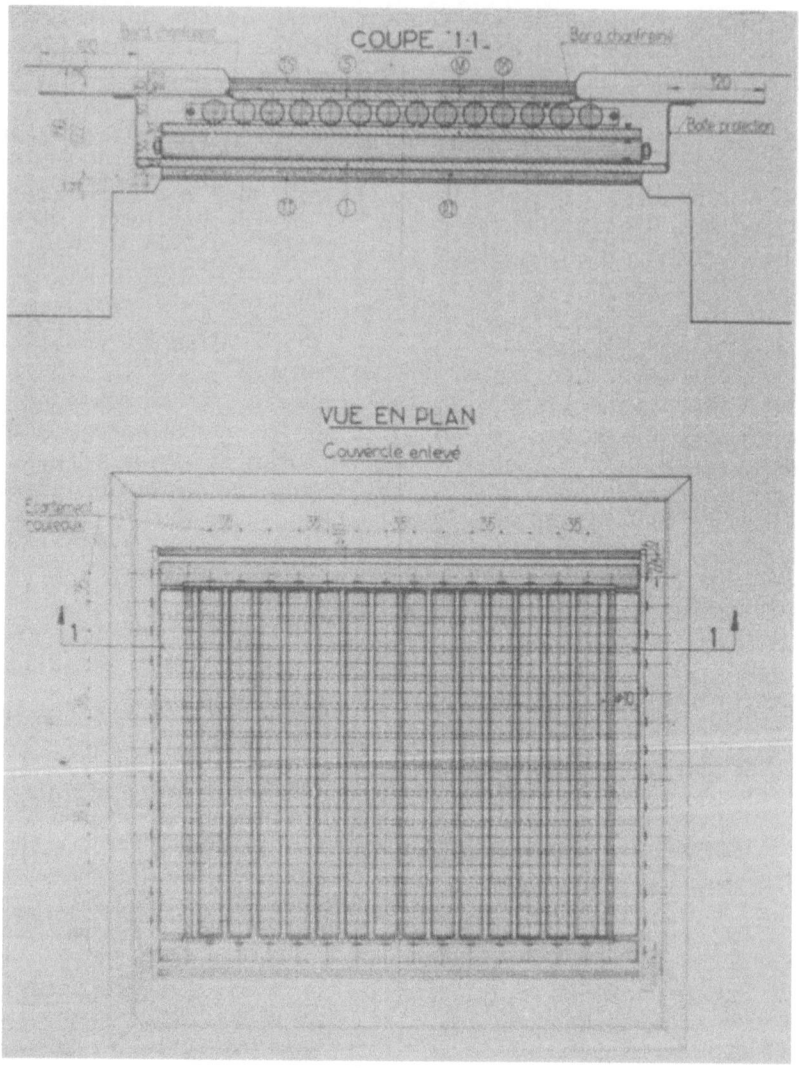

3. Microclimate
4. Social space for the new population.

In the chapter consecrated to the geology and the mining in-
dustry, we can read :
"The Burenville district is wholly situated in the so-called Liège
Coal Basin. It is a syncline oriented to N 60° E. There are nume-
rous workable seams, parallel and slightly dipping to the South-
East. They are cut by the St-Gilles fault".

At the time of the study, the mining industry was still very
active and there were more than 20 active collieries in the only
Liège basin. Moreover, consulting the plans of the drains, we find
14 blind pits ("bures") in the triangle of the rue St-Nicolas,
rue du Calvaire and rue St-Laurent. The name of Burenville evokes
its origine : "Bure el ville". The Burenville district is divided
into three parts, belonging respectively to :
West : S.A. des Charbonnages de l'Espérance et Bonne Fortune;
East : S.A. des Charbonnages de Bonne-Espérance, Batterie, Bonne-
       Fin et Violette;
South : S.A. des Charbonnages de Gosson-Kessales.

These different companies work the seams at 50 to 800 m depth.

An agreement between the first and the third quoted allows
the St-Nicolas pit of the first quoted to extend its activities
to seams which can still be worked in the concession of the third
quoted.

The working forecasts of the studied zone related to a total
seam thickness of about 5 to 7 m at the time, after June 1959,
while up to this date, a considerable quantity of coal had already
been won at this place. The net production of the three companies
quoted above in the zone comprising Burenville was for 1959 :

$$372.000 + 420.000 + 348.200 = 1.140.200 \text{ tons.}$$

With an average volumetric cleanness of 78 % (ratio between
the solid before the working and the total volume of the won seam)
and a gravimetric cleanness of 69 % (coal density : 1.35, stone
density : 2.25) 22 m^3 rock had to be got for 78 m^3 coal, i.e.
450 kg dirt per ton net production.

In the Burenville report, the damage to the dwellings, observed in the district in 1959, has been noted. We can read : "The
zone which has particularly suffered the disaster can be easily
bounded to the surroundings of the Place St-Nicolas. The explanation of such a concentration of damage is very simple". "At the
boundary of two joined concessions, a section of ground which has
not yet been worked forms a real loose wall between two subsidence

basins which lean on it. As a result there are important tensions perpendicularly to the pillar and in the immediate surroundings".

The geologist's observations were confirmed by the Direction of Public Utility Services :
- telephone cables more tense than when they are laid in;
- numerous breakings in the water system.

The report ended as follows : "Wherever we build, we have to think of the mine deformations ... (many drawing out zones as compared to the simple subsidence zones and this because of the pillars between the concessions and the workings)".

So, the architects were asked to construct :
- modest dwelling-places (limited number of houses);
- squat (not or almost not elongated base section, limited height);
- homogenous resistance (no annex, no loggia, no roof dislevelment);
- belted;
- on pillars or on flags according to the case;
- oriented to have a better behaviour during sollicitations on the rim of basins;
- always on slopes less than 10°.

The architects tried to take this advice into account considering :
- the ground surface and conformation belonging to the Society;
- the prescriptions of the communication network (the lay-out of the Boulevard Ste-Beuve was imposed);
- the climatic constraints of a windy district;
- the use of existing communal equipment;
- the social aspect of the feeling of belonging to the district;
- the aesthetic aspect;
- the supplementary costs of the building reinforcement.

The earthquake of 8 November 1983 exacerbated the sollicitations and accelerated the collapse process. It left in the Liège region (lato sensu) traces of horizontal and vertical movements, tensions in all directions, which severely damaged hundreds of houses. The group of "La Maison Liégeoise" did not escape the tremors provoking the fall of objects and the cracking of walls inside the buildings. But the main structures resisted in a way that contrasts with the environment.

We think that we may ascribe this satisfying behaviour to the respect of some pieces of advice concerning the implantation and the conception of the buildings. In this respect the architects chose :
- simplicity ,
- almost symmetrical volumes,
- a limited elongation of the parallelepipedic shapes,

- regularity for the elevation lines,
- limitation of the height,
- anti-twist peripheral stiffening,
- armed basement flags where differential settling was feared.

If there had been money to do more, may be paraseismic solutions would have been chosen for the linking of the non-structural elements (partitions) with the framework. The solutions proposed at the time went out of the working plan bounds while offering no real guarantees.

Economically speaking, it is too early to strike the balance. The supplementary costs of the construction are probably well paid by the reduction of repairing costs in the "main walls" item.

It's a matter to be followed.

REFERENCES

Fonder, A. et Plumier, A. - Calcul et conception des ouvrages
    soumis aux effets sismiques. Université de Liège, Faculté
    des Sciences Appliquées. 118 p., Déc. 1980.

Loi ...
    Loi du 12 juillet 1976 relative à la réparation de certains
    dommages causés à des biens privés par des calamités natu-
    relles. Moniteur Belge du 13 août 1976, pp. 10149 à 10165.

Ministère des Travaux Publics
    Projet de plan de secteur de Liège.
    Cartes 42/1 à 3 et 5 à 7. Bruxelles.

Bureau d'Etudes Louis Jacquet.
    Aménagement de Burenville. Liège, Rapport inédit, 137 p.,
    1959.

Arrêté ...
    Arrêté Royal du 23 novembre 1983, considérant comme une cala-
    mité publique, les dégâts provoqués par le tremblement de
    terre du 8 novembre 1983 dans plusieurs communes des Pro-
    vinces de Liège et du Limbourg et délimitant l'étendue géo-
    graphique de cette calamité. Moniteur Belge du 8 décembre
    1983, pp. 15174-15175.

Venter, J., Pirnay, L. et Lesage, G. - Protection des bâtiments
    contre les mouvements de terrains. Annales des Mines de
    Belgique, Tome III, 4ème livr., juillet 1954, pp. 511-527.

Kratzsch, H. - Mining Subsidence Engineering. Springer Verlag,
    Berlin, Heidelberg, New York, 1983, 543 p., 380 fig.

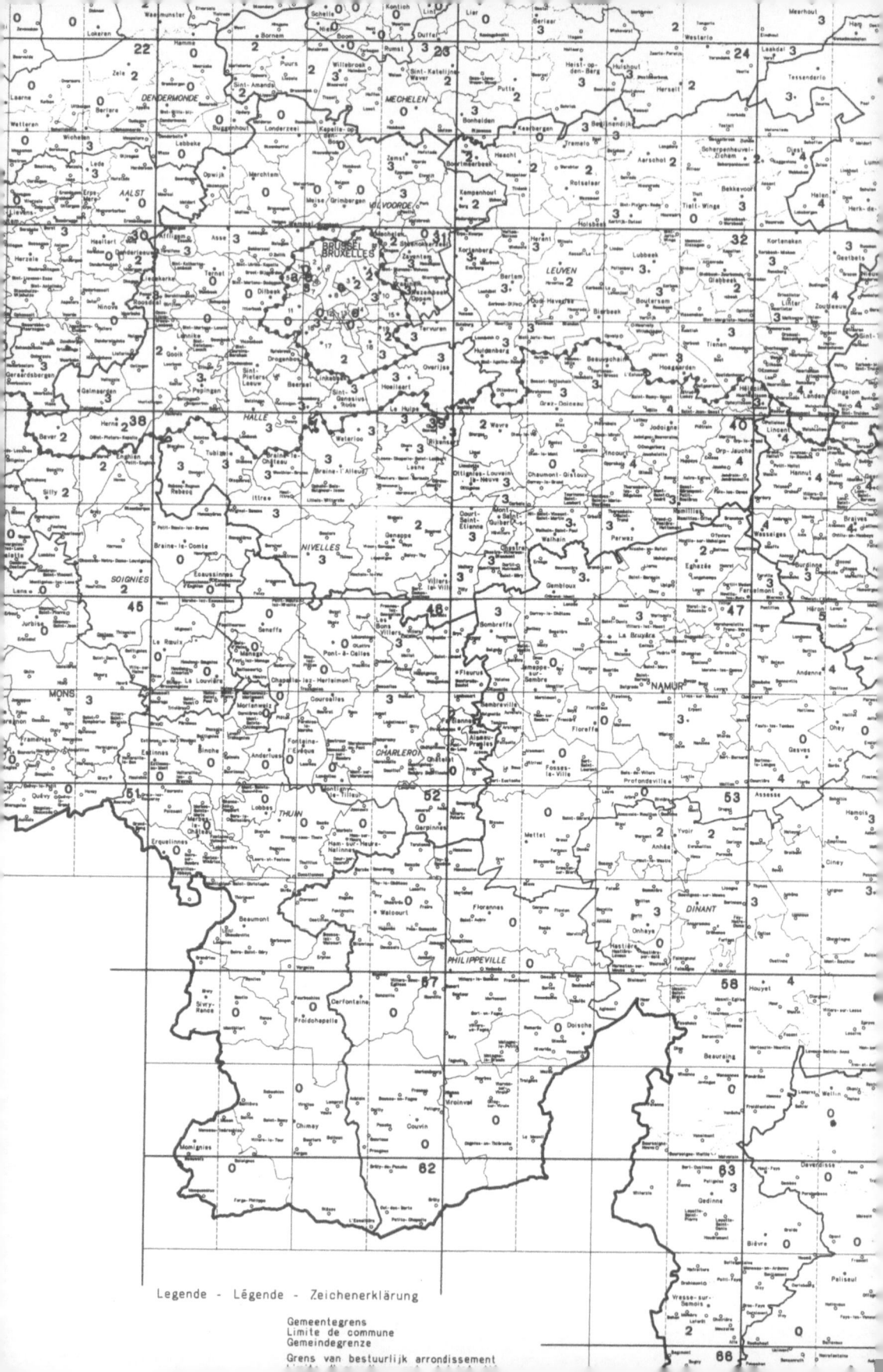

Legende - Légende - Zeichenerklärung

Gemeentegrens
Limite de commune
Gemeindegrenze

Grens van bestuurlijk arrondissement